Functional Nanomaterials for Sensors

Because of their novel chemical and physical properties, functional nanomaterials have found increasing industrial applications in nanoelectronics, energy science, and biological applications. *Functional Nanomaterials for Sensors* surveys advances in functional nanomaterials and their use in sensing. It covers their properties, synthesis, design, fabrication, and applications, including for chemical, biological, and gas sensing, environmental remediation, fuel cells, catalysis, electronic devices, and biotechnology.

FEATURES:

- Describes how nanomaterial functionalization is being used to create more effective sensors
- Discusses various synthesis procedures, characterization techniques, and which nanomaterials should be used for sensing applications
- Provides an in-depth look into oxide nanostructures, carbon nanostructures, and two-dimensional (2D) material fabrication
- Explores the challenges of using nanoscale sensors for large-scale industrial applications

This book is aimed at materials, chemical, biotech, and electronics researchers and industry professionals working on sensor design and development.

Emerging Materials and Technologies

Series Editor: Boris I. Kharissov

The *Emerging Materials and Technologies* series is devoted to highlighting publications centered on emerging advanced materials and novel technologies. Attention is paid to those newly discovered or applied materials with potential to solve pressing societal problems and improve quality of life, corresponding to environmental protection, medicine, communications, energy, transportation, advanced manufacturing, and related areas.

The series takes into account that, under present strong demands for energy, material, and cost savings, as well as heavy contamination problems and worldwide pandemic conditions, the area of emerging materials and related scalable technologies is a highly interdisciplinary field, with the need for researchers, professionals, and academics across the spectrum of engineering and technological disciplines. The main objective of this book series is to attract more attention to these materials and technologies and invite conversation among the international R&D community.

Nanotechnology Platforms for Antiviral Challenges: Fundamentals, Applications and Advances
Edited by Soney C George and Ann Rose Abraham

Carbon-Based Conductive Polymer Composites: Processing, Properties, and Applications in Flexible Strain Sensors
Dong Xiang

Nanocarbons: Preparation, Assessments, and Applications
Ashwini P. Alegaonkar and Prashant S. Alegaonkar

Emerging Applications of Carbon Nanotubes and Graphene
Edited by Bhanu Pratap Singh and Kiran M. Subhedar

Micro to Quantum Supercapacitor Devices: Fundamentals and Applications
Abha Misra

Application of Numerical Methods in Civil Engineering Problems
M.S.H. Al-Furjan, M. Rabani Bidgoli, Reza Kolahchi, A. Farrokhian, and M.R. Bayati

Advanced Functional Metal-Organic Frameworks: Fundamentals and Applications
Edited by Jay Singh, Nidhi Goel, Ranjana Verma and Ravindra Pratap Singh

Nanoparticles in Diagnosis, Drug Delivery and Nanotherapeutics
Edited by Divya Bajpai Tripathy, Anjali Gupta, Arvind Kumar Jain, Anuradha Mishra and Kuldeep Singh

Functional Nanomaterials for Sensors
Suresh Sagadevan and Won-Chun Oh

For more information about this series, please visit: www.routledge.com/Emerging-Materials-and-Technologies/book-series/CRCEMT

Functional Nanomaterials for Sensors

Edited by
Suresh Sagadevan and Won-Chun Oh

CRC Press
Taylor & Francis Group
Boca Raton London New York

CRC Press is an imprint of the
Taylor & Francis Group, an **informa** business

First edition published 2023
by CRC Press
6000 Broken Sound Parkway NW, Suite 300, Boca Raton, FL 33487–2742

and by CRC Press
4 Park Square, Milton Park, Abingdon, Oxon, OX14 4RN

CRC Press is an imprint of Taylor & Francis Group, LLC

ISBN: 978-1-032-20495-6 (hbk)
ISBN: 978-1-032-20496-3 (pbk)
ISBN: 978-1-003-26385-2 (ebk)

DOI: 10.1201/9781003263852

Typeset in Times
by Apex CoVantage, LLC

Contents

Preface

Recently, with the improvement of human quality of life, many efforts are being made to live healthy, pleasant, and comfortable lives. Sensors are one of the essential factors for improving the quality of life, and high-performance intelligent sensors are required, so demand for them is expected to increase further in the future. Recently, functional materials that give functions to sensors are becoming more important. In particular, nanofunctional materials existing in various components and forms are attracting more attention because they have unique physical, chemical, mechanical, and optical properties differentiated from existing bulk materials. Recently, research on nanotubes including carbon nanotubes (CNTs) and nanocomposites has been very active to be further developed in existing nanoparticle research and applied as sensors. This book examines the latest nanosensor research and development (R&D) trends such as gas sensors, water quality sensors, biosensors, light sensors, and physical sensors, and it is predicted that high functionality and miniaturization of sensors by nanomaterials will be used as a very important part.

Research on nanosensors using nanomaterials such as metals, inorganic, organic, biological, and composite materials having a shape of less than or equal to 100 nm in various compositions and forms such as nanoparticles, nanowires, and nanotubes has been actively reported. This is due to the unique physical, chemical, mechanical, and optical properties of nanomaterials. For example, when a nanosensing material having a large specific surface area is used in a gas sensor, a low concentration air pollution source may be detected even at a low operating temperature, thereby lowering power consumption and ultimately miniaturizing a sensor, and thus, a material having a very large specific surface area such as nanoparticles, nanowires, and graphene has been used. This book describes two points. The first point is about how to manufacture nanostructured materials such as nanoparticles and nanowires, and the second point is about the latest nanosensor technology trends using nanomaterials. In fact, many nanosensor research results using nanomaterials are being published, so the contents covered in this book will be part of them, and it can be used as a reference for researchers and developers who want to conduct related research and development.

Trends in sensor R&D technology using various nanomaterials were described. Although the vast amount of research worldwide deals with only a few research cases, nanomaterials are developed in various ways in the form of nanoparticles, nanowires, nanotubes, and nanocomposites, and it is confirmed that they are expanding to various applications such as gas sensors, water sensors, biosensors, light sensors, and physical sensors. From this, furthermore, it can be predicted that future state-of-the-art sensor technologies will act as more essential factors for high functionality and miniaturization of nanomaterials. However, sensors using nanowires have the advantage of being able to manufacture nanowires of a wide variety of compositions and structures compared to the top-down method, but more research is needed on the technology for assembling and aligning nanowires in two dimensions as in the top-down method. In addition, although research has already been conducted at advanced research institutes in various countries, research on the stability of ultrafine nanomaterials—such as toxicity to the human body and the environment—is expected to develop into a technology that can be sustainable for R&D and industrialization.

Assoc. Prof. Dr. Suresh Sagadevan
Fellow of the Royal Society of Chemistry (FRSC),
London, United Kingdom

Prof. Dr. Won-Chun Oh
Department of Advanced Materials Science &
Engineering, Hanseo University, Seosan-si,
Chungnam, Korea

Editors

Dr. Suresh Sagadevan is an associate professor at the Nanotechnology & Catalysis Research Centre, University of Malaya. He has published more than 350 research papers in the ISI top-tier journals & Scopus. He has authored 12 international books series and 40 book chapters. He was selected as one among the top 2% of scientists worldwide by Stanford University consecutively in 2020 and 2021. In 2021, he was recognized for his outstanding contribution in the research activities with the award as Fellow of the Royal Society of Chemistry (FRSC). He was a guest editor and editorial board member of many reputed ISI journals. He is a member of many professional bodies at the national and international levels. He was a recognized reviewer for many reputed journals. Indeed, he is working in various fields such as nanofabrication, functional materials, crystal growth, graphene, polymeric nanocomposite, glass materials, thin films, bio-inspired materials, drug delivery, tissue engineering, supercapacitors, optoelectronics, photocatalytic, green chemistry, and biosensor applications.

Dr. Won-Chun Oh is a full professor in the Department of Advanced Materials and Engineering at Hanseo University in Korea and the School of Materials Science and Engineering at Anhui University of Science and Technology in China, and he is guest professor in universities in China, Thailand, and Indonesia. He obtained the 'Research Front' award from Korean Carbon Society, and obtained the 'Yangsong' award from Korea Ceramic Society, the 'Excellent Paper Award' from Korea Journal of Material Research, and the 'Best Paper Award' from the *Journal of Industrial and Engineering Chemistry* for his pioneering work, and an 'Award of Appreciation' from ICMMA2011, ICMMA2014, and ICMMA2019. He is an Indian Chemical Manufacturers and Merchants Association (ICMMA) committee board member and has been appointed as one of the conference chairman and vice chairmen from 2007 to the present year. He was appointed as one of the 'Top 100 Scientist in the World' at IBC, UK and 'Top 2% Scientists in the World' at Stanford University. He is the author or a coauthor of 890 papers published in domestic and international journals and speeches on conference as special lectures, plenary lectures, and as a keynote lecture speaker. He serves as the editor-in-chief of the *Journal of Multifunctional Materials and Photoscience, Asian Journal of Materials Chemistry,* and is an Advisory Board member of the *Asian Journal of Chemistry and Nanomaterials.*

Contributors

M. M. Abdullah
Advanced Materials and
Nano-Research Centre
Department of Physics
Faculty of Science and
Arts Najran University
Najran, Saudi Arabia

Himanshu Aggarwal
Asian Journal of Chemistry
New Delhi, India

Waqar Ahmed
Takasago i-Kohza
Malaysia-Japan International
Institute of Technology
Universiti Teknologi Malaysia
Kuala Lumpur, Malaysia

Shumaila
Department of Physics
Jamia Millia Islamia
New Delhi, India

Syed Wazed Ali
Department of Textile and
Fibre Engineering
Indian Institute of Technology
New Delhi, India

Isha Arora
Department of Chemistry
Amity Institute of Applied Sciences
Amity University
Noida, Uttar Pradesh, India

Amrish Chandra
Amity Institute of Pharmacy
Amity Universit
Noida, Uttar Pradesh, India

Harshita Chawla
Department of Chemistry,
Amity Institute Of Applied Sciences
Amity University
Noida, Uttar Pradesh, India

Bhasker Pratap Choudhary
Department of Applied Sciences
Chandigarh Engineering College
Jhanjeri, Mohali, India

Daniele Dondi
Department of Chemistry
University of Pavia
Viale Taramelli and
INFN, Sezione di Pavia
Via Agostino Bassi
Pavia, Italy

Ganjar Fadillah
Department of Chemistry
Universitas Islam
Yogyakarta, Indonesia

Is Fatimah
Department of Chemistry
Universitas Islam
Yogyakarta, Indonesia

Seema Garg
Department of Chemistry
Amity Institute of Applied Sciences
Amity University
Noida, Uttar Pradesh, India

Erwann Guenin
Laboratoire TIMR UTC-ESCO
Centre de recherche de Royallieu
rue du docteur Schweitzer
Compiègne, Cedex, France

Abu Hashem
Nanotechnology and
Catalysis Research Centre
Institute for Advanced Studies
Universiti Malaya
Kuala Lumpur, Malaysia
and
Microbial Biotechnology Division
National Institute of Biotechnology
Ganakbari, Ashulia, Savar, Dhaka

Volker Hessel
School of Chemical Engineering and
Advanced Materials
The University of Adelaide
Adelaide, Australia

Manuel Varon Hoyos
School of Chemical Engineering and
Advanced Materials
The University of Adelaide
Adelaide, Australia

M. A. Motalib Hossain
Nanotechnology and
Catalysis Research Centre
Institute for Advances Studies
Universiti Malaya
Kuala Lumpur, Malaysia

Mohd Rafie Johan
Nanotechnology and Catalysis
Research Centre
Institute for Advanced Studies
Universiti Malaya
Kuala Lumpur, Malaysia

Chariya Kaewsaneha
School of Integrated Science and
Innovation
Sirindhorn International Institute of
Technology (SIIT)
Thammasat University
Pathum Thani, Thailand

Ravinder Kale
Fibres & Textile Processing
Technology Institute of Chemical
Technology
Mumbai, India

Bakht Mand Khan
Department of Applied Physical and
Material Sciences
University of Swat
Khyber Pakhtunkhwa, Pakistan

Sunny Khan
Department of Physics
Jamia Millia Islamia
New Delhi, India

Estelle Leonard
Laboratoire TIMR UTC-ESCO
Centre de Recherche de Royallieu
rue du docteur Schweitzer
Compiègne, Cedex, France

Magdalena Luty-Błocho
Faculty of Non-Ferrous Metals
AGH University of Science and
Technology
Krakow, Poland

Mohammad Al Mamun
Nanotechnology and Catalysis
Research Centre,
Institute for Advanced Studies
Universiti Malaya
Kuala Lumpur, Malaysia
and
Department of Chemistry
Jagannath University
Dhaka, Bangladesh

Ab Rahman Marlinda
Nanotechnology and Catalysis
Research Centre (NANOCAT)
Universiti Malaya
Kuala Lumpur, Malaysia

Sugandha Gupta
Bhagwan Parshuram Institute of
Technology
Indraprastha University
New Delhi, India

Won-Chun Oh
Department of Applied Physical and
Material Sciences,
University of Swat
Khyber Pakhtunkhwa, Pakistan

Pakorn Opaprakasit
School of Integrated Science and
Innovation, Sirindhorn International
Institute of Technology (SIIT)
Thammasat University
Pathum Thani, Thailand

Gani Purwiandono
Department of Chemistry
Universitas Islam
Yogyakarta, Indonesia

Suresh Sagadevan
Catalysis Research Centre
University of Malaya
Kuala Lumpur, Malaysia

Mehmood Shahid
School of Integrated Science and
Innovation, Sirindhorn
International Institute of
Technology (SIIT)
Thammasat University
Pathum Thani, Thailand

Ajeet Singh
Molecular Spectroscopy and
Biophysics Lab
Department of Physics
Deva Nagari College
Uttar Pradesh, India

N. B. Singh
Department of Chemistry and
Biochemistry School of Basic
Science & Research and
Research and Development
Sharda University
Greater Noida, India

Preeti Singh
Fibres & Textile Processing Technology,
Institute of Chemical Technology
Mumbai, India
and
Department of Textile and
Fibre Engineering,
Indian Institute of Technology
New Delhi, India

Kefayat Ullah
Department of Applied Physical and
Material Sciences
University of Swat
Khyber Pakhtunkhwa, Pakistan

Ravish Kumar Uppadhayay
Molecular Spectroscopy and
Biophysics Lab
Department of Physics
Deva, Nagari College
Uttar Pradesh, India

Ahmed Usman
Institute of Physics
Academia Sinica Academia Road
Nangang, Taipei, Taiwan

Dhanalakshmi Vadivel
Department of Chemistry
University of Pavia
Viale Taramelli and
INFN, Sezione di Pavia
Via Agostino Bassi
Pavia, Italy

Yasmin Abdul Wahab
Nanotechnology and Catalysis
Research Centre
Institute for Advanced Studies
Universiti Malaya
Kuala Lumpur, Malaysia

Wiyogo Prio Wicaksono
Department of Chemistry
Universitas Islam
Yogyakarta, Indonesia

Wiyogo Prio Wicaksono
Department of Chemistry
Faculty of Mathematics and
Natural Sciences
Univerisitas Islam Indonesia
Kampus Terpadu
Sleman, Yogyakarta, Indonesia

Om Prakash Yadav
Department of Physics
K. S. Saket P. G. College
Ayodhya, India

Praveen Yadav
Department of Physics
K. S. Saket P. G. College
Ayodhya, India

M. Zulfequar
Department of Physics
Jamia Millia
Islamia
New Delhi, India

1 Functional Nanomaterials Processing Methods

Ganjar Fadillah, Gani Purwiandono,
Wiyogo Prio Wicaksono, and Is Fatimah

CONTENTS

1.1 INTRODUCTION

The rapid development of industrialization, urbanization, and community development has led to a decrease in the quality of the environment and human health. Monitoring various pollutants is currently the main task in maintaining environmental quality. Various analytical techniques and materials have been developed to monitor environmental quality, one of which is sensor devices. It is undeniable that the development of sensors currently attracts much attention because it has an extensive and applicable application in industry and field analysis. In addition, various materials have also been developed, such as carbon-based [1,2], metal oxide nanocomposites [3,4], metal-organic frameworks (MOFs) [5], precious metals/alloys [6], polymers [7], etc. However, one of the most critical parameters in material preparation for sensors is the material processing technique.

In general, material processing techniques for sensor applications can be classified into two types: thick-film and thin-film processing. Thin films have thicknesses on the order of 0.1 μm (micrometers) or less, while thick films are thousands of times thicker, typically in the 10–50 μM range. Previous studies reported that differences in the characteristics of the resulting layer could affect the material characteristics and the responsiveness of the material as a sensor [8,9]. Thick and thin films can be

DOI: 10.1201/9781003263852-1

1

produced from differences in the fabrication and deposition of materials on a substrate. Compared to a thick film, the thin film material is more widely used for sensor applications because thin-film material produces a higher sensitivity value than thick film and bulk materials [10]. In addition, the thin nature of the sensor surface provides a relatively high surface area so that it can be deformed more easily. Vaishnav et al. (2015) developed the thin-film sensor base indium tin oxide for toluene sensing [11]. The thin-film sensor was prepared by thermal evaporation technique and the results showed that the prepared sensor enhances the sensitivity and selectivity for toluene with the detection range from 10–1000 mg L^{-1}. In a related study presented by Yadav et al. (2020), the authors studied the CdS-SnO$_2$ thick-film sensor for monitoring toluene gas [12]. They revealed that the sensitivity of the sensor was directly dependent on the crystal and grain size of the material. Small crystal size will increase the volume ratio so that the sensitivity value increases. This proves that thin film is more suitable for sensor applications because it has a small particle size and large surface area. Therefore, this chapter will discuss several methods for producing thick and thin films, especially those applied to sensors. Finally, this chapter discusses the basic principles, advantages, and disadvantages, and potential for future research, especially in sensor application.

1.2   THIN-FILM PROCESSING

1.2.1   Physical Vapor Deposition (PVD)

Physical vapor deposition (PVD) is a cost-effective and excellent method for depositing materials onto surfaces to form various morphologies of thin films or nanostructures that could be prepared by condensation of a vaporized material with a highly controllable vacuum techniques such as sputter deposition and thermal evaporation [13]. Besides sputtering dan thermal evaporation, PVD could employ laser ablation, electron beam, pulse laser, and vacuum arc-based emission to evaporate the materials [14]. Fundamentally, PVD is a typical vacuum coating process whereby atoms or molecules are vaporized from a liquid or solid source, then the vapor is transported in the vacuum or low-pressure condition, followed by a condensation process on a substrate to form a solid deposit as described in Figure 1.1.

Moreover, PVD offers superiority over solution-based processing techniques like electrodeposition, including direct control of aspect ratio, pure-high vacuum processing, ability of most of the inorganic and organic materials to be deposited, and being a greener method, as well as oriented attachment to the substrate [15]. PVD techniques are also excellent in reproducibility and demonstrate good stability in the sensing performance of the fabricated nanostructure [16]. Despite the photocatalytic and solar-cell technologies, the recent development of PVD techniques has been extensively explored for the fabrication of the functional micro-nano structures for the sensors. Besides the advantages, PVD techniques still face several disadvantages, including complexity in the fabricated nanostructure shape, relative expense, and the complexity of the process [17].

Cu et al. (2013) has employed the hot filament physical vapor deposition (HFPVD) techniques to fabricate the nanostructured carbon films that consist of nanosheets

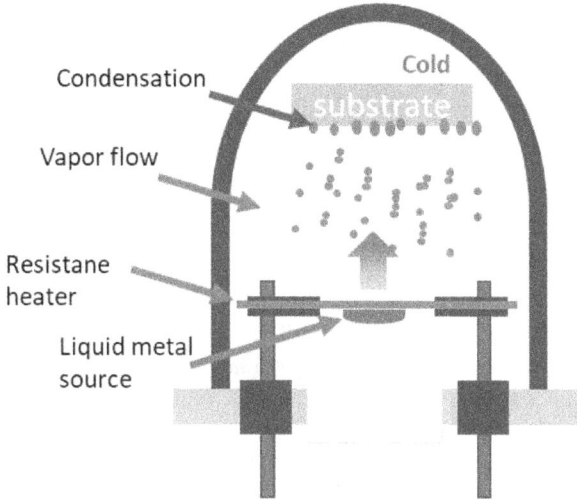

FIGURE 1.1 The schematic illustration of physical vapor deposition.

Source: (Created by Contributors; adapted from [13])

and nanohoneycombs structures on Si (100) substrates under the methane and in vacuum conditions, followed by Au deposition using sputtering techniques, producing the modified four-point electrodes for resistivity based humidity sensors [16]. The different conditions in the transportation of the vapored materials, such as in methane and vacuum condition, led to the formation of different nanostructures. The carbon nanosheets were successfully formed under the methane condition, while the carbon nanohoneycombs were grown in the vacuum condition. Furthermore, the performance of the HFPVD prepared carbon nanostructures were evaluated for the relative humidity (RH) sensing with the result as detailed in Figure 1.2B and 1.2C. It shows that the nanosheet structure performed better humidity sensing compared to nanohoneycomb that may be due to the small particles size as the site that enhances the adsorption of water vapor molecules. Durmanov et al. (2018) have employed the electron beam PVD (EBPVD) under vacuum conditions in the fabrication of thin silver (Ag) film with the folded surface structure containing pore-like nanoscale cavities and indentations on mica substrate for non–label selective virus detection [18]. The structure and particle size of the nanostructure materials prepared by EBPVD techniques could be variated by controlling the temperature, voltage, and exposure time. The SERS enhancement has been successfully achieved by the prepared Ag nanostructure compared to the aluminum (Al) plate, revealing the semi-regular array pore-like structure with a rough surface in between allowing the surface-enhanced Raman scattering (SERS) activity. The combination of PVD with the other deposition methods, such as glancing angle deposition (GLAD) techniques, has been studied by Horprathum et al. (2014) [19]. The combination of the techniques (PVD-GLAD) offers the fabrication of a well-controlled multi-dimensional morphology of the nanostructural thin film of silver (Ag) and oxide compounds, i.e., tungsten oxide

(WO_3), titanium dioxide (TiO_2), and indium tin oxide (ITO), by controlling the composition aspect ratio, porosity, and shape. Interestingly, the controlled morphology and its structure on both p-type silicon (100 mm) wafers and glass slides substrates show the different optical, electrical, and mechanical properties that are beneficial in the enhancement of the sensing activities. The fabricated WO_3 exhibited various nanostructures with different GLAD angles and rotation speeds. It shows that below 65°, the dense film WO_3 was formed, while above that temperature, columnar films were produced. For the rotation speed, fast speed led to the formation of vertically aligned nanorods, whereas spiral-shaped nanostructures were produced in the slow speed rotation. The evaluation performance of the different WO_3 nanostructures has been investigated in the NO_2 gas sensing. The result showed that the WO_3 nanorods yielded much-improved performance due to a larger effective surface area.

PVD techniques show good prospects in the future due to their advantages over their drawbacks. It is challenging to extensively investigate the potency of various substrate types and transportation conditions of vapored materials, deeply understand the relationship of the PVD process parameters to the morphology of the thin film or nanostructure, as well as provide an easier process.

1.2.2 ATOMIC LAYER DEPOSITION (ALD)

Atomic layer deposition (ALD) is a thin film deposition technique based on a chemical process in which a material is exposed to a reactant in the gas phase sequentially to form a thin layer. In this technique, the precursor will react with the material surface sequentially, one by one repeatedly, until all reactive sites on the surface are saturated and form a thin layer on the material's surface. Many studies have reported that the maximum amount of material deposited on the surface after a single exposure process (the so-called ALD cycle) is affected by surface–precursor interactions [20,21]. Varying the number of cycles during the deposition process makes it possible to grow material of uniform size and shape (Figure 1.2); therefore, the ALD method is considered one of the best deposition methods to produce fragile films by controlling the thickness and composition of the films at the atomic level. In addition, the advantages of this method are that it is possible to grow multilayer structures, and some thermally unstable precursors are still possible to be used for the deposition process because, in this method, the decomposition process is carried out at relatively low temperatures in the presence of a catalyst [22]. Therefore, this method has a broader application than chemical deposition because it can cause deposition on organic or biological substrates that generally degrade at high temperatures. However, this method has many limitations; namely, the substrate must have high purity, the reaction is slow, and it is necessary to remove excess precursors at the end of the process until the percentage of impurities is below 1%.

Zhao et al. (2022) fabricated the iron (Fe)-based MOF (PCN-333) thin film for selective and sensitive dopamine sensors by ALD [23]. They combined the gas and liquid approaches for obtaining the sensor. The electrochemical sensor test showed that the developed material has a high sensitivity for dopamine detection (4637 μA mM^{-1} cm^{-2}) with a wide linear range and low limit of detection at 0.50–140 μM and 0.14 μM, respectively. The ALD method provides a uniform dispersion size of

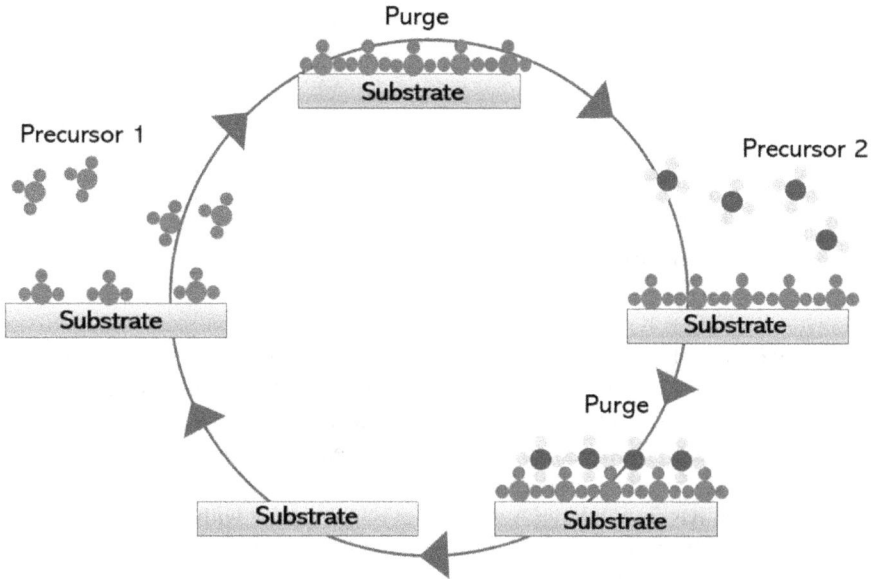

FIGURE 1.2 The repeating cycle of atomic layer deposition for controlling the size and shape of deposited materials.

Source: (Created by Authors)

material so it can improve the surface-active sites and transfer electron rate. Another result presented by Sun et al. (2021) prepared the flexible strain sensor based on ZnO/poly(vinylidene fluoride) (ZnO/PVDF) [24]. The ALD method can produce flexible sensors based on functional polymers due to its fabrication in a low-temperature process. Mainly some polymers which have low activity like PVDF it will be difficult to deposit onto the substrate; however, with the ALD methods, it is possible to carry out because the methods will trigger the activation of the functional group of polymer structure so that it will become easier to deposit onto the surface's substrate. Based on the previous studies, several challenges are still required to develop in the future such as a new precursor for selective and specific application of the sensor, fabricating two-dimensional structure with innovative ALD, and development of substrate materials for improving the amount of deposited film [22].

1.2.3 CHEMICAL DEPOSITION (CD)

Chemical deposition (CD) is one of the fabrication techniques for modifying the surface of a sensor. The material to be coated is reacted with different chemicals that allow a reaction to occur and form a layer as shown in Figure 1.3. Compared to physical deposition, chemical deposition has many advantages, such as simplicity, controlled composition, directional variations, up-scaling capability, and ultra-thin film growth. In this technique, the thickness of the layer Is strongly influenced by the deposition time. Kumar et al. (2021) synthesized the ZnS/p-Si for an ultraviolet

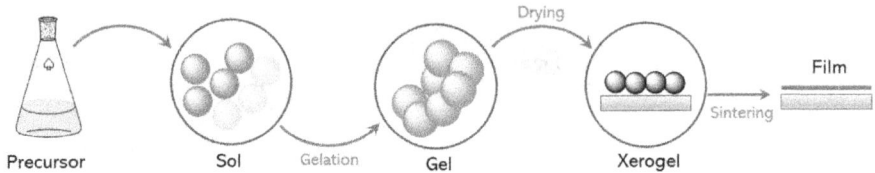

FIGURE 1.3   Illustration of chemical deposition by sol-gel methods.

Source: (Created by Authors)

sensor by chemical bath deposition [25]. The film was carried out on the surface of a silicon glass substrate with zinc acetate as the precursor at a constant temperature. The microscopic observations showed that the formed thin layer has an excellent homogeneity and crystallinity with grain boundaries and grains free of voids and is solid throughout the film. In addition, the UV sensor test described that the prepared sensor material has fast and high responses at 68.98 mA/W without biasing results. Although the CD has been widely reported and showed that the prepared film by this method has good sensitivity performance in various sensor applications, the CD techniques still have limitations such as low ability to control thin films and high density of structural defects. Reyes-Vallejo et al. (2021) studied the effect of deposition time and complexing agents on morphological, structural, electrical, and optical properties of Cu_2O (copper oxide) thin films [26]. Their study found that the long deposition time and increasing concentration of TEA (triethanolamine) complexing agents produced the thick layer film. Increasing the concentration of TEA can trigger a slowdown in the reaction rate because TEA can form complexes with copper ions, thereby promoting the growth of a thicker film and preventing Cu_2O from being deposited on the substrate surface. Due to limitations in regulating the thickness of the resulting film, other deposition methods have been developed, namely chemical vapor deposition (CVD) and electrochemical deposition (ED).

Several other methods for adjusting the layer thickness have been reported; for example, using spray-assisted layer by layer chemical deposition. Chen et al. (2021) developed spray-assisted chemical deposition for controlling the thickness of the copper (Cu) layer on the carbon fiber fabrics (CFFs) [27]. The thickness of the coating can be adjusted by regulation of the spraying process and the type of copper used. This deposition process showed that increasing spraying time could form micron-sized flocculant-coating consisting of Cu nanoparticles uniformly covering the surface. Furthermore, the thickness has been demonstrated to directly affect the performance of a sensor material with an increasing conductivity value up to 5.01×10^{-3} Ω and electron transfer rate.

1.2.3.1   Chemical Vapor Deposition (CVD)

In material for sensor applications, chemical vapor deposition (CVD) is a process by which chemical components in the vapor phase react to form a thin film layer on a surface. Due to the requirement that the reactants begin in the vapor phase, a chemical reaction is essential for thin-film development. When it comes to controlling the qualities of thin films created, our capacity to manage the gas-phase constituents, gas-phase physical conditions, solid surface, and the envelope that surrounds them all comes down to one thing: controllability. It is a sequential process that begins with the initial gas phase, goes through several quasi-steady–state subprocesses, and forms a solid film with a particular microstructure in its final state [28]. Figure 1.4 depicts a schematic representation of this sequence: (1) the initial step is gaseous surface diffusion; (2) adsorption of reactive species onto surfaces frequently happens after migration, and the reactants react on the surface, generally catalyzed; (3) surface desorption and dispersion of reaction byproducts; and the last (4) incorporation of condensed solid product into growing film microstructure [29,30].

CVD is the favored technique in many circumstances because of its several significant advantages. The following are some of the most important. Unlike with sputtering and other techniques, it is not limited to deposition in the direct line of sight (as is the case with evaporation and other processes). The coating in CVD can be applied to deep recesses, holes, and other complex three-dimensional shapes relatively quickly. For example, CVD tungsten can fill through holes in integrated circuits with an aspect ratio of 10:1. The deposition rate is excellent, and it is possible to generate thick coatings quite quickly (in some cases, centimeteres thick). The procedure is generally competitive with and, in some cases, more cost-effective than other processes in most situations. CVD equipment does not often require an ultrahigh vacuum and can be configured to accommodate a wide range of process variations [29]. This material's versatility allows for several compositional modifications during deposition, and it is easily capable of achieving co-deposition of elements or compounds.

On the other hand, CVD is flexible at 600°C and above; many substrates are not thermally stable at these temperatures. However, the introduction of plasma-CVD and metallo-organic CVD has helped mitigate this issue to some extent. The demand

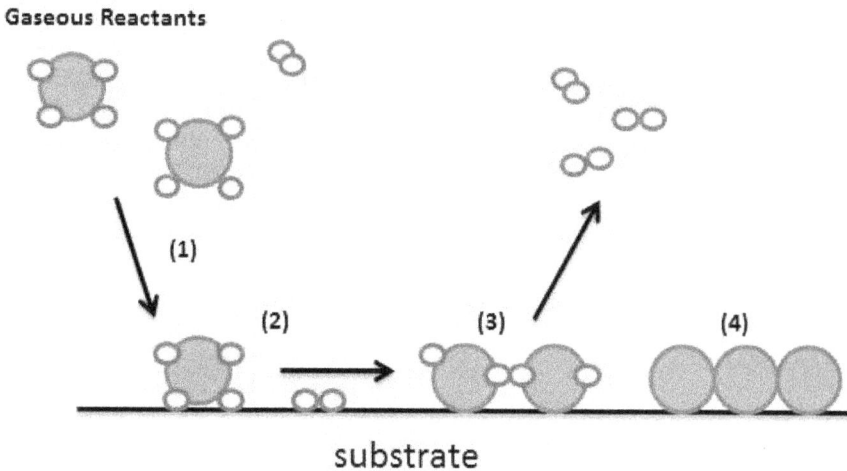

Gaseous Reactants

substrate

FIGURE 1.4 The growth pathways on the substrate surface during thermal CVD.

Source: (Created by Authors)

for chemical precursors (the beginning chemicals) with high vapor pressure—which is frequently hazardous and, in some cases, exceedingly toxic—is another downside of the process. Furthermore, the byproducts of the CVD reactions are poisonous and corrosive, and they must be neutralized, which can be a time-consuming and expensive procedure [29].

CVD has been applied in the development of material for sensor applications. Kuznetsov et al. (2021) used metal-organic chemical vapor deposition to manufacture a chemically aggressive fiber-optic sensor (MOCVD) using single-mode optical fiber with chemically etched cladding coated with thin tin dioxide layer (SnO_2). The separation of transverse electric (TE) and transverse magnetic (TM) lossy mode resonances (LMRs) was detected for resonances up to third-order at 345°C deposition temperature. The SnO_2-coated fiber-optic sensor is chemically resistant to acids (including aqua regia) and alkalis. The sensor's sensitivities to sulfuric, hydrochloric, and nitric acids were 34.8, 35.2, and 32.0 nm/M, respectively [31]. Luan et al. (2021) also created high-temperature conductive films sensors using SiBCN ceramics coated with polymer-derived ceramics (PDC) and CVD. The PDC plus CVD method produced a film with minimal flaws and strong conductivity at 1000°C. The overall decrease in conductivity is attributed to increased graphitization of free carbon in PDC-SiBCN [32]. Their research shows that CVD may create a protective and electrically conductive application for sensors.

1.2.3.2 Electrochemical Deposition (ED)

Electrochemical deposition (electrodeposition, ED) is a technique for producing thin and tight coating of conducting/semiconducting materials (metal, metal oxides, salt, organic, or metal-organic framework) onto the conductor/semiconductor substrate by simple electrolysis deposition reaction of the metal ions [33,34]. The technique requires the imposition of a direct current (DC) electrical field to drive the redox reaction in the electrodeposition cell set up, as shown in Figure 1.5. The cell may consist of two electrodes—that is, anode/counter electrode and cathode/working electrode—or three electrodes: counter, working, and the reference electrode [35]. Fundamentally, the deposition layer could be produced by the reduction reaction of the metal ions from the electrolyte onto the surface of the cathode (substrate). If the anode is the active electrode, the oxidation reaction will occur, whereas if the inert metal such as platinum is employed as the anode, the circuit will complete without oxidizing itself. The redox reaction in the electrodeposition process uses an anode with an active electrode as follows (Eq. 1–2).

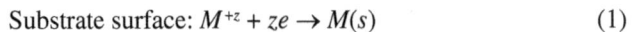

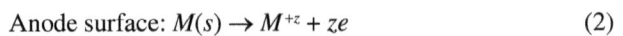

$$\text{Substrate surface: } M^{+z} + ze \rightarrow M(s) \tag{1}$$

$$\text{Anode surface: } M(s) \rightarrow M^{+z} + ze \tag{2}$$

Electrodeposition offers several advantages, such as the ability to produce hundreds of microns of conductive film thicker than compared to PVD and CVD, its low cost, its simplicity, its ability to be applied in the simultaneous coating process of large samples number as well as large size of the sample, and its ability to fabricate kinds of deposit structures while controlling the preparation parameters [23]. However,

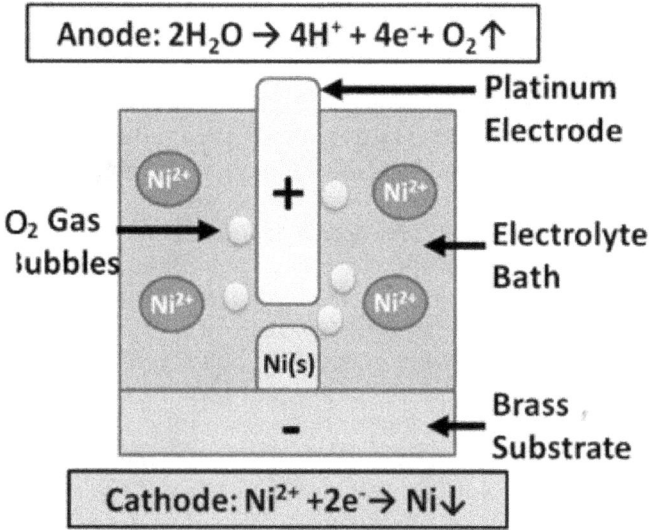

FIGURE 1.5   The illustration of electrochemical deposition of nickel.

Source: (Reprinted with Copyright permission from Elsevier B.V.) [36]

some disadvantages still obsess, such as requiring the absence of impurity in the electrolyte solution and being limited only for the conductive substrate.

Several operation parameters—including deposition current density, applied potential, temperature, pH, concentration, and composition of the solution, as well as electrodeposition time—could be controlled to fabricate a typical shape, thickness, and structure growth of deposits [22,37,38]. Recently, the electrodeposition technique extensively employed in the development of functional nanomaterials for sensors application, particularly for the fabrication of various nanostructured materials (NPs). Fundamentally, the NPs could be formed through nucleation and growth step from the metallic ion's reduction reaction driven by DC power supply. Interestingly, with this technique, the preparation of the NPs does not require a capping agent/surfactant/any dispersion agent [39]. The electrodeposit nanomaterials have been used for electrocatalysis, electrochemical sensing/biosensing, and surface plasmon resonance-based sensing [38]

Nia et al. (2017) have performed one-step simultaneous electrodeposition of mixture solution of graphene oxide (GO) and copper chloride (CuCl$_2$) to synthesize the composite of copper nanoparticles and reduced graphene oxide (CuNPs-rGO) as the electrocatalyst for non-enzymatic hydrogen peroxide sensor [40]. Before the electrodeposition step, the pre-treatment polishing process using alumina slurry toward bare glassy carbon electrode (GCE) was applied. Two electrodeposition routes have been employed for the fabrication of the modified bare GCE with CuNPs-rGO. The first route was using cyclic voltammetry (CV) with three cycles from 0--1.5 V vs saturated calomel electrode (SCE) at a scan rate of 10 mV/s. Second was an amperometry technique at a constant potential of −0.4 V vs. SCE was employed for 420 s. Interestingly, the electrodeposition route using the CV technique exhibited a higher reduction current in the H_2O_2 electrochemical sensing compared to the prepared modified electrode with amperometry technique due to the CV technique could produce the well distributed and smaller particle size of CuNPs. The CV-prepared CuNPS-RCG/GPE electrode provided high conductivity as well as a large surface area promoting the synergistic effect that affected the high electrocatalytic performance toward H_2O_2 reduction.

Siampour et al. (2020) have developed a new approach by using a pre-seeding step to improve the sensing performance of electrodeposition of gold (Au) nanostructure on fluorine-doped tin oxide (AuNSs/FTO) electrode for non-enzymatic glucose detection. The simple pre-seeding was carried out by vacuum thermal deposition of Au thin film, followed by annealing at 500°C [41]. The pre-seeding step on the substrate could influence the morphology of the Au growth fabricated by the electrodeposition technique, leading to increase of the electrochemical sensing performance. The pre-seeding process produced a dense array of semi-spherical Au nanoparticles, leading to a unique tiny hole of AuNSs was successfully fabricated after the electrodeposition step. The morphology could be variated by controlling the deposition potential, deposition time, and concentration of the electrolyte/substrate, and employing the surfactant.

In the future, electrodeposition has prospective as the technique for the mass fabrication of the modified electrode for various sensing applications, such as clinical diagnostics and environmental monitoring, as well as for industrial process control due to the simple procedure, could produce controllable thickness of deposit, as well as possible in preparing various unique morphology both by controlling the electrodeposition conditions and or applying the pre-treatment step, producing high sensitivity toward various analytes.

1.3   THICK-FILM PROCESSING

1.3.1   SCREEN-PRINTING

Screen-printing is a thick-film technology (TFT) that is widely used to fabricate sensors, especially in the development of biosensors. This method is generally carried out by deposition of a polymer, dielectric insulator, or metal conductor on an inorganic or polymeric substrate to increase sensitivity and selectivity. In general, there are two methods of screen-printing: (1) off-contact printing, a method is most used with one side of the part on the coated substrate—in this method, there is a gap between the substrate surface and the plate; and (2) contact printing, during which the whole screen is in contact with the media during the complete squeegee. The thick imprinting surface is possible to form several layers on a substrate with a similar serigraphic shape. Depending on the electrode application, several techniques for deposition of a sensitive layer on the surface of the sensor substrate have been reported, such as membrane immobilization, physical adsorption, covalent bonding between functional groups on the substrate surface, polymerization, etc. [42]. Surface molecule printing is generally carried out concerning the polymerization reaction on a solid substrate surface, as shown in Figure 1.6. Xia et al. (2021)

fabricated an electrochemical sensor-based surface molecularly imprinted polymer to determine 5-hydroxytryptamine [43]. The imprinting surface was prepared on the surface of MWCNTs@IL-Au with the composition of 5-hydroxytryptamine (5-HT), azobisisobutyronitrile (AIBN), and ethylene glycol dimethyl acrylate (EGDMA) as template, initiator, and cross-linking agent, respectively. The synthesized sensor showed that the modification with surface imprinting could improve the sensitivity and selectivity of the sensor for the detection of 5-HT due to its high electrostatic and hydrogen interaction between the analyte and the surface electrode. Guo et al. (2019) synthesized the capacitive pressure sensor-based carbon nanotubes (CNTs) layers [44]. The sensor was prepared by traditional etching with mixing solution between CNTs and polydimethylsiloxane (PDMS). Their study revealed that the composition between CNTs and PMDS played a crucial factor in improving sensitivity. Reported studies have proven that the sensitivity of electrodes prepared by screen-printing is strongly influenced by the composition of the coated material and the printing method on the substrate surface.

Surface imprinting is developed to detect several other biomolecular compounds and can also be used to detect some heavy metals. Bakhshpour and Denizli (2020) synthesized imprinting Cd(II) ions based surface plasma resonance (SPR) sensors [45]. The thick-film sensor was fabricated using poly(hydroxyethyl methacrylate) and Au nanoparticles (Au NPs) as modifier imprinting reagents with N-methacrylolyl-L-cysteine functional groups. Their study proved that the prepared sensors have an excellent sensitivity with the limit of detection (LOD) at 0.01 μg L^{-1}. Several screen-printing technologies have been developed, which help in increasing sensitivity and selectivity. In addition, this technology can reduce the cost of production and effectiveness for monitoring pollutants in the environment. Solid substrates based on carbon and gold nanoparticles such as graphene, graphite, fullerenes, and quantum dots have been reported and successfully printed on their surfaces. However, surface

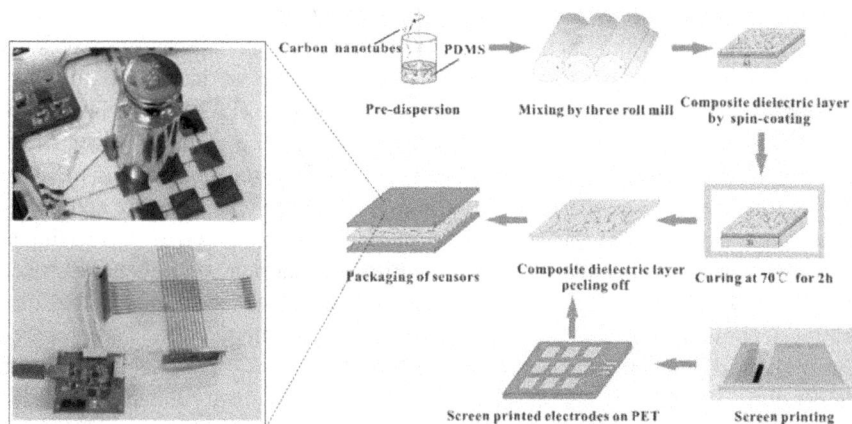

FIGURE 1.6   Illustration of screen-printing technique for capacitive sensor.

Source: (Reprinted with permission with MDPI Publisher) [44]

printing still has some limitations such as low stability and reproducibility, especially polymer-based surface printing [46]. Therefore, fabrication techniques are still an interesting topic to be developed in the future: (1) controlling the composition and thickness of the impinging layer on specific surface electrodes, (2) high reproducibility; and (3) potential to all types of solid substrates.

1.3.2 DIP-COATING

The dip-coating method, also known as the saturation or impregnation method, is one of the most popular methods to produce a uniform thin layer on a flat surface substrate by immersing the substrate in a coating material solution then lifting the substrate at a controlled speed, followed by a drying process to remove residual solvent. This dip-coating method has been widely developed for research purposes, especially in developing sensor materials, because of its easy and low-cost preparation [47]. However, this method still has weaknesses, such as the low consistency of the resulting layer thickness. Generally, this method consists of five main stages: immersion, start-up, deposition, drainage, and evaporation. In addition, several studies have reported that several parameters directly affect the thickness of the resulting film, such as the concentration and viscosity of the coating solution, immersion time, substrate withdrawal speed, and several cycles. Therefore, changes in the characteristics of the resulting layer will cause differences in sensor performance. Alrammous et al. (2019) fabricated the humidity sensor-based GO sheets on paper substrate [48]. The sensors were prepared by immersing paper-based substrates into GO suspensions solution with various concentrations and immersion times followed by drying at room temperature. The results showed that the prepared sensors have a different response to humidity. After comparison, the sensor with a porous structure, prepared by immersing in 2 mg mL^{-1} GO solution for 10 minutes, showed the highest sensing response to humidity until reaching 90% RH.

In contrast to other coating methods, this method has a broader application because it can be used for various substrates, such as plain (flat) and irregular (cylindrical) substrates. Therefore, this method is widely used for sensor fabrication, especially in fiber-based gas sensors. For example, Ke et al. (2021) fabricated the ammonia sensor-based polyaniline on carboxylated polyacrylonitrile (PAN) fabric [49]. PAN, which is hydrolyzed by alkaline conditions on the surface of the fiber substrate, can increase the loading of aniline on the surface. The sensing performance test showed that the prepared sensor has good sensitivity at room temperature and a wide linear range between 1.5–1500 ppm. Shen et al. (2018) fabricated SWCNT (single-walled carbon nanotube) and MWCNT (multi-walled carbon nanotube) fiber-based gas sensors decorated with ZnO QDs [50]. Carbon nanotubes were modified by dip-coating technique with the help of ultrasonic at room temperature. The prepared sensor showed good stability and response for NH_3, HCHO, and C_2H_5OH at room temperature. In addition, the developed sensor exhibited better mechanical resistance. Although this technique has shown good performance in sensor applications, the method still has a significant limitation: the weak interaction between the substrate and the coating materials [51]. Therefore, developing strategies to increase the adhesive power between coating materials and substrates is still an interesting topic to be developed. However, several strategies that have been developed include the manufacture of a superamphiphobic polymer substrate, a two-step coating process, and a combination of spray and dip-coating techniques [51,52].

1.3.3 SPIN-COATING

The spin-coating process is widely used to create uniform films with thicknesses of several micrometers. Because of its simplicity, the relatively low cost of the equipment utilized, and the excellent results obtained, it has risen to become one of the choices for most research centers across a wide range of disciplines, especially in material for the sensor [53]. Centrifugal deposition is a technique used in this process. It is carried out in three stages: deposition, spin-up, and evaporation, as shown in Figure 1.7. The material is deposited on the turntable in the first stage. Then the turntable is spun up and spun down in series, with the evaporation stage throughout the operation. Centrifugal force is used to spread the solution applied to the turntable. Following this stage, the applied layer is allowed to dry completely. It is feasible to achieve uniform evaporation of the solvent due to rapid rotation. Because of evaporation or simply drying, high volatile components of a solution are removed from the surface of the substrate, while the low volatile components of the solution stay on the surface's substrate [54]. The coating solution's viscosity and the coating machine's rotation speed determine the thickness of a deposited layer.

The advantages of the spin-coating approach include lower loss of materials and the fact that it is a very inexpensive method. A spinning plate is significantly less expensive than a vacuum deposition technique, and it is also easier to deposit the layer. It should be noted that this spin-coating technique has some drawbacks; for example, the difficulty in creating multiple layers (three or more layers), the presence of contaminants (such as traces of solvent or oxygen), and the inability to create fragile films (less than 10 nm) [55].

Based on its application, the spin-coating technique is frequently utilized in the material sensor to create a thick material layer for use in sensors. Ayyala and Covington (2021) used a photolithography-assisted spin-coated process to synthesize

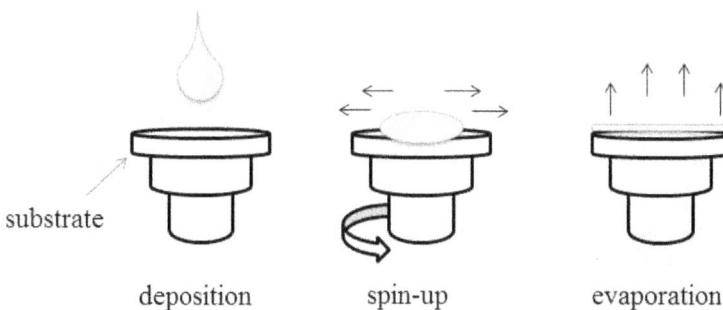

FIGURE 1.7   Illustration of the thick-film material sensor using a spin-coating process.

Source: (Created by authors)

a rapid and stable chemo-resistive nickel oxide (NiO)-based sensor for a wide variety of volatile organic chemicals. According to the results of the SEM investigation, the sensors had a thickness of 5–10 μm. It was discovered that the sensor response for the hydroxyl and carbonyl functional groups at the studied concentrations was significantly higher than the response for the other vapors [56]. Another study of $Cs_3Bi_2I_6Br_3$ for an ammonia-based sensor was synthesized by Jiao et al. (2021) using a simple liquid spin-coating technique. The sensor performance testing has shown the device's responses to ammonia with a thickness of 10 μm. The sensitivity response of ammonia up to 11.8–500 ppm, and the recovery process, repeatability, and enhanced stability after 14 days [57]. Koli et al. (2021) used the spin-coating method to produce a film sensor using sol-gel $LaCrO_3$ perovskite for environmental applications. The morphology of prepared films revealed that the thickness of the film was about 5 μm. The $LaCrO_3$ sensor is susceptible to CO and CO_2 gases at 200°C–300°C. Additionally, the modified $LaCrO_3$ sensor was shown to be effective in terms of gas response, selectivity, humidity sensing performance, response recovery, and reusability performance, among other things. When it came to sensing qualities, the $LaCrO_3$ sensor performed exceptionally well for all of those stated reasons. The relative humidity experiment was conducted at various humidity levels ranging from 10–90% [58]. Their study revealed a potential method for synthesizing material-based-sensor applications.

1.4   CONCLUSION AND FUTURE PROSPECTIVE

The sensor's sensitivity is directly affected by the dimensions and thickness of the layer of material to be used as devices. Determination of the type of thickness of the layer depends on the sensor application. However, most current thin films are widely used as sensors because they have better sensitivity due to a decrease in particle size and an increase in the surface area of the material. As a result, thick and thin film fabrications have been widely reported, such as physical and chemical vapor deposition, electrochemical deposition, atomic layer deposition, screen-printing, dip-coating, spin-coating, etc. Each of these fabrication techniques has some advantages and disadvantages. Some challenges in future research are still interesting to be studied; for example, improving the chemical interaction between coating material with a substrate and improving the selectivity, sensitivity, and stability of sensor materials.

REFERENCES

[1] J.V. Maciel, G.D. da Silveira, A.M.M. Durigon, O. Fatibello-Filho, D. Dias, Talanta 236 (2022) 122881.
[2] G. Fadillah, S. Triana, U. Chasanah, T.A. Saleh, Sensing and Bio-Sensing Research 30 (2020) 100391.
[3] T.A. Saleh, G. Fadillah, TrAC Trends in Analytical Chemistry 120 (2019) 115660.
[4] G. Fadillah, W.P. Wicaksono, I. Fatimah, T.A. Saleh, Microchemical Journal 159 (2020) 105353.
[5] J. Zhang, L. Gao, Y. Zhang, R. Guo, T. Hu, Microporous and Mesoporous Materials 322 (2021) 111126.

[6] F.-C. Chang, J.-W. Huang, C.-H. Chen, Journal of Alloys and Compounds 843 (2020) 155885.

[7] S. Matindoust, G. Farzi, M.B. Nejad, M.H. Shahrokhabadi, Reactive and Functional Polymers 165 (2021) 104962.

[8] M. Moznuzzaman, M.R. Islam, I. Khan, Sensing and Bio-Sensing Research 32 (2021) 100419.

[9] S.K. Al-Hayali, A.M. Salman, A.H. Al-Janabi, Measurement 170 (2021) 108703.

[10] B.C. Sertel, N.A. Sonmez, M.D. Kaya, S. Ozcelik, Ceramics International 45 (2019) 2917.

[11] V.S. Vaishnav, S.G. Patel, J.N. Panchal, Sensors and Actuators B: Chemical 210 (2015) 165.

[12] P. Yadav, A.K. Sharma, S.K. Yadav, A.K. Vishwakarma, L. Yadava, Materials Today: Proceedings 38 (2021) 2792.

[13] P.A. Pandey, G.R. Bell, J.P. Rourke, A.M. Sanchez, M.D. Elkin, B.J. Hickey, N.R. Wilson, Small 7 (2011) 3202.

[14] S.M. Rossnagel, Journal of Vacuum Science & Technology A 21 (2003) S74.

[15] F. Yu, J. Xu, H. Li, Z. Wang, L. Sun, T. Deng, P. Tao, Q. Liang, Progress in Natural Science: Materials International 28 (2018) 28.

[16] J. Chu, X. Peng, P. Feng, Y. Sheng, J. Zhang, Sensors and Actuators B: Chemical 178 (2013) 508.

[17] A.S.H. Makhlouf, in: A.S.H. Makhlouf, I. Tiginyanu (Eds.), Nanocoatings and Ultra-Thin Films, Woodhead Publishing, Oxford, UK, 2011, p. 3.

[18] N.N. Durmanov, R.R. Guliev, A.V. Eremenko, I.A. Boginskaya, I.A. Ryzhikov, E.A. Trifonova, E.V. Putlyaev, A.N. Mukhin, S.L. Kalnov, M.V. Balandina, A.P. Tkachuk, V.A. Gushchin, A.K. Sarychev, A.N. Lagarkov, I.A. Rodionov, A.R. Gabidullin, I.N. Kurochkin, Sensors and Actuators B: Chemical 257 (2018) 37.

[19] M. Horprathum, P. Eiamchai, J. Kaewkhao, C. Chananonnawathorn, V. Patthanasettakul, S. Limwichean, N. Nuntawong, P. Chindaudom, AIP Conference Proceedings 1617 (2014) 7.

[20] L.M. Mohlala, T.-C. Jen, P.A. Olubambi, Procedia CIRP 93 (2020) 9.

[21] K.-J. Qian, S. Chen, B. Zhu, L. Chen, S.-J. Ding, H.-L. Lu, Q.-Q. Sun, D.W. Zhang, Z. Chen, Applied Surface Science 258 (2012) 4657.

[22] H. Xu, M.K. Akbari, S. Kumar, F. Verpoort, S. Zhuiykov, Sensors and Actuators B: Chemical 331 (2021) 129403.

[23] Z. Zhao, Y. Kong, G. Huang, C. Liu, C. You, Z. Xiao, H. Zhu, J. Tan, B. Xu, J. Cui, X. Liu, Y. Mei, Nano Today 42 (2022) 101347.

[24] C. Sun, J. Zhang, Y. Zhang, F. Zhao, J. Xie, Z. Liu, J. Zhuang, N. Zhang, W. Ren, Z.-G. Ye, Applied Surface Science 562 (2021) 150126.

[25] A. Kumar, M. Kumar, V. Bhatt, S. Mukherjee, S. Kumar, H. Sharma, M.K. Yadav, S. Tomar, J.-H. Yun, R.K. Choubey, Sensors and Actuators A: Physical 331 (2021) 112988.

[26] O. Reyes-Vallejo, J. Escorcia-García, P.J. Sebastian, Materials Science in Semiconductor Processing 138 (2022) 106242.

[27] D. Chen, C. Liu, Z. Kang, Colloid and Interface Science Communications 40 (2021) 100365.

[28] M.S. Hammond, Chemical Vapor Deposition 2 (1996) 210.

[29] H.O. Pierson, in: H.O. Pierson (Ed.), Handbook of Chemical Vapor Deposition (CVD) (Second Edition), William Andrew Publishing, Norwich, NY, 1999, p. 36.

[30] J.-H. Park, T. Sudarshan, Chemical vapor deposition, ASM International, Los Angeles, USA, 2001.

[31] P.I. Kuznetsov, D.P. Sudas, E.A. Savelyev, Sensors and Actuators A: Physical 321 (2021) 112576.

[32] X. Luan, S. Gu, Q. Zhang, L. Cheng, R. Riedel, Sensors and Actuators A: Physical 330 (2021) 112824.

[33] X. Zhang, K. Wan, P. Subramanian, M. Xu, J. Luo, J. Fransaer, Journal of Materials Chemistry A 8 (2020) 7569.

[34] S. Prakash, J. Yeom, in: S. Prakash, J. Yeom (Eds.), Nanofluidics and Microfluidics, William Andrew Publishing, Scarborough, ON, Canada, 2014, p. 87.

[35] G. Fadillah, F. Ariani, AIP Conference Proceedings 2370 (2021) 050001.

[36] N. Manukyan, A. Kamaraj, M. Sundaram, Procedia Manufacturing 34 (2019) 197.

[37] A.M. Tarditi, M.L. Bosko, L.M. Cornaglia, in: M.S.J. Hashmi (Ed.), Comprehensive Materials Finishing, Elsevier, Oxford, 2017, p. 1.

[38] D. Tonelli, E. Scavetta, I. Gualandi, Sensors 19 (2019) 1186.

[39] L. Feng, X. Sun, S. Yao, C. Liu, W. Xing, J. Zhang, in: W. Xing, G. Yin, J. Zhang (Eds.), Rotating Electrode Methods and Oxygen Reduction Electrocatalysts, Elsevier, Amsterdam, 2014, p. 67.

[40] P. Moozarm Nia, P.M. Woi, Y. Alias, Applied Surface Science 413 (2017) 56.

[41] H. Siampour, S. Abbasian, A. Moshaii, K. Omidfar, M. Sedghi, H. Naderi-Manesh, Scientific Reports 10 (2020) 7232.

[42] G. Fadillah, O.A. Saputra, T.A. Saleh, Trends in Environmental Analytical Chemistry 26 (2020) e00084.

[43] Y. Xia, Y. Wang, M. Zhang, F. Zhao, B. Zeng, Microchemical Journal 160 (2021) 105748.

[44] Z. Guo, L. Mo, Y. Ding, Q. Zhang, X. Meng, Z. Wu, Y. Chen, M. Cao, W. Wang, L. Li, Micromachines 10 (2019) 715.

[45] M. Bakhshpour, A. Denizli, Microchemical Journal 159 (2020) 105572.

[46] A.M. Musa, J. Kiely, R. Luxton, K.C. Honeychurch, TrAC Trends in Analytical Chemistry 139 (2021) 116254.

[47] C. Zhou, N. Shi, X. Jiang, M. Chen, J. Jiang, Y. Zheng, W. Wu, D. Cui, H. Haick, N. Tang, Sensors and Actuators B: Chemical 353 (2022) 131133.

[48] R. Alrammouz, J. Podlecki, A. Vena, R. Garcia, P. Abboud, R. Habchi, B. Sorli, Sensors and Actuators B: Chemical 298 (2019) 126892.

[49] F. Ke, Q. Zhang, L. Ji, Y. Zhang, C. Zhang, J. Xu, H. Wang, Y. Chen, Composites Communications 27 (2021) 100817.

[50] Z. Gao, Z. Lou, S. Chen, L. Li, K. Jiang, Z. Fu, W. Han, G. Shen, Nano Research 11 (2018) 511.

[51] P. Muthiah, B. Bhushan, K. Yun, H. Kondo, Journal of Colloid and Interface Science 409 (2013) 227.

[52] H. Wang, H. Zhou, A. Gestos, J. Fang, T. Lin, ACS Applied Materials & Interfaces 5 (2013).

[53] B.S. Yilbas, A. Al-Sharafi, H. Ali, in: B.S. Yilbas, A. Al-Sharafi, H. Ali (Eds.), Self-Cleaning of Surfaces and Water Droplet Mobility, Elsevier, Cambridge, USA, 2019, p. 45.

[54] A. Mishra, N. Bhatt, A.K. Bajpai, in: P. Nguyen Tri, S. Rtimi, C.M. Ouellet Plamondon (Eds.), Nanomaterials-Based Coatings, Elsevier, Oxford, UK, 2019, p. 397.

[55] A. Boudrioua, M. Chakaroun, A. Fischer, in: A. Boudrioua, M. Chakaroun, A. Fischer (Eds.), Organic Lasers, Elsevier, Oxford, UK, 2017, p. 49.

[56] S.K. Ayyala, J. Covington, Chemosensors 9 (2021) 247.

[57] W. Jiao, J. He, L. Zhang, Journal of Alloys and Compounds 895 (2022) 162561.

[58] P.B. Koli, K.H. Kapadnis, U.G. Deshpande, U.J. Tupe, S.G. Shinde, R.S. Ingale, Environmental Challenges 3 (2021) 100043.

2 Functional Nanomaterials for Potential Applications

Isha Arora, Seema Garg, Harshita Chawla,
Amrish Chandra, Suresh Sagadevan,
and M.M. Abdullah

CONTENTS

2.1   INTRODUCTION

Advancement in industrialization, leading to toxic byproducts, has altered the environment by unleashing a distinct variety of toxins and emission of hazardous gases into the atmosphere. In this regard, conventionally implemented strategies such as immobilization, biological and chemical oxidation, and incineration have been largely employed to treat several types of organic and toxic industrial contaminants. In parallel, the massive

DOI: 10.1201/9781003263852-2

development of nanotechnology has sparked over the years. Nanomaterials (NMs), due to their nanoscale sizes, have augmented a new prospect for their several industrial and environmental applications, including sewage treatment and removal of hazardous contaminants. Confronting the subject of treatment of hazardous contaminants from the environment, NMs delivering unique optical, magnetic, or electrical properties become a fundamental key for environmental remediation.

Lately, the emphasis has largely shifted ahead from bulky to NMs, leading to substantial advances in nanoscience and technology for the development of NMs. It is important for us to understand that there is a difference in bonding in a small sized metal and that of a bulk semiconductor cluster. Therefore, the distinct physicochemical material characteristics, that were unattainable in the conventional bulk matrix, drew several research scholars and scientists to elevate their attention toward the field of nanoscience and technology (Abdullah et al., 2017; Chenab et al., 2020; Mansoori & Soelaiman, 2005). Due to the presence of superior surface to volume ratio, NMs offer much elevated, effective, and heightened physical attributes than that of bulk materials.

There are the following three major reasons for distinct superior properties in NMs than that of their bulk materials.

1. *Much greater surface area–to-volume ratio than their conventional forms:* The surface area–to-volume ratio of NMs is extraordinarily high, leading to greater surface area. As a result, the characteristics of bulk materials can be possibly improved or modified.
2. *Quantum effect:* The nanometer sizes of NMs also have spatial confinement effect on the materials. The energy band structure and charge carrier density in the materials can be modified quite differently from their bulk and in turn will modify the electronic and optical properties of the materials.
3. *Reduced imperfections:* nanostructures and NMs favor a self-purification process in that the impurities and intrinsic material defects move near the surface on thermal annealing which in turn increases the perfection in materials and thus affects properties of NMs.

In today's world, nanotechnology, NMs, and functional NMs are completely in charge for sustainability of humankind. The utility of NMs and functionalized NMs are currently being studied with majestic scientific and academic rigor. It has been highly observed that despite conventional ways for addressing environmental challenges, optimizing physical attributes with diverse functionalization, provides remarkable advancements to the pristine NMs. Additional desirable physical characteristics—including size, morphology, porosity, shape, and chemical composition—can all be established by further functionalizing or optimizing NMs. Not only are physical features improved, but NMs also possess the ability to exploit the target contaminant's specific surface chemistry by the means of functionalizing or anchoring the functional groups, as compared to conventional approaches, focused for remediation processes (Aashima & Mehta, 2020). Functionalization of a nanoparticle (NP) can be simply defined as the surface modification of the NP, such as conjugation of chemicals or biomolecules onto the surface of bio-NP—such as folic acid, biotin molecules, oligo nucleotides, peptides, antibodies, etc.—to improve characteristics and strike the degradable target with high accuracy.

Moreover, the functionalized NM also introduces unique physical characteristics, including anti-corrosion, anti-agglomeration, and non-invasive qualities (Thiruppathi et al., 2017). NMs must be functionalized to obtain and utilize their hidden potential in distinct industrial sectors. Several unique hidden features of NPs can be successfully elevated via simple surface functionalization or modification, including hydrophilicity, hydrophobicity, conductivity, and corrosion resistance. In several scientific works, it has been observed that NMs with functional properties can be specifically employed based on 'lab in a chip' strategy, which in turn ensures help with authentic evaluation of human health, as well as several research studies related to eco-friendly energy production technology (Crutchley, 2016; Karim, 2020). To improve performance and reduce toxic effects, chemically functionalized NPs—viz. TiO_2, Al_2O_3, Fe_2O_3, etc.—have been successfully added by researchers into organic/epoxy coatings. Additionally, scientists have also achieved improved precision in targeted medication delivery and active cellular drug absorption by employing multifunctional NMs in clinical investigations. These multifunctional NMs can be achieved by simple conjugation of NM with the chemical species. Presently, several researchers have turned their attention to the polymer nanocomposites. NMs, homogeneously dispersed in a polymer matrix and organic solvents, can successfully produce new range of functionalized nanocomposites offering better efficacy and improved thermal, mechanical, optical, and electrical properties, and can exploit the full potential of NMs for a successive range of applications including paints, ceramics, coatings, drug delivery, etc. (Domun et al., 2015; Fernando, 2009; Kumar & Sinha Ray, 2018; Mout et al., 2012).

Functionalized NMs, by showcasing their prestigious advantages for human health and environmental remediation, offer a numerous range of applications in variety of sectors including the catalysis industry, wastewater treatment plants, soil pollution remediation, air pollution remediation, the agricultural sector, the agri-food industry, the pharmaceutical industry, the biomedical sector, electronics, the energy sector, electrical applications, environmental remediation, the surfaces and coating industries, the textile industries, cosmetics, the aerospace sector, the sports sector, and many more.

This chapter will specifically focus on distinct general characteristics of functionalized NMs and will answer several questions including how an NM can be functionalized by the help of various advanced technologies/strategies, and what is the current industrial overview and future aspects of functionalized NM. Furthermore, we will also discuss wide range of applications offered by functionalized NMs.

2.2 DIFFERENCE BETWEEN FUNCTIONAL AND STRUCTURAL NMS BASED ON THEIR GENERAL CHARACTERISTICS

Based on their distinct qualities, solid nanoparticles can be categorized into structural and functional NMs.

1. *Functional NMs* have always possessed special features—which include electrical, thermal, magnetic, or photonic capabilities—where as *structural NMs*' implementation objective is focused more on their strength and durability.
2. *Functional NMs* generally depend on the electronic configuration and electron rotation of the material.

3. *Structural NMs* are highly influenced by atom bonding, such as metal bonds, ionic bonds, covalent bonds, hydrogen bonds, and microstructures such as crystal structures, grain size, morphology, dislocation substructure, and second phase properties (Pan et al., 2020).

2.3 STRATEGIES TO FUNCTIONALIZE A NANOMATERIAL (NM)

To extract the maximum hidden capability of an NM for biomedical, nanocomposite, sensing, bioimaging, energy conversion, electrical, and other applications, optimization of NPs, dispersion in complex matrices and interactions among NPs with other biomolecules or chemical molecules is vastly suggested. This task can be accomplished by adequately surface functionalizing and modifying the structure of NMs, which optimizes their engagement with their surroundings. The colloidal stability, dispersion, and controlled assembly of nanostructures are all regulated by their functionalization. Attributes including hydrophobicity, hydrophilicity, conductivity, and corrosivity can be easily introduced into NP surfaces by altering the surface surroundings (Kumar & Sinha Ray, 2018; Mout et al., 2012), while integrating surface functionality to NMs through in situ formation and post-formulation amendments is a difficult task. Hence, as a result, researchers are presently equipped with the advanced technologies and toolkits for performing the functionalization technique. Advanced strategies that are highly introduced nowadays include incorporation of surfactants, polymers, small molecules, polymer, dendrimers, biomolecules, and inorganic materials into NM to functionalize them. The subject of NM interface modification and functionalization has received much attention and inquiry in recent years (Kumar & Sinha Ray, 2018).

Numerous established synthesis strategies for functionalization of NMs include the following.

2.3.1 CHEMISORPTION

The use of chemisorption such as incorporation of thiol groups at the surface of NPs, and physisorption viz. milling and/or mixing.

2.3.2 ELECTROSTATIC INTERACTIONS

Negatively charged NPs attract positively charged nanostructured molecules, or vice-versa.

2.3.3 COVALENT INTERACTIONS

Covalent functionalization permits distinct chemical species to be coupled to NMs by forming covalent interactions. Biomolecules, tiny organic molecules, polymers, and inorganic compounds are commonly covalently bonded to the surface of NMs using this strategy to provide improved dispersion, hydrophobicity, reactivity, high colloidal stability, and diverse characteristics. This strategy is a highly appealing technique as it modifies common NMs, such as metal oxides nanostructures, two-dimensional (2D) NMs viz. graphene oxide (GO), carbon nanotubes (CNTs), and nanoclays to fine tune materials for a range of applications, including bioimaging, environmental remediation, packaging, cosmetics, catalysis, and many more. Using heterobifunctional cross-linker molecules, several functionalization techniques have been established. The most widely used instances of cross-linker molecules for covalent functionalization are organofunctional alkoxysilanes (3-aminopropyl)-triethoxysilane (APTES), (3-glycidoxypropyl)-dimethyl-ethoxysilane (GPMES), (3-mercaptopropyl)-trimethoxysilane (MPTMS), and others), glutaraldehyde (GA), N-hydroxysuccinimide (NHS), and 1-ethyl-3- (EDC) chemistry (Stamov et al., 2011; Treccani et al., 2013). The covalent functionalization strategy includes few basic reactions.

1. Reactions for successful immobilization of biomolecules in case of salinization and GA include the following steps:
   i. Oxide surface salinization using APTES.
   ii. Amine-terminated substrate reaction with GA.
   iii. Incorporating biomolecules via covalent bonds.
   iv. Imine bond generation among biomolecule and substrate.

2. Reactions for successful immobilization of biomolecules in case of EDC/NHS methods include the following steps:
   i. Reaction of EDC with carboxyl group of the biomolecule.
   ii. Formation of unstable intermediate (O-acylisourea).
   iii. Reaction of NHS with intermediate.
   iv. Formation of stable amine reactive NHS ester.
   v. Reaction of amine-terminated surface with NHS ester.
   vi. Formation of stable amine bond.
   vii. Biomolecule immobilization.

These techniques incorporating the cross-linker biomolecules can be used for permanent functionalization of peptides, oligonucleotides, and several other NPs on activated surfaces via covalent interactions between amine-terminated/aldehyde substrates and carboxyl/amine/hydroxyl groups of guest molecules.

2.3.4 NON-COVALENT INTERACTIONS

Non-covalent interactions are specifically based on receptor-ligand systems and weak interactions. Non-covalent functionalization requires both attractive and repulsive forces between molecules. Chemisorption and/or physisorption can be used to adsorb guest molecules onto the surface of NMs. Several inorganic and organic systems used for identification, association, and monitoring contain non-covalent interactions. The reversibility of non-covalent functionalization, which leads to a self-correcting mechanism, is an appealing characteristic. Mechanisms with great reactivity to physical factors such as pH and temperature can be constructed using non-covalent interactions.

Non-covalent functionalization employs both the surface charge of the NM and functional mediety. The primary step of beginning with non-covalent functionalization strategy is the inculcation of the functional material over the surface of the NP by the backing of non-covalent interactions including van der Waals interactions, hydrogen bonding, and electrostatic interactions (Kanth et al., 2020). Overall, NMs with increased reactivity, miscibility, mechanical stability, bio-compatibility, catalytic performance, and sensing behavior can be successfully made by non-covalent functionalization (Georgakilas et al., 2016; Gupta et al., 2002; Lehn, 2002).

2.3.5 INTRINSIC SURFACE ENGINEERING INCLUDING HETEROATOM INCORPORATION AND DEFECT ENGINEERING

Intrinsic surface engineering can be defined as variations to the crystal lattice of NMs at the atomic scale. Heteroatom integration or defect engineering can be used to accomplish this. The electronic, catalytic, and morphological properties of NMs are altered by doping of electron-rich and electron-deficient elements. Doping is another useful strategy for altering the characteristics of an NM by intentionally doping an NM with another substance. Outgrowing the hidden potential of an NP by the doping strategy can be done by two methods. The first method is to mount the doping ingredient throughout the entire particle during the synthesis of NM. However, doping a substance during the synthesis of NM does not involve surface modification entirely, because some of the dopant gets accumulated over the surface of the NP; this in turn could modulate the characteristics of newly synthesized NP as compared to the pristine one. The second method of doping is to dope a tiny amount of substance over the surface of prevailing NP without producing a new formal coating (Deng et al., 2009; Hunt, 2020; Tong et al., 2008). Dopants also cause crystal lattice disruption and variations in band widths and lengths, leading to the formation of several binding sites. Dopants can be incorporated into crystal structures through cation and anion replacement or covalent interaction with external atoms. Numerous in situ and post-treatment synthesis techniques for introducing heteroatoms into NMs have been devised (Kong et al., 2014; Kumar et al., 2016).

2.4 APPLICATIONS

2.4.1 WASTEWATER TREATMENT: PHOTODEGRADATION OF ORGANIC POLLUTANTS

Photoadsorption and photodegradation are two major processes that are employed on larger scale for wastewater treatment. With the emerging field of nanotechnology, functionalization of NMs to produce new advanced and smart NPs are highly applicable in both adsorption and degradation processes for removal of contaminants from wastewater. Photodegradation process is generally considered to possess better benefits than the adsorption process. In recent decades, researchers have faced a tremendous number of challenges for removal of heavy metal to dye degradation. To overcome such downsides, solar/artificial light harvesting strategy for photocatalytic degradation of contaminants in water sources has lately proven to be an effective method. Contaminant transport from the aqueous phase to the photocatalytic surface, dye adsorption, production of reactive oxygen radicals, surface oxidation of adsorbed molecules, and eventually desorption of reaction products, is all part of the photocatalytic degradation of contaminant removal from wastewater. Photocatalytic process offers several benefits, as well as disadvantages. One of the most observed downsides is ultrafast recombination of charge carrier species and consequent loss of photoactive sites, which further leads to a significant decrease in dye degradation effectiveness. During the deactivation of active sites, as the intermediate byproducts accumulate due to partial oxidation of organic molecules, the process significantly limits the photon transit toward the surface of the catalyst due to physical obstruction of active sites. Periodic repetition of the photocatalyst deactivation concludes the NMs to lack their efficiency and functionality. However, it has been observed that regeneration of photoactive sites can be achieved by exposing the NMs to the ultraviolet (UV) spectrum and water-rich environments concomitantly to rehydrate the surface of photocatalyst, or atmospheric oxygen injection to produce a greater number of reactive radicals (Boulamanti et al., 2008; Chenab et al., 2020; Chiam et al., 2020; Kaewgun & Lee, 2010; Kumar, 2017; Moradi et al., 2017; Pandey et al., 2020; Wu et al., 2017). Dyes have been established as among the most serious water pollutant concerns, and their unprocessed sewage effluents into the surroundings have caused widespread controversy for wastewater degradation. The methods for eliminating dyes, currently in use, can be divided into the following three categories.

1. *Physical approaches* = surfactants and adsorption.
2. *Chemical approaches* = hydrogen peroxide oxidation, Fenton's method.
3. *Biological approaches* = algae and bacteria.

There has been a surge of attention in launching innovative nanoadsorbents exhibiting excellent adsorption capability in recent years. As an instance, Figure 2.1 depicts that pursuant to the functionalization of β-cyclodextrin, the adsorption capability of zero valent iron NPs (Fe^0) processed by Ferula persica extract, for the elimination of crystal violet (CV) was reported to be 99.8%, as compared to the conventional chemically synthesized Fe^0 NPs. The spike in the proportion of active sorption sites of nano adsorbent is thought to be responsible for such a high decolorization efficiency (Nasiri et al., 2019).

Concluding the water pollution remediation concern, numerous advanced functionalized nanophotocatalysts—viz. carbon nanotube and carbon nanorod-based (Awfa et al., 2019; Azzam et al., 2019; Bhati et al., 2016; Fathy et al., 2017; Vadivel et al., 2016), graphene oxide and reduced graphene oxide, and graphene quantum dot-based (Anjum et al., 2019; Pan et al., 2015; Pérez-Ramírez et al., 2016; Vadivel et al., 2014; Yang et al., 2015)—have been immensely fabricated and shown tremendous results in photocatalytic degradation of dyes, pigments, or other organic contaminants from wastewater.

2.4.2   AIR POLLUTION

Air pollution is generally caused by physical as well as chemical contaminants emitted into the natural atmosphere. According to various scientific studies, air contaminants are associated with numerous ailments such as lung- and heart-related

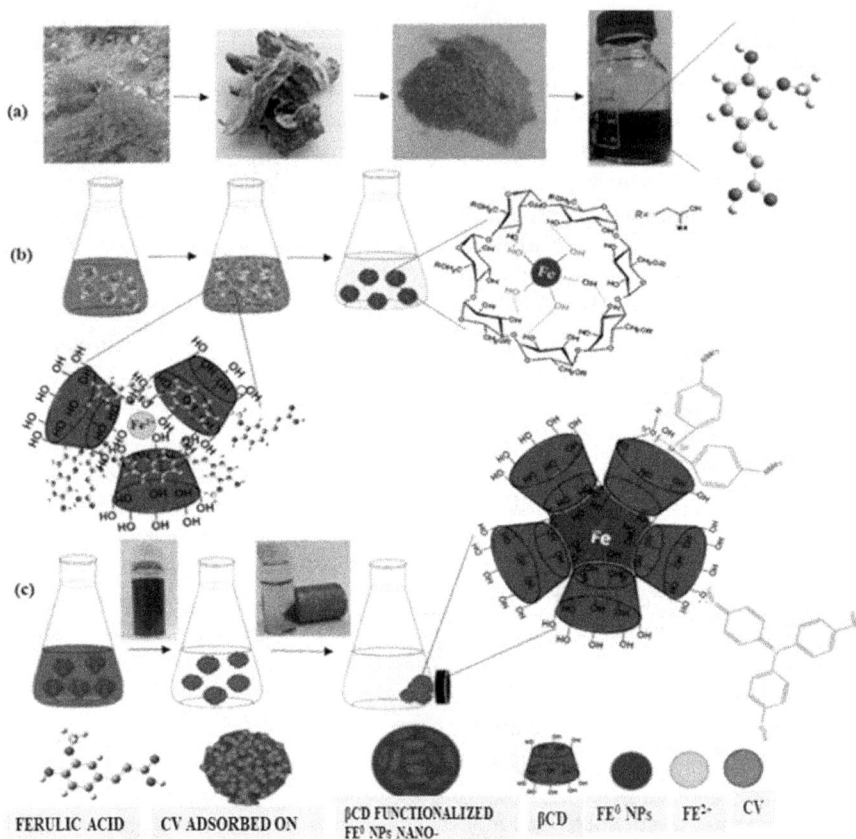

FIGURE 2.1 Schematic illustration for (A, B) synthesis of biogenic β-cyclodextrin function-alized Fe0 NPs (5G-Fe0 NPs/βCD) using Ferula persica root extract (C) removal of CV (Nasiri et al., 2019).

diseases (Aashima & Mehta, 2020; Buoli et al., 2018). Sulfur oxides, ozone, nitrogen oxides, carbon dioxide, carbon monoxide, polycyclic aromatic hydrocarbons, and volatile organic compounds are all contaminants linked to poor air quality in the atmosphere. Among all stated causes of air pollution, benzene (a volatile organic compound, or VOC) is the most abundant and hazardous contaminant that causes pollution in indoor air and the petrochemical industries. Presently, control of industrial benzene emissions as an utmost eliminated byproduct is a fundamental matter of concern for environmental remediation. Hence, innovative strategies for monitoring, interpreting, and addressing air pollutants at even lower contamination concentrations must be devised. As an instance, titania-ceria nanohybrid under xenon lamp illumination (shown in Figure 2.2) has been observed to be highly catalytic active for the gas-phase benzene oxidation.

Under symbiotic effect circumstances between photocatalysis of titania and thermocatalysis of ceria, the carbon dioxide rate generation of titania-ceria nanohybrids with Ti:Ce molar ratio 0.108 is 36.4 times higher than the kinetics recorded in

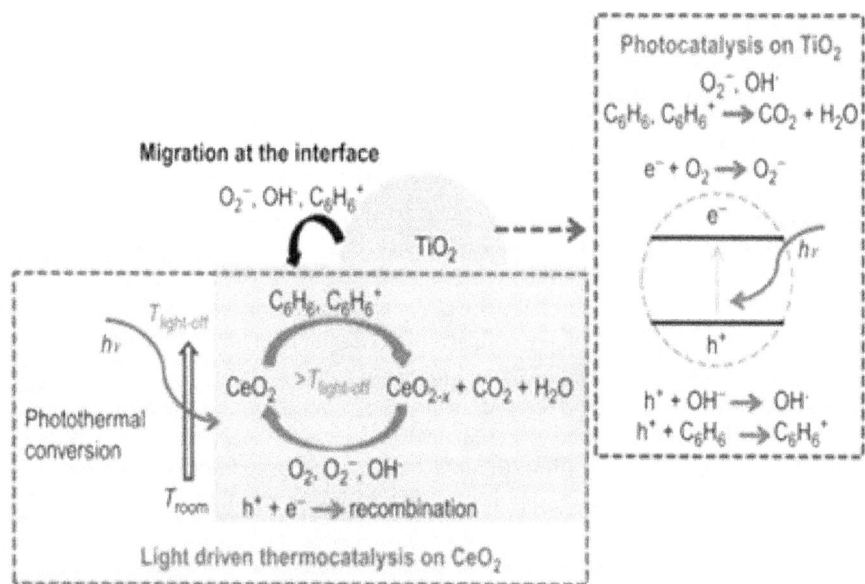

FIGURE 2.2 Schematic illustration of solar-light-driven thermocatalysis and the synergetic effect between the photocatalysis (Bellardita et al., 2020).

standard photothermal catalytic conditions at room temperature (Aashima & Mehta, 2020; Zeng et al., 2015). In recent reports, clay-based titania photocatalyst has been demonstrated as a novel treatment of air pollution. For the photocatalytic treatment of toluene, a benchtop catalytic flow reactor with a clay-based titania system also exhibited tremendous results. While constructing p-type gas sensors for the sensing of ethanol, Au-NiO core shell NPs have been proven to be more effective than pristine NiO NPs. In case of Au-NiO nano composite, this behavior is caused by an increase in charge carriers. Similarly, several other NMs that functionalized or modified by doping or heterocomposite fabrication are found to be tremendously efficient for the treatment of air pollution, as compared to the pristine NPs (Aashima & Mehta, 2020; Abadi et al., 2020; Kibanova et al., 2009; Majhi et al., 2018).

2.4.3 Soil Pollution

According to estimates, around 60% of soil from earth's crust is deteriorated, and the cause of deterioration is probably the elevation in the concentration of metals, disinfectants, pesticides, microorganisms, antibiotics, and contamination from underground water storage reservoirs; installation of fertilizer, oil, and fuel retention; waste percolation from landfills; and industrial wastes direct dumping into the soil. Apart from water and air pollution, soil pollution is also a deliberate environmental issue that arises mostly because of the occurrence of detrimental chemicals, and contaminants in the soil. Functionalization of ligand NM by the association of two nanojunctions can result in more robust soil treatment technologies (Tripathi et al., 2015).

Several recent past research reports have showcased few relevant instances related to ligand functionalization for soil remediation, which are as follows:

1. For the removal of hydrodechlorination of polychloroethanes, bimetallic lead/iron NPs demonstrated efficient results (Lien & Zhang, 2005).
2. For extraction of distinct metal ions, silica-coated maghemite functionalized with diethylenetriaminepentaacetic acid (DTPA). These types of functionalized NPs have resulted in nearly 80% metal ion removal from contaminated soil samples (Hughes et al., 2018).
3. Another report demonstrated the decreased and extractable concentrations of Pb, Cu, and Zn from the soil, after the treatment of soil with the calcium phosphate NPs (CPNs) (Arenas-Lago et al., 2016).
4. Aerobic oxidation of 4-chlorophenol was done by Fe_2O_3 nanowires containing zero valent iron (Ai et al., 2013).
5. For degradation of carbon tetrachloride to methane, biogenic Fe_3O_4 and Fe_2O_3 NPs, were investigated to give efficient results (McCormick & Adriaens, 2004).
6. For the removal of hexachlorocyclohexane from polluted soil slurry, Bimetallic Pd/Fe systems stabilized via carboxymethyl cellulose. The association of stabilizer with bimetallic NPs has enhanced the efficacy of NPs with 100% removal of pollutants when compared with pristine ones. The liberation of hydrogen gas by Pd accomplishes the suggested mechanism, which began with the corrosion of iron. As a result, a powerful reducing agent emerges, which destroys the pollutant (Singh et al., 2012).
7. As illustrated in Figure 2.3, The degradation of a growing contaminant, methylparaben, was achieved using a nano-hetro assemblage of superparamagnetic Fe_3O_4 and bismuth vanadate assembled within the Pinus roxburghii,

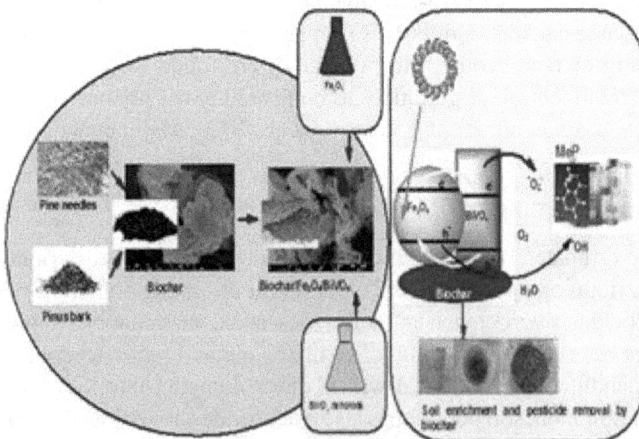

FIGURE 2.3 Nano-hetro assembly of superparamagnetic Fe_3O_4 and bismuth vanadate for pesticide removal (Kumar et al., 2017).

generated from biochar (MeP). The degradation reported was 97.4% in just two hours (Kumar et al., 2017).

8. Calcium peroxide (CaO_2) NPs can be used to explicitly disintegrate 2,4-DCP (Qian et al., 2016).
9. The disintegration of polychlorinated biphenyls (PCBs) from polluted soil slurry employing the gum stabilized Pd/Fe NPs was mainly due to the hydrophobicity of PCBs. Surfactants inculcated in the functionalization shown an enhancement in the dissolution of PCBs and an increment in their elimination from soil slurry (Fan et al., 2013).
10. Li et al. conveyed that the entire dechlorination and elimination of nearly 0% of pentachlorophenol (PCP) from polluted soil slurry using Pd/Fe NPs is conceivable. PCP is converted into phenol by using synthesized NPs and removed thorough electroosmosis from soil (Li et al., 2011).

2.4.4 BIOSENSING

During the past several years, distinct social sectors—including clinical diagnostics, food safety, energy production, and environmental remediation—have become highly aware of the urgent requirement of the ultrasensitive detection of biomolecules. For the ultrasensitive detection of biomolecules, signal amplification strategies (shown in Figure 2.4), such as mass spectroscopy or polymerase chain amplification, can be effectively used. The delicate or even single molecule have sparked a lot of interest in the field of nanotechnology for biosensing applications (Chiang et al., 2011; Taniguchi et al., 2009). Despite their hypersensitivity, signal amplification approaches are detrimental and frequently struggle from challenging output deduction, exorbitant prices, and sophisticated operation. With the help of emerging nanoscience and technology

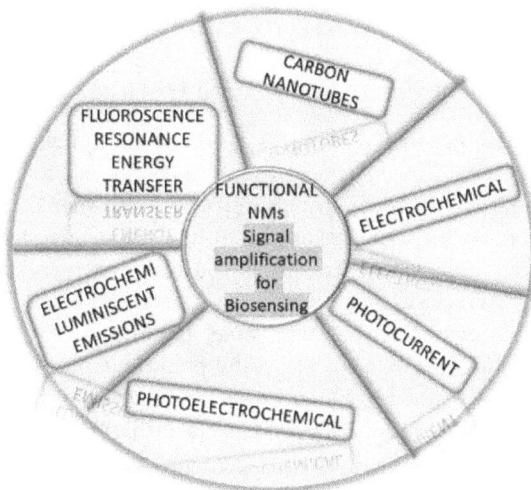

FIGURE 2.4 Schematic illustration of signal amplification strategies using bio functional NMs in biosensing.

advances, signal amplification techniques—due to their speedy analytical strategy and facile downsizing—show tremendous potential in obtaining high sensitivity and selectivity for ultrafast detection of the biomolecules. In this regard, biofunctionalization of NPs has introduced a viable and long-lasting synergic effect in the catalysis industry to speed up signal transduction, resulting in extremely lower detection limits, and widened linear correlation range. Nanoscaled materials can be efficiently used as chemical and biological sensors for single-molecule biomolecule detection. To achieve a high detection sensitivity in biosensing, an appropriate strategy for functionalizing nanostructures with biomolecules as signal prompting elements must be established. For immobilization of biomolecules over the NM, several strategies can be employed including covalent/non-covalent interactions with the functional groups onto the surface of NP, doping of NP with a relevant substance, and physio-chemical adsorption of biomolecules all over the NP (Veiseh et al., 2010). In the signal amplification technique, nanostructured material usually acts as a trigger element for detecting the signal tags. Additionally, various other affinity interactions including barnase-barstar or avidin-biotin systems are highly efficient for vigorous association of multifunctional nanoframeworks. The progress of conjugation of NM with the biomolecule is dependent on the detection of target molecule and on the signal amplification rate. Furthermore, photonic nanoparticle probes, including fluorescence power transmission nanobeads and quantum dots (QDs), outperform natural dyes and fluorescent proteins in terms of signal luminance, light absorption stability, and chromatic light emission. Surface-extended Raman scattering and Rayleigh scattering recognition, and fluorescent and photoacoustic imaging, have all been used by employing nanoprobes. Colorimetric, chemiluminescent, and optical analysis of bioanalytes, due to their cost effectiveness, high selectivity and sensitivity, high conductivity, high catalytic stability, and easy contraction, have recently captivated huge attention and significance in employing the functional NMs.

2.4.5 CATALYSIS INDUSTRY

2.4.5.1 METAL-BASED FUNCTIONAL NMS FOR HYDROGEN GAS EVOLUTION

The emphasis on fossil fuel–based resources as a key source of energy has culminated in environmental disaster, as well as negative effects on the environment. For sustainability of humankind and the environment, a shift toward eco-friendly sources of energy is highly in demand. Hydrogen fuel evolution from water splitting has become one of the most viable, eco-friendly, and sustainable sources of energy. Industrially, several techniques such as natural gas steam reforming process, photoelectrochemical water splitting, nuclear powered water splitting, water electrolysis, and photonic thermochemical water splitting, are used for successful production of hydrogen gas. Among all these stated methods for hydrogen fuel evolution, water electrolysis is the most economically efficient method. Water electrolysis setup entails one anode and one cathode submerged in an aqueous media. Hydrogen evolution reaction takes place at cathode in which catalysts used are platinum based and oxygen evolution reaction takes place at anode where catalysts are IrO_2- or RuO_2-based. The elemental forms of Pt-, Ir-, and Ru-based catalysts used for water-splitting reactions are rarely abundant on the earth's crust and highly expensive. Hence, it becomes a drawback for scalable water splitting in industries. Water electrolysis process necessitates a voltage of 1.23 V vs. RHE. A potential greater than 1.23 V (overpotential) is obligatory to complete the reaction at a respectable rate. Therefore, as a substitute to Pt-, Ru-, or Ir-based catalysts, electrocatalysts have been cultivated by scientists for hydrogen or oxygen evolution reactions at an industrial scale, as they possess a characteristic to lower the overpotential and thereby increasing the efficiency of the process. There are certain characteristics necessary for an electrocatalyst to be introduced in water splitting which, including the following.

1. Cost-effective in nature.
2. High specific activity.
3. High active surface area.
4. Better electron transport efficiency.
5. Low overpotential.
6. Eco-friendly.
7. High current density.

Several economically viable strategies—such as hydrothermal, solvothermal, chemical vapor deposition, and electrochemical deposition methods—are used to functionalize the nanoelectrocatalysts. By omitting polymer binders, nanofunctionalization permits water-splitting electrocatalysts to maximize active surface area, inject disorder and permeability to the crystallites, improve electrochemical performance, acquire enhanced stability, and produce synergistic effects by doping nanomaterials with metals. All of these changes contribute to the development of cost-effective, durable, and efficient nanoelectrocatalysts (Rajakaruna & Ariyarathna, 2020).

2.4.5.2 Graphene-Based Functional NM for Enhanced Catalytic Properties

Graphene-based NMs offer a comprehensive array of catalytic applications, such as photocatalysis, electrocatalysis, organic synthesis, carbocatalysis, and environmental remediation (Santosh Bahadur Singh & Hussain, 2020). Graphene-based NMs possess excellent competence to catalyze the redox-based reactions. To functionalize/modify nanographene, the electronic structure must be adjusted by modifying the dispersion of electronic states and expanding the distribution of states (DOS) around the Fermi level. The nature of the materials employed in functionalization and the process of production are the two major factors that affect the functionalization of nanographene (Huang et al., 2018; Zhao et al., 2018; Zhu et al., 2018). Functionalization of nanographene incorporating organics, proteins, molecular linkers, and other soft matter or solid synthetic metals, semiconductors, and other hard matter are the most commonly used precursors for constructing highly capable nanographene-based catalytic materials. In order to functionalize nanographene, several distinct methodologies are used which includes thermal, hydrothermal, mechanical, and chemical functionalization of nanographene (Jia et al., 2017). In comparison to uncatalyzed procedures, catalyzed reactions are ecological, fuel efficient, low cost, and environmentally sustainable. As a result, catalysis aids in the conservation of natural resources. Organic synthesis, pharmaceuticals, energy, sensors, oil refineries, and natural resource cleanup are some of the applications of catalysts. Researchers have recently become enthusiastic in nanographene-based catalysts as heterogeneous, metal-free, and environmentally friendly catalysts (Fang et al., 2018; Li et al., 2018; Luo et al., 2018; Ren et al., 2018; Shi & Zhang, 2016; Zheng et al., 2017). Graphene-based catalysis does indeed have a promising future since it is more economical and long-lasting than metal-based catalysts. Doping with hetero atoms including nitrogen, boron, or sulfur can control the catalytic activity of graphene-based catalysts by providing effective active sites (Xiong et al., 2018). Due to their flexible structure, large surface area, high electrical and thermal conductivities, charge carrier mobility at room temperature, and high chemical stability, graphene-based nanomaterials—in conjunction with well-known conventional photocatalytic materials like titanium dioxide—have introduced new horizons in the field of photocatalyzed environmental cleanup (Voiry et al., 2013).

2.5 CURRENT INDUSTRIAL OVERVIEW OF FUNCTIONAL NANOMATERIALS

In today's world, modern science, research, and engineering authentication, as well as broad scientific purpose, are essential components of sustainability of energy and industrial resources for humankind. The necessities of human technological insight and clairvoyance are massive and far-reaching nowadays. In the context of comprehensive scientific aspiration and scientific perseverance, functionalized NMs are the smart materials of present and future requirements. Due to a high surface-to-volume ratio and corresponding high reactivity on a continuum that ranges between 1 nm and a few hundred nanometers that are invisible at a microscopic level, novel solutions to various environmental challenges have arisen presently in several research works. Surface engineered or functionalized NMs are more likely to become involved with humans and the environment because of intense breakthroughs in nanomedicine, scientific diagnostics, and biomedical sciences and engineering. Nanoscience and technology have introduced strong anticipation among numerous scientists and research scholars to work to establish newly and advanced chemical strategies to functionalize NMs for the purpose of advanced applications in cancer therapy, medical diagnostics, tissue engineering, eco-friendly fuel production, wastewater treatment, environmental engineering, smart food packaging for commercial use, bioimaging, chemical sensing, therapeutics, photocatalysis, and much more. Nanoparticle functionalization offers an appropriate interface from substances to healthcare and vice-versa, and it is becoming increasingly relevant in the exploration and design of novel therapeutic carriers and diagnostic equipment. This technique connects the biochemical basis involved in physiology to drug targets and other biomedical engineering applications such as imaging, drug delivery science, and tissue engineering by utilizing various types of NMs and their functionalization. Recently, several scientific research works have been primarily related to fabrication of functionalized carbonaceous NMs, as they acquire low-dimension carbon allotropes, enhance the immobilization success, and increase biomolecules utilization potential. There are several recent advances on formulation of single-walled functionalized carbon nanotubes for the synthesis of enzymes carbon NM conjugates and peptides. Along with the advanced synthesis and implementation of functionalized NMs for industrial scalability, there exist a few downsides of functionalized nanoscaled material, which restricts several applications useful for environmental remediation (Darwish & Mohammadi, 2018; "Handbook of Functionalized Nanomaterials for Industrial Applications," 2020; Oliveira et al., 2015; Predoi et al., 2017; Subbiah et al., 2011).

2.6 FUTURE TRENDS AND CHALLENGES OF FUNCTIONAL NANOMATERIALS

- Functional NMs must be distinguished according to their characteristic capabilities and toxicity extent rather than their magnitude, as this will lead to an easier assessment of toxicological jeopardy (Bierkandt et al., 2018).
- After looking at the unique characteristics of active nanographene for catalytic applications, more study on nanographene functionalization is required to boost its manufacturing approach, yields, electrical and thermal conductivity, and other qualities without jeopardizing their conventional favorable intrinsic properties (Santosh Bahadur Singh & Hussain, 2020).
- Nanoscaled treatment is replacing traditional remediation approaches due to the selectivity, sensitivity, efficacy, quick elimination duration, and renewability it provides for treatment. These characteristics of nanoremediation have reinforced the optimistic viewpoint for environment remediation. Hence, nanocontamination necessitates a long-term review, and new NMs should undergo a complete risk assessment before being used in treatment processes.
- A deep knowledge of the strategies that are utilized at the nanometer scale, acquired through in situ synchrotron radiation and neutron trial and error, will indeed be vital in accumulating the expertise required for the existing empirical evidence of nanostructured materials in engineering disciplines, and will be a fascinating study to pursue for future environment sustainability (Aashima & Mehta, 2020).
- Nanofunctional damage pathways should be highly comprehended as input to distinct modeling techniques (Aashima & Mehta, 2020).
- The composition of an NM surface serves an essential impact on toxicity, bioavailability, and ecological consequence. The surface of an NM can be altered to provide tremendous scientific improvements; however, caution must be exercised to guarantee that this will not raise the particular chemical hazard, exposure, or threat; hence, a relatively secure design approach to research is recommended (Hunt, 2020).
- The REACH laws of the European Union have by far the most stringent guidelines for surface functionalized NMs, and these regulations are anticipated to expand in the future. However, if surface modification alters the NMs hazards, supplementary national regulatory requirements might well be necessary (Hunt, 2020).
- The path and consequences of NMs once discharged into the atmosphere are still largely unexplored and challenging to anticipate. As a result, thorough design and optimization of NMs is required to extract optimum profits.

2.7 SUMMARY

With massive development in the field of nanoscience and technology, organic NMs are widespread in the environment, and while ancient civilizations such as Rome, Greece, China, and India needed to know how to control and manipulate and implement several of these NPs in numerous practices to ensure specific objectives, the purposeful and decisive use of nanostructured materials in diverse commodities is a trend of the twenty-first century. Despite typical approaches to managing environmental concerns, enhancing physical features with diversified functionalization has been shown to give significant improvements to pristine NMs. In contrast to the traditional approaches centered on remedial procedures, NMs have the potential to exploit the target contaminant's specific surface chemistry by functionalizing or attaching the functional groups, allowing them to harness the target contaminant's distinctive surface chemistry. Functional NMs hold a lot of promise and can be used in a variety of ways. Functionalization of an NM can be readily implemented by distinct strategies—including covalent/non-covalent interactions, chemisorption, physisorption, and intrinsic surface engineering—which are deeply rephrased in this chapter. This chapter has also shown remarkable advantageous sustainable applications whereby a functional NM showcases the better characteristics of an NM with enhanced adsorption kinetics, catalytic activity, anti-corrosion, and anti-agglomeration. To conclude, we can say that the functionalization strategy remarkably exploits the hidden capabilities of a pristine NM, such as hydrophilicity, hydrophobicity, conductivity, and corrosion resistance. However, toxic kinetics is yet a matter of concern for environmental remediation and yet to be understood, pending future research.

REFERENCES

Aashima, & Mehta, S. K. (2020). Chapter 18: Impact of functionalized nanomaterials towards the environmental remediation: Challenges and future needs. In Hussain, C. M. (Ed.), *Handbook of Functionalized Nanomaterials for Industrial Applications*. Elsevier, pp. 505–524. https://doi.org/10.1016/B978-0-12-816787-8.00018-1

Abadi, M. B. H., Shirkhanloo, H., & Rakhtshah, J. (2020). Air pollution control: The evaluation of TerphApm@MWCNTs as a novel heterogeneous sorbent for benzene removal from air by solid phase gas extraction. *Arabian Journal of Chemistry, 13*(1). https://doi.org/10.1016/j.arabjc.2018.01.011

Abdullah, N. H., Shameli, K., Etesami, M., Chan Abdullah, E., & Abdullah, L. C. (2017). Facile and green preparation of magnetite/zeolite nanocomposites for energy application in a single-step procedure. *Journal of Alloys and Compounds, 719*. https://doi.org/10.1016/j.jallcom.2017.05.028

Ai, Z., Gao, Z., Zhang, L., He, W., & Yin, J. J. (2013). Core-shell structure dependent reactivity of Fe@Fe$_2$O$_3$ nanowires on aerobic degradation of 4-chlorophenol. *Environmental Science and Technology, 47*(10). https://doi.org/10.1021/es4005202

Anjum, M., Miandad, R., Waqas, M., Gehany, F., & Barakat, M. A. (2019). Remediation of wastewater using various nano-materials. *Arabian Journal of Chemistry, 12*(8). https://doi.org/10.1016/j.arabjc.2016.10.004

Arenas-Lago, D., Rodríguez-Seijo, A., Lago-Vila, M., Couce, L. A., & Vega, F. A. (2016). Using Ca$_3$(PO$_4$)$_2$ nanoparticles to reduce metal mobility in shooting range soils. *Science of the Total Environment, 571*. https://doi.org/10.1016/j.scitotenv.2016.07.108

Awfa, D., Ateia, M., Fujii, M., & Yoshimura, C. (2019). Novel magnetic carbon nanotube-TiO_2 composites for solar light photocatalytic degradation of pharmaceuticals in the presence of natural organic matter. *Journal of Water Process Engineering, 31.* https://doi.org/10.1016/j.jwpe.2019.100836

Azzam, E. M. S., Fathy, N. A., El-Khouly, S. M., & Sami, R. M. (2019). Enhancement the photocatalytic degradation of methylene blue dye using fabricated $CNTs/TiO_2/AgNPs/$ Surfactant nanocomposites. *Journal of Water Process Engineering, 28.* https://doi.org/10.1016/j.jwpe.2019.02.016

Bellardita, M., Fiorenza, R., Palmisano, L., & Scirè, S. (2020). Photocatalytic and photothermocatalytic applications of cerium oxide-based materials. In Scirè, S., & Palmisano, L. (Eds.), *Cerium Oxide (CeO$_2$): Synthesis, Properties and Applications.* Elsevier, pp. 109–167. https://doi.org/10.1016/b978-0-12-815661-2.00004-9

Bhati, A., Singh, A., Tripathi, K. M., & Sonkar, S. K. (2016). Sunlight-induced photochemical degradation of methylene blue by water-soluble carbon nanorods. *International Journal of Photoenergy, 2016.* https://doi.org/10.1155/2016/2583821

Bierkandt, F. S., Leibrock, L., Wagener, S., Laux, P., & Luch, A. (2018). The impact of nanomaterial characteristics on inhalation toxicity. *Toxicology Research, 7*(3). https://doi.org/10.1039/c7tx00242d

Boulamanti, A. K., Korologos, C. A., & Philippopoulos, C. J. (2008). The rate of photocatalytic oxidation of aromatic volatile organic compounds in the gas-phase. *Atmospheric Environment, 42*(34). https://doi.org/10.1016/j.atmosenv.2008.07.016

Buoli, M., Grassi, S., Caldiroli, A., Carnevali, G. S., Mucci, F., Iodice, S., Cantone, L., Pergoli, L., & Bollati, V. (2018). Is there a link between air pollution and mental disorders? *Environment International, 118.* https://doi.org/10.1016/j.envint.2018.05.044

Chenab, K. K., Sohrabi, B., Jafari, A., & Ramakrishna, S. (2020). Water treatment: Functional nanomaterials and applications from adsorption to photodegradation. *Materials Today Chemistry, 16.* https://doi.org/10.1016/j.mtchem.2020.100262

Chiam, S. L., Pung, S. Y., & Yeoh, F. Y. (2020). Recent developments in MnO_2-based photocatalysts for organic dye removal: A review. *Environmental Science and Pollution Research, 27*(6). https://doi.org/10.1007/s11356-019-07568-8

Chiang, C. K., Chen, W. T., & Chang, H. T. (2011). Nanoparticle-based mass spectrometry for the analysis of biomolecules. *Chemical Society Reviews, 40*(3). https://doi.org/10.1039/c0cs00050g

Crutchley, R. J. (2016). Preface to functional nanomaterials. *Coordination Chemistry Reviews, 320–321.* https://doi.org/10.1016/j.ccr.2016.04.016

Darwish, M., & Mohammadi, A. (2018). Functionalized nanomaterial for environmental techniques. *Nanotechnology in Environmental Science, 1–2.* https://doi.org/10.1002/9783527808854.ch10

Deng, L., Wang, S., Liu, D., Zhu, B., Huang, W., Wu, S., & Zhang, S. (2009). Synthesis, characterization of Fe-doped TiO_2 nanotubes with high photocatalytic activity. *Catalysis Letters, 129*(3–4). https://doi.org/10.1007/s10562-008-9834-5

Domun, N., Hadavinia, H., Zhang, T., Sainsbury, T., Liaghat, G. H., & Vahid, S. (2015). Improving the fracture toughness and the strength of epoxy using nanomaterials-a review of the current status. *Nanoscale, 7*(23). https://doi.org/10.1039/c5nr01354b

Fan, G., Cang, L., Qin, W., Zhou, C., Gomes, H. I., & Zhou, D. (2013). Surfactants-enhanced electrokinetic transport of xanthan gum stabilized nanoPd/Fe for the remediation of PCBs contaminated soils. *Separation and Purification Technology, 114.* https://doi.org/10.1016/j.seppur.2013.04.030

Fang, Z., Peng, L., Qian, Y., Zhang, X., Xie, Y., Cha, J. J., & Yu, G. (2018). Dual tuning of Ni-Co-A (A = P, Se, O) nanosheets by anion substitution and holey engineering for efficient hydrogen evolution. *Journal of the American Chemical Society, 140*(15). https://doi.org/10.1021/jacs.8b01548

Fathy, N. A., El-Shafey, S. E., & El-Shafey, O. I. (2017). Synthesis of a novel MnO_2@carbon nanotubes-graphene hybrid catalyst (MnO_2@CNT-G) for catalytic oxidation of basic red 18 dye (BR18). *Journal of Water Process Engineering, 17.* https://doi.org/10.1016/j.jwpe.2017.03.010

Fernando, R. H. (2009). Nanocomposite and nanostructured coatings: Recent advancements. *ACS Symposium Series, 1008.* https://doi.org/10.1021/bk-2009-1008.ch001

Georgakilas, V., Tiwari, J. N., Kemp, K. C., Perman, J. A., Bourlinos, A. B., Kim, K. S., & Zboril, R. (2016). Noncovalent functionalization of graphene and graphene oxide for energy materials, biosensing, catalytic, and biomedical applications. *Chemical Reviews, 116*(9). https://doi.org/10.1021/acs.chemrev.5b00620

Gupta, P., Vermani, K., & Garg, S. (2002). Hydrogels: From controlled release to pH-responsive drug delivery. *Drug Discovery Today, 7*(10). https://doi.org/10.1016/S1359-6446(02)02255-9

Handbook of Functionalized Nanomaterials for Industrial Applications. (2020). *Handbook of Functionalized Nanomaterials for Industrial Applications.* Elsevier. https://doi.org/10.1016/c2018-0-00341-2

Huang, X. Y., Wang, A. J., Zhang, L., Fang, K. M., Wu, L. J., & Feng, J. J. (2018). Melamine-assisted solvothermal synthesis of PtNi nanodentrites as highly efficient and durable electrocatalyst for hydrogen evolution reaction. *Journal of Colloid and Interface Science, 531.* https://doi.org/10.1016/j.jcis.2018.07.051

Hughes, D. L., Afsar, A., Laventine, D. M., Shaw, E. J., Harwood, L. M., & Hodson, M. E. (2018). Metal removal from soil leachates using DTPA-functionalised maghemite nanoparticles, a potential soil washing technology. *Chemosphere, 209.* https://doi.org/10.1016/j.chemosphere.2018.06.121

Hunt, N. J. (2020). Chapter 29: Handbook of surface-functionalized nanomaterials: Safety and legal aspects. In Hussain, C. M. (Ed.), *Handbook of Functionalized Nanomaterials for Industrial Applications.* Elsevier. https://doi.org/10.1016/B978-0-12-816787-8.00029-6

Jia, J., Xiong, T., Zhao, L., Wang, F., Liu, H., Hu, R., Zhou, J., Zhou, W., & Chen, S. (2017). Ultrathin N-doped Mo_2C nanosheets with exposed active sites as efficient electrocatalyst for hydrogen evolution reactions. *ACS Nano, 11*(12). https://doi.org/10.1021/acsnano.7b06607

Kaewgun, S., & Lee, B. I. (2010). Deactivation and regeneration of visible light active brookite titania in photocatalytic degradation of organic dye. *Journal of Photochemistry and Photobiology A: Chemistry, 210*(2–3). https://doi.org/10.1016/j.jphotochem.2009.12.018

Kanth, P. C., Verma, S. K., & Gour, N. (2020). Chapter 10: Functionalized nanomaterials for biomedical and agriculture industries. In Hussain, C. M. (Ed.), *Handbook of Functionalized Nanomaterials for Industrial Applications.* Elsevier. https://doi.org/10.1016/B978-0-12-816787-8.00010-7

Karim, M. E. (2020). Chapter 30: Functional nanomaterials: Selected legal and regulatory issues. In Hussain, C. M. (Ed.), *Handbook of Functionalized Nanomaterials for Industrial Applications.* Elsevier. https://doi.org/10.1016/B978-0-12-816787-8.00030-2

Kibanova, D., Cervini-Silva, J., & Destaillats, H. (2009). Efficiency of clay - TiO_2 nanocomposites on the photocatalytic elimination of a model hydrophobic air pollutant. *Environmental Science and Technology, 43*(5). https://doi.org/10.1021/es803032t

Kong, X. K., Chen, C. L., & Chen, Q. W. (2014). Doped graphene for metal-free catalysis. *Chemical Society Reviews, 43*(8). https://doi.org/10.1039/c3cs60401b

Kumar, A. (2017). A review on the factors affecting the photocatalytic degradation of hazardous materials. *Material Science & Engineering International Journal, 1*(3). https://doi.org/10.15406/mseij.2017.01.00018

Kumar, A., Shalini, Sharma, G., Naushad, M., Kumar, A., Kalia, S., Guo, C., & Mola, G. T. (2017). Facile hetero-assembly of superparamagnetic Fe_3O_4/$BiVO_4$ stacked on biochar for solar photo-degradation of methyl paraben and pesticide removal from soil. *Journal of Photochemistry and Photobiology A: Chemistry, 337.* https://doi.org/10.1016/j.jphotochem.2017.01.010

Kumar, N., Mittal, H., Parashar, V., Ray, S. S., & Ngila, J. C. (2016). Efficient removal of rhodamine 6G dye from aqueous solution using nickel sulphide incorporated polyacrylamide grafted gum karaya bionanocomposite hydrogel. *RSC Advances, 6*(26). https://doi.org/10.1039/c5ra24299a

Kumar, N., & Sinha Ray, S. (2018). Synthesis and functionalization of nanomaterials. In Sinha Ray, S. (Ed.), *Springer Series in Materials Science* (Vol. 277). Springer. https://doi.org/10.1007/978-3-319-97779-9_2

Lehn, J. M. (2002). Toward complex matter: Supramolecular chemistry and self-organization. *Proceedings of the National Academy of Sciences of the United States of America, 99*(8). https://doi.org/10.1073/pnas.072065599

Li, Y., Yin, K., Wang, L., Lu, X., Zhang, Y., Liu, Y., Yan, D., Song, Y., & Luo, S. (2018). Engineering MoS_2 nanomesh with holes and lattice defects for highly active hydrogen evolution reaction. *Applied Catalysis B: Environmental, 239.* https://doi.org/10.1016/j.apcatb.2018.05.080

Li, Z., Yuan, S., Wan, J., Long, H., & Tong, M. (2011). A combination of electrokinetics and Pd/Fe PRB for the remediation of pentachlorophenol-contaminated soil. *Journal of Contaminant Hydrology, 124*(1–4). https://doi.org/10.1016/j.jconhyd.2011.03.002

Lien, H.-L., & Zhang, W. (2005). Hydrodechlorination of chlorinated ethanes by nanoscale Pd/Fe bimetallic particles. *Journal of Environmental Engineering, 131*(1). https://doi.org/10.1061/(asce)0733-9372(2005)131:1(4)

Luo, Y., Li, X., Cai, X., Zou, X., Kang, F., Cheng, H. M., & Liu, B. (2018). Two-dimensional MoS_2 confined $Co(OH)_2$ electrocatalysts for hydrogen evolution in alkaline electrolytes. *ACS Nano, 12*(5). https://doi.org/10.1021/acsnano.8b00942

Majhi, S. M., Naik, G. K., Lee, H. J., Song, H. G., Lee, C. R., Lee, I. H., & Yu, Y. T. (2018). Au@NiO core-shell nanoparticles as a p-type gas sensor: Novel synthesis, characterization, and their gas sensing properties with sensing mechanism. *Sensors and Actuators, B: Chemical, 268.* https://doi.org/10.1016/j.snb.2018.04.119

Mansoori, G. A., & Soelaiman, T. A. F. (2005). Nanotechnology – An introduction for the standards community. *Journal of ASTM International, 2*(6). https://doi.org/10.1520/JAI13110

McCormick, M. L., & Adriaens, P. (2004). Carbon tetrachloride transformation on the surface of nanoscale biogenic magnetite particles. *Environmental Science and Technology, 38*(4). https://doi.org/10.1021/es030487m

Moradi, M., Haghighi, M., & Allahyari, S. (2017). Precipitation dispersion of Ag–ZnO nanocatalyst over functionalized multiwall carbon nanotube used in degradation of acid orange from wastewater. *Process Safety and Environmental Protection, 107.* https://doi.org/10.1016/j.psep.2017.03.010

Mout, R., Moyano, D. F., Rana, S., & Rotello, V. M. (2012). Surface functionalization of nanoparticles for nanomedicine. *Chemical Society Reviews, 41*(7). https://doi.org/10.1039/c2cs15294k

Nasiri, J., Motamedi, E., Naghavi, M. R., & Ghafoori, M. (2019). Removal of crystal violet from water using β-cyclodextrin functionalized biogenic zero-valent iron nanoadsorbents synthesized via aqueous root extracts of *Ferula persica. Journal of Hazardous Materials, 367.* https://doi.org/10.1016/j.jhazmat.2018.12.079

Oliveira, S. F., Bisker, G., Bakh, N. A., Gibbs, S. L., Landry, M. P., & Strano, M. S. (2015). Protein functionalized carbon nanomaterials for biomedical applications. *Carbon, 95.* https://doi.org/10.1016/j.carbon.2015.08.076

Pan, D., Jiao, J., Li, Z., Guo, Y., Feng, C., Liu, Y., Wang, L., & Wu, M. (2015). Efficient separation of electron-hole pairs in graphene quantum dots by TiO_2 heterojunctions for dye degradation. *ACS Sustainable Chemistry and Engineering, 3*(10). https://doi.org/10.1021/acssuschemeng.5b00771

Pan, L., Liu, X., & Zhu, G. (2020). *Functional Nanomaterial for Photoenergy Conversion.* World Scientific Publishing Co Pte Ltd. https://doi.org/10.1142/11887

Pandey, S., Do, J. Y., Kim, J., & Kang, M. (2020). Fast and highly efficient catalytic degradation of dyes using κ-carrageenan stabilized silver nanoparticles nanocatalyst. *Carbohydrate Polymers, 230.* https://doi.org/10.1016/j.carbpol.2019.115597

Pérez-Ramírez, E. F., Luz-Asunción, M. de la, Martínez-Hernández, A. L., & Velasco-Santos, C. (2016). Graphene materials to remove organic pollutants and heavy metals from water: Photocatalysis and adsorption. In *Semiconductor Photocatalysis — Materials, Mechanisms and Applications.* Edited by Wenbin Cao, IntechOpen. https://doi.org/10.5772/62777

Predoi, D., Motelica-Heino, M., Guegan, R., & Le Coustumer, P. (2017). Advances in functionalized materials research 2016. *Journal of Nanomaterials, 2017.* https://doi.org/10.1155/2017/1269319

Qian, Y., Zhang, J., Zhang, Y., Chen, J., & Zhou, X. (2016). Degradation of 2,4-dichlorophenol by nanoscale calcium peroxide: Implication for groundwater remediation. *Separation and Purification Technology, 166.* https://doi.org/10.1016/j.seppur.2016.04.010

Rajakaruna, R. M. P. I., & Ariyarathna, I. R. (2020). Chapter 4: Functionalized metal-based nanoelectrocatalysts for water splitting. In Hussain, C. M. (Ed.), *Handbook of Functionalized Nanomaterials for Industrial Applications.* Elsevier. https://doi.org/10.1016/B978-0-12-816787-8.00004-1

Ren, X., Wei, Q., Ren, P., Wang, Y., & Chen, R. (2018). Synthesis of flower-like $MoSe_2@MoS_2$ nanocomposites as the high efficient water splitting electrocatalyst. *Materials Letters, 231.* https://doi.org/10.1016/j.matlet.2018.08.049

Santosh Bahadur Singh, & Hussain, C. M. (2020). Chapter 5: Functionalized nanographene for catalysis. In Hussain, C. M. (Ed.), *Handbook of Functionalized Nanomaterials for Industrial Applications.* Elsevier. https://doi.org/10.1016/B978-0-12-816787-8.00005-3

Shi, Y., & Zhang, B. (2016). Recent advances in transition metal phosphide nanomaterials: Synthesis and applications in hydrogen evolution reaction. *Chemical Society Reviews, 45*(6). https://doi.org/10.1039/c5cs00434a

Singh, R., Misra, V., Mudiam, M. K. R., Chauhan, L. K. S., & Singh, R. P. (2012). Degradation of γ-HCH spiked soil using stabilized Pd/Fe^0 bimetallic nanoparticles: Pathways, kinetics and effect of reaction conditions. *Journal of Hazardous Materials, 237–238.* https://doi.org/10.1016/j.jhazmat.2012.08.064

Stamov, D. R., Khoa Nguyen, T. A., Evans, H. M., Pfohl, T., Werner, C., & Pompe, T. (2011). The impact of heparin intercalation at specific binding sites in telopeptide-free collagen type I fibrils. *Biomaterials, 32*(30). https://doi.org/10.1016/j.biomaterials.2011.06.031

Subbiah, R., Veerapandian, M., & Yun, K. S. (2011). Nanoparticles: Functionalization and multifunctional applications in biomedical sciences. *Current Medicinal Chemistry, 17*(36). https://doi.org/10.2174/092986710794183024

Taniguchi, K., Kajiyama, T., & Kambara, H. (2009). Quantitative analysis of gene expression in a single cell by qPCR. *Nature Methods, 6*(7). https://doi.org/10.1038/nmeth.1338

Thiruppathi, R., Mishra, S., Ganapathy, M., Padmanabhan, P., & Gulyás, B. (2017). Nanoparticle functionalization and its potentials for molecular imaging. *Advanced Science, 4*(3). https://doi.org/10.1002/advs.201600279

Tong, T., Zhang, J., Tian, B., Chen, F., & He, D. (2008). Preparation of Fe^{3+}-doped TiO_2 catalysts by controlled hydrolysis of titanium alkoxide and study on their photocatalytic activity for methyl orange degradation. *Journal of Hazardous Materials*, *155*(3). https://doi.org/10.1016/j.jhazmat.2007.11.106

Treccani, L., Yvonne Klein, T., Meder, F., Pardun, K., & Rezwan, K. (2013). Functionalized ceramics for biomedical, biotechnological and environmental applications. *Acta Biomaterialia*, *9*(7). https://doi.org/10.1016/j.actbio.2013.03.036

Tripathi, V., Fraceto, L. F., & Abhilash, P. C. (2015). Sustainable clean-up technologies for soils contaminated with multiple pollutants: Plant-microbe-pollutant and climate nexus. *Ecological Engineering*, *82*. https://doi.org/10.1016/j.ecoleng.2015.05.027

Vadivel, S., Theerthagiri, J., Madhavan, J., Santhoshini Priya, T., & Balasubramanian, N. (2016). Enhanced photocatalytic activity of degradation of azo, phenolic and triphenyl methane dyes using novel octagon shaped BiOCl discs/MWCNT composite. *Journal of Water Process Engineering*, *10*. https://doi.org/10.1016/j.jwpe.2015.12.001

Vadivel, S., Vanitha, M., Muthukrishnaraj, A., & Balasubramanian, N. (2014). Graphene oxide-BiOBr composite material as highly efficient photocatalyst for degradation of methylene blue and rhodamine-B dyes. *Journal of Water Process Engineering*, *1*. https://doi.org/10.1016/j.jwpe.2014.02.003

Veiseh, O., Gunn, J. W., & Zhang, M. (2010). Design and fabrication of magnetic nanoparticles for targeted drug delivery and imaging. *Advanced Drug Delivery Reviews*, *62*(3). https://doi.org/10.1016/j.addr.2009.11.002

Voiry, D., Yamaguchi, H., Li, J., Silva, R., Alves, D. C. B., Fujita, T., Chen, M., Asefa, T., Shenoy, V. B., Eda, G., & Chhowalla, M. (2013). Enhanced catalytic activity in strained chemically exfoliated WS_2 nanosheets for hydrogen evolution. *Nature Materials*, *12*(9). https://doi.org/10.1038/nmat3700

Wu, S., Xiong, J., Sun, J., Hood, Z. D., Zeng, W., Yang, Z., Gu, L., Zhang, X., & Yang, S. Z. (2017). Hydroxyl-dependent evolution of oxygen vacancies enables the regeneration of BiOCl photocatalyst. *ACS Applied Materials and Interfaces*, *9*(19). https://doi.org/10.1021/acsami.7b01701

Xiong, L., Bi, J., Wang, L., & Yang, S. (2018). Improving the electrocatalytic property of CoP for hydrogen evolution by constructing porous ternary CeO_2-CoP-C hybrid nanostructure via ionic exchange of MOF. *International Journal of Hydrogen Energy*, *43*(45). https://doi.org/10.1016/j.ijhydene.2018.09.117

Yang, B., Tian, Z., Zhang, L., Guo, Y., & Yan, S. (2015). Enhanced heterogeneous fenton degradation of methylene blue by nanoscale zero valent iron (nZVI) assembled on magnetic Fe_3O_4/reduced graphene oxide. *Journal of Water Process Engineering*, *5*. https://doi.org/10.1016/j.jwpe.2015.01.006

Zeng, M., Li, Y., Mao, M., Bai, J., Ren, L., & Zhao, X. (2015). Synergetic effect between photocatalysis on TiO_2 and thermocatalysis on CeO_2 for gas-phase oxidation of benzene on TiO_2/CeO_2 nanocomposites. *ACS Catalysis*, *5*(6). https://doi.org/10.1021/acscatal.5b00292

Zhao, Z., Liu, H., Gao, W., Xue, W., Liu, Z., Huang, J., Pan, X., & Huang, Y. (2018). Surface-engineered PtNi-O nanostructure with record-high performance for electrocatalytic hydrogen evolution reaction. *Journal of the American Chemical Society*, *140*(29). https://doi.org/10.1021/jacs.8b04770

Zheng, T., Sang, W., He, Z., Wei, Q., Chen, B., Li, H., Cao, C., Huang, R., Yan, X., Pan, B., Zhou, S., & Zeng, J. (2017). Conductive tungsten oxide nanosheets for highly efficient hydrogen evolution. *Nano Letters*, *17*(12). https://doi.org/10.1021/acs.nanolett.7b04430

Zhu, X. Y., Zhang, L., Yuan, P. X., Feng, J. J., Yuan, J., Zhang, Q. L., & Wang, A. J. (2018). Hollow Ag44Pt56 nanotube bundles with high electrocatalytic performances for hydrogen evolution and ethylene glycol oxidation reactions. *Journal of Colloid and Interface Science*, *532*. https://doi.org/10.1016/j.jcis.2018.08.016

3 Functional Nanomaterials for Characterization Techniques

*Is Fatimah, Gani Purwiandono,
Ganjar Fadillah, and Wiyogo Prio Wicaksono*

CONTENTS

3.1   INTRODUCTION

Sensors and biosensors have received considerable attention from various circles, including researchers. The demands and continuous development of industry trigger the rapid growth of pollution and contamination, and requirements for product quality control and human health. This development has been driven by rapid advances in semiconductor material technology in recent decades [1, 2]. This semiconductor material-based sensor provides many advantages for continuous development and is easy to integrate with biosensor systems. However, the development of materials for sensors is limited to inorganic semiconductor materials; several responsive polymer materials

DOI: 10.1201/9781003263852-3

have also been developed, especially in medicine [3, 4]. Sensors and biosensors must have several special requirements to be reliable in industrial application such as high sensitivity, short time response, good selectivity, wide range area for required concentration, low production cost, reusability, excellent stability, etc., as shown in Figure 3.1. Therefore, many types of materials have been developed and reported. These materials have various properties and characteristics. For example, graphene oxide–based sensors doped with metal oxide semiconductor materials are widely developed as gas sensors because they have a large surface area and bridge in electron transfer when redox reactions occur on the surface of the sensor material [5, 6]. Therefore, the material characteristics need to be studied first before being applied as a sensor.

In general, material characteristics for sensor applications can be studied by basic characterization to determine the surface properties of the material and advanced characterization to study the specific characteristics of materials such as impedance, electrochemical, etc. Because the properties of materials are critical to determining the aims of the sensor, this chapter discusses some characterization techniques explicitly in the application of sensors and biosensors. The determination of the characterization technique to be used is very dependent on the sensor application to be applied. Therefore, this chapter also discusses the classification of sensors based on the characterizations and industrial applications used.

3.2 CLASSIFICATION OF SENSOR-BASED CHARACTERIZATION

A *chemical sensor* is a device that converts chemical information into a signal either qualitatively or quantitatively for analytical purposes. Chemical information from a sensor can come from a chemical reaction or the physical properties of a target material or analyte that will determine the compositions. Various types of sensors have been developed with various types of responsive materials for use in various applications such as environmental, pharmaceutical, food, gas and oil industries, etc.

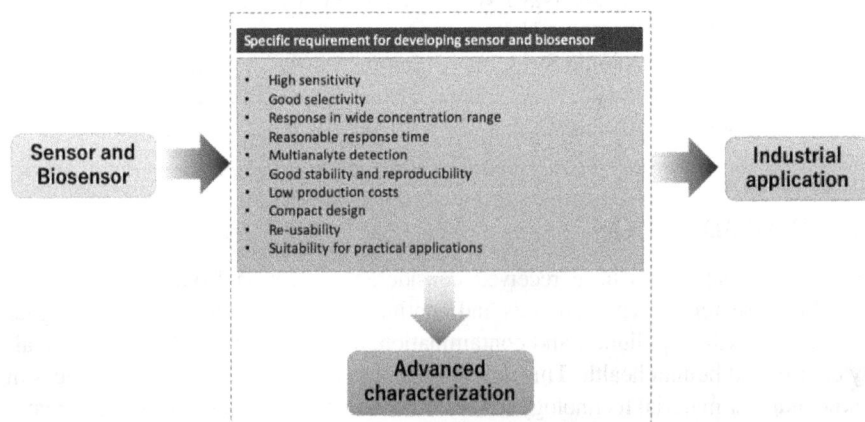

FIGURE 3.1   The specific requirement properties of sensor and biosensor.

Source: (Created by authors)

A sensor has four main processes: sampling, sample transportation, signal processing, and data processing. The signal processing section is an essential part of the development of a sensor. Therefore, many sensors are generally developed with high sensitivity and selectivity properties for a specific response.

Material characterization for sensors is generally based on the type of sensor being developed. For example, Jiang et al. (2021) developed the imprinted electrochemical sensor for the detection of *Salmonella* [7]. They used cyclic voltammetry (CV), different pulse voltammetry (DPV), and electrochemical impedance spectroscopy (EIS) to determine the electrochemical properties. López et al. (2021) designed the optical sensor-based imprinted polymer for the determination of ion Pb^{2+} [8]. For this case, spectroscopy reflectance is used as a robust characterization for determining the sensor's performance. Material characterization is usually designed based on the function of the sensor that has been developed. Therefore, the classification of various types of sensors is fundamental to know in designing the instrumentation of a sensor. Based on the principle of the operating system, chemical sensors can be classified into several types, as shown in Figure 3.2.

An optical sensor is based on changes in optical phenomena resulting from the interaction between the analyte and the receptor material. Optical-based sensors have two basic methods, namely label-free and label-based methods. Optical sensors are currently being developed in various fields and can be used with various other spectroscopic techniques such as Raman spectroscopy, reflectometric interference spectroscopy (RIF), photoluminescence spectroscopy, surface plasmon resonance spectroscopy (SPR), and infrared spectroscopy. This type of sensor generally uses optical fibers known as optodes in various configurations so that their applications are vast. Fang et al. (2021) summarized the sensor-based optic for monitoring several pesticides compounds in foods [9]. The type of optical sensor widely used is the fluorescent type; this sensor has been reported to have good sensitivity with a low limit of detection (LoD) value. The interaction between the material and the target analyte changes the FL intensity through enhancing or quenching, with the intensity value being correlated with the target concentration.

The electrochemical sensor converts interaction between the analyte and surface material into a signal that can represent the concentration of the analyte in quantification. The reaction that occurs can be stimulated by giving a current or potential to cause a spontaneous reaction. Some electrochemical-based sensors include potentiometry, voltammetry, polarography, chemically sensitized field-effect transistors, potentiometric solid-gas sensors, etc. Li et al. (2021) fabricated the electrochemical sensor-based Zn/Ni-ZIF-8/XC-72/Nafion hybrid materials for rapid detection of Cu(II) and Pb(II) [10]. The basic characterization used for this material are like SEM, TEM, XPS, and FT-IR. At the same time, the advanced characterizations are CV, DPV and EIS. Furthermore, the prepared electrode material showed an excellent sensitivity for simultaneously detecting metal ions in actual samples. In this type of sensor, the development of a material with a spontaneous interaction response is the main factor in obtaining a sensitive sensor.

The mass-sensitive device generates a frequency signal in response to the mass change caused by the accumulation of the analyte. This sensor can measure a range of low analyte concentrations up to the nanogram level but is not suitable for

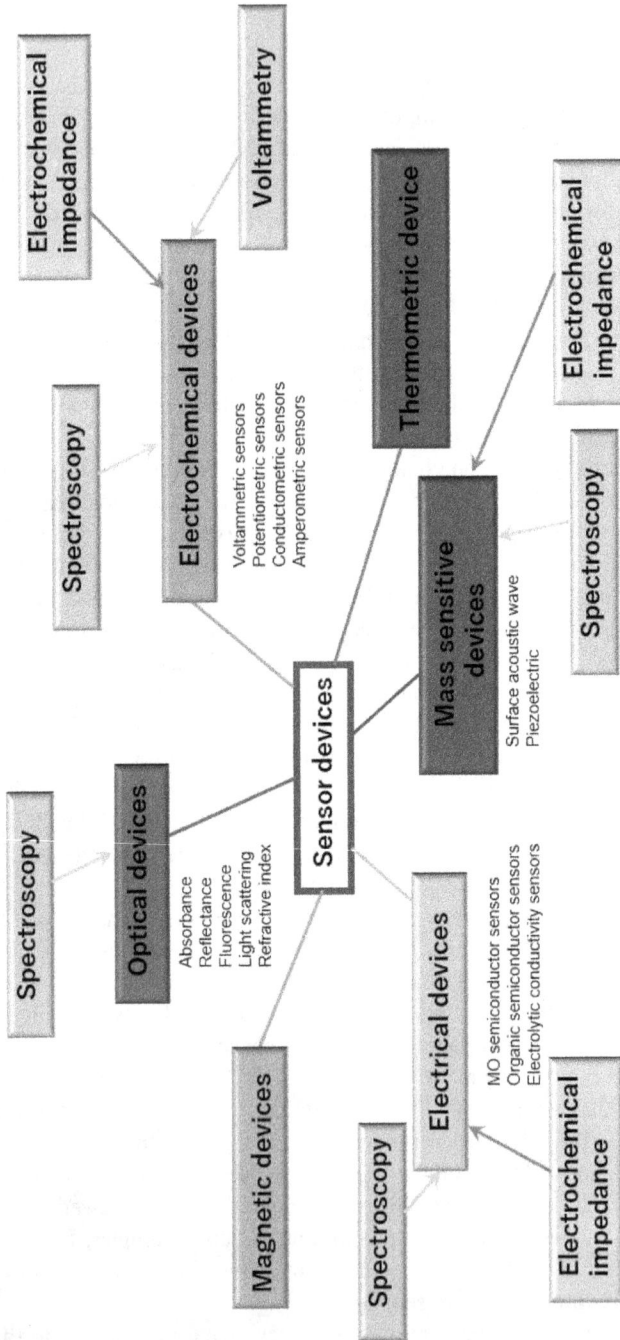

FIGURE 3.2   Classification of sensor-based principle and characterization required.

Source: (Created by authors)

application as a routine sensory device. Piezoelectric devices and surface acoustic waves are examples of applications of this type of sensor. In addition, this sensor has been widely used as a gas sensor and biochemical recognition [11]. Another device is a magnetic and thermometric sensor. These two types of sensors are generally used based on changes in the material's physical properties, such as changes in the paramagnetic properties of the gas being analyzed or changes in heat resulting from a chemical reaction between the analyte and material that is responsive to temperature.

3.3   BASIC CHARACTERIZATION

3.3.1   X-Ray Diffraction (XRD)

X-ray diffraction analysis is a very basic analytical method in the preparation of materials for sensor application. XRD analysis is primary utilized to identify the phase of crystalline material and the real structure of synthesized material over various possibilities in defects and deviation. The basic principle of XRD analysis is the interaction between crystalline material with X-rays; as the incident X-rays impinge the building block of a crystalline material, they will be scattered at different planes of the material. The X-ray beams are generated by the electron hot to metal anode having high energy, such Cr, Fe, Co, Cu, Mo, or Ag.

The scattered pattern results in diffracted X-rays with different optical path lengths to travel. The magnitude of the path length represents the distance between the incident angle of the X-ray beam and the crystal planes, which stated in the famous Bragg Equation (Eq. 1):

$$n(\lambda) = 2d\ sin\theta \qquad (1)$$

Where d is the spacing between diffracting planes, θ is the incident angle, n is an integer, and λ is the beam wavelength

The specific directions appear as spots on the diffraction pattern called reflections [12].

Constructive interference occurs only if the path difference (given by $2d\ sin\theta$) is a multiple (n = 1, 2, . . .) of the used wavelength of the X-ray beam. By the Bragg equation, the distance between lattice planes of the crystalline can be determined, and represented as the plot of X-ray intensity versus the angle between the incident and the diffracted beam (denoted as 2θ) or called a diffractogram (Figure 3.3).

As each material has a typical diffraction pattern, the identity of the crystalline phase of the measured sample can be identified by comparing with references, and one of these is known as the Joint Committee on Powder Diffraction Standards (JCPDS) standard. By the specific pattern of reflection of material, identification of different phases in a synthesized material can be detected by the combination of the peak. The interplanar distance d(hkl) of the reflection is related to the lattice parameter regarding to the unit cell, a, b, and c of unit structure. Particularly, the identification for a specific procedure in the synthesis can also be performed by examination of the characteristic peaks; for example, in the synthesis of reduced-graphene oxide (rGO) from graphite via graphene oxide (GO) presented in Figure 3.4.

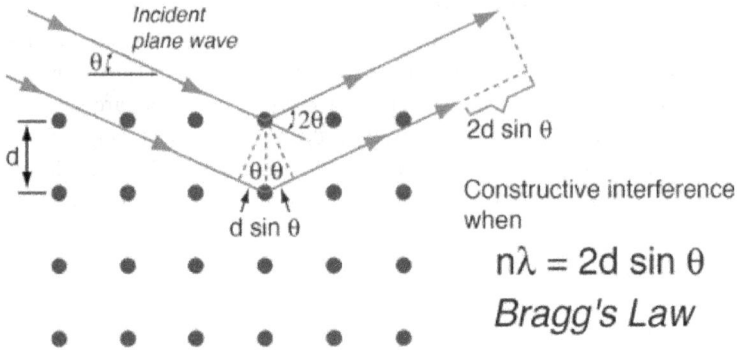

FIGURE 3.3   X-ray diffraction principle.

Source: (Created by authors)

The characteristic diffraction peak of the GO is reflection associated with (002) plane observed at 2θ of around $10°$ with an average d-spacing of around 0.79 nm, which indicates the interlayer spacing. This is also proof of an extended space due to the intercalation of the oxygen-containing functional groups into the layers of graphite, which the characteristic peaks for the (002) reflection are observed at 2θ around 26.53° decreased in intensity. Characteristically, as a proof of the reduction to the oxygen-containing layers, the pattern of rGO displayed a shifting of d-spacing at approximately $2\theta = 24.5°$ associated with the distance layer of around 0.36 nm. By comparing the intensity of the peak, the degree of graphitization can also be determined, as the higher intensity is attributed to the higher degree of graphitization [13].

The intensity of peak quantitatively reflects the degree of crystallinity, so it can be utilized for either the identification of defect or chemical change of a crystalline material or the crystallite growth of the material; for example, in crystallinity control of ZrO_2 by atomic layer deposition experiment [14]. Evaluation of the effect of deposition temperatures to the formation of the monoclinic crystalline phase was performed by monitoring the peak corresponding to the specific peak [15]. By using the corresponding peaks, the crystallite size of the phase can be calculated based on the Scherer equation (Eq. 2):

$$d = \kappa\lambda \, / \, B\cos\theta) \tag{2}$$

Where d is the mean crystalline size of the NPs, λ is the wavelength of radiation (1.5406 Å), θ is the angle of selected reflection, and B is the intensity of full width at half maximum (FWHM) of the selected reflection

Some drawbacks are faced in the use of conventional XRD for thin-film analysis. Conventional X-rays with large glancing angles of incidence will go through a few to several hundred micrometers inside the material depending on its "radiation" density. The problem will be related to the beam penetration depth being much greater than the thickness of the thin film. In addition, the detailed analysis for sub-micrometric layers of a thin-film specimen is more difficult since the reflection will appropriately represent a complete structure after deposition, while a large number of overlapping peaks

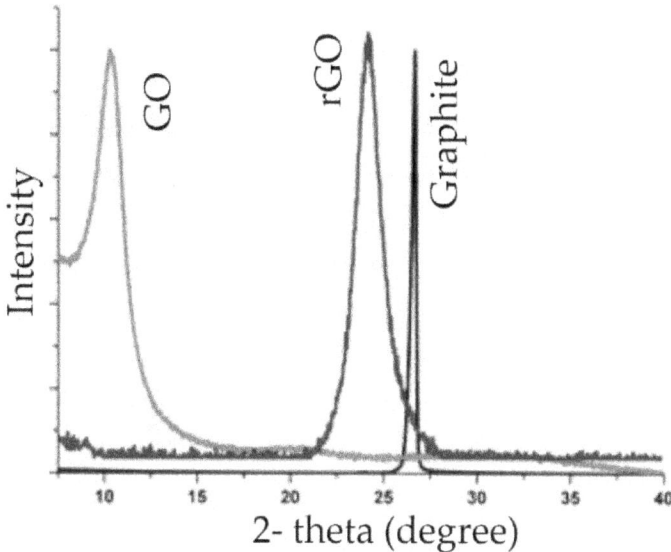

FIGURE 3.4   Diffractogram of reduced-graphene oxide (rGO), graphene oxide (GO), and graphite.

Source: (Reprinted with copyright permission from Hindawi Publisher)

from the different crystallographic phases cannot be well represented. To overcome this problem, grazing incidence configurations in XRD (GIRXD) measurements have been developed. The basic idea of this method is a modification of the Bragg-Brentano geometry for producing an "asymmetric" diffraction result. The modification allows access to small depths in the sample by varying the incidence angle. A parallel mono-chromatic X-ray beam falls on the sample surface at a fixed, low glancing angle, α (larger than the critical angle for total reflection, αc, but usually smaller than 10°), and the diffraction profile is recorded by detector scan only (Figure 3.5).

Figure 3.6 represents the comparison of conventional XRD and GIXRD patterns from analysis of TiO_2. By GIXRD, the species of titanium oxides such as Ti_2O and Ti_2O_3 are detected.

3.3.2   X-Ray Photoelectron Spectroscopy (XPS)

XPS is a method for determining the composition and chemical bonding state on the material's surface. In XPS, electrons in the material absorb a particular form of photon and further emerge from the material. The emitted photoelectrons represent a graph of intensity versus electron energy [17]. The electron's kinetic energy (E_K) is the experimental quantity measured by the spectrometer and depends on the photon energy of X-rays employed. The electron's binding energy (E_B) is the parameter that identifies the electron specifically, both in terms of its parent element and atomic energy level. The relationship between the parameters involved in the XPS experiment is described as the following Eq. 3.

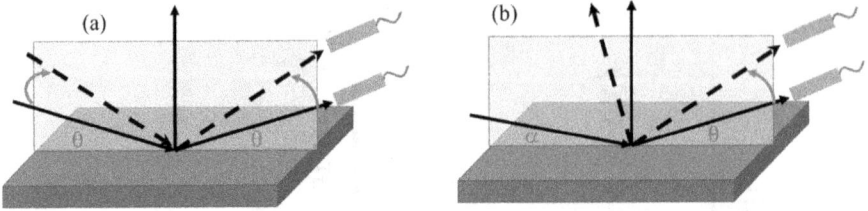

FIGURE 3.5 Schematic diagrams of (a) symmetric θ–2θ, and (b) asymmetric glancing angle XRD geometry.

Source: (Reprinted with copyright permission [16])

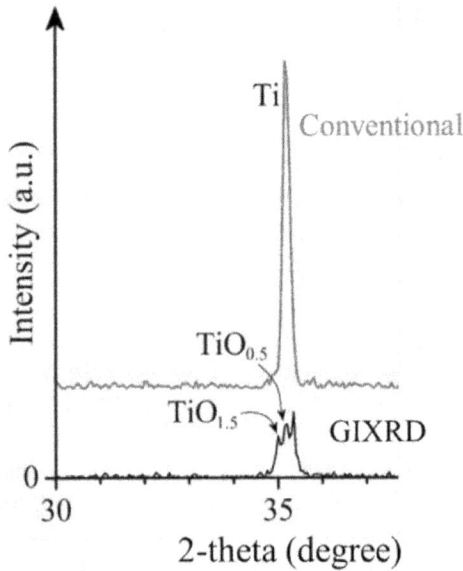

FIGURE 3.6 Comparison on conventional and GIXRD pattern of TiO_2 thin film.

Source: (Reprinted with copyright permission from Scientific Report) [16].

$$E_B = h\nu - E_K - W \tag{3}$$

Where hν is the photon energy, E_K is the electron's kinetic energy, and W is the spectrometer work function

The process of photoemission forms when an electron from the K shell is ejected from the atom. The photoelectron spectrum will reproduce the electronic structure of an element with a specific binding energy. The specific binding energy of the photoemission is valuable chemical information about an atom. In the characterization of material for sensor application, the interaction between the surface of the material sensor with chemical compounds is expressed in physical, chemical, and biological bandings [18]. The easiness of the material surface-chemical compounds interaction expresses the sensitivity. In a material-based sensor, the sensitivity of the sensors is correlated with the change of energy. The lower energy indicates the high sensitivity of the sensor. In a solid-based sensor, the origin of the sensitivity might be explained from the view of energetics. The dissociation energy of the bonding between the material and sample explains how the interaction was formed. The higher binding energy means the interaction between materials and the sample is stable. The energetics data show a simple explanation for why the solid-based sensor shows high sensitivity with low power consumption [19].

3.3.3 SCANNING ELECTRON MICROSCOPY (SEM) AND TRANSMISSION ELECTRON MICROSCOPY (TEM)

SEM is one of the versatile methods for the identification of topographic and topographic materials in the microstructure scale. By combining SEM with energy dispersive X-ray spectrophotometry (SEM-EDS), the analysis produces not only the image but also quantitative information about material composition. By the specific image obtained from the analysis, nanomaterial can be assigned as a specific identification such as nanobelt, nanoneedle, nanoflower, etc. In the preparation and characterization of sensor material, SEM-EDS analysis is usually employed for the detection of surface morphology, determination of fracture, detection of impurities or foreign phases, and detection of the defect. For such soluble nanoparticles or quantum dots of metal or metal oxide, the particle size distribution can also be determined from the image. Figure 3.7 shows the presentation of the SEM image of Au nanorings.

SEM analysis is based on the detection of high-energy electrons emitted from the surface of the sample from the interaction of electrons from an electron gun, and an objective lens focuses on a small spot (Figure 3.8).

Some analysis variables such as the accelerating voltage and the distance between the sample and electron gun (working distance) are crucial parameters that need to be optimized to achieve the best quality images. Technically, the images from the analysis can be observed from the secondary electron (SE) and back-scattered electron (BSE) detectors. SE can collect surface topography of material produced by electron-electron scattering from the surface of the specimen. In contrast, BSE has resulted from electron-nuclei scattering, and it reports the contrast of composition (COMPO) of the specimen [21]. Typically, images obtained from SE have a strong signal, and in contrast, BSE imaging is characterized as low or limited signal, and low-loss electron (LLE) imaging with an inherently noisier subset of all BSEs. However, LLE and BSE exhibit the advantages such as they are more readily modeled than the SEs as higher energy scattering processes and higher energy electrons are less influenced by charging. Figure 3.6 presents the difference between SE and BSE techniques.

The essential characteristic of material for sensor applications is the surface analyzer. The physical and chemical composition on the surfaces determines the nature of the interactions. Transmission electron microscopy (TEM) and high-resolution transmission electron microscopy (HR-TEM) analyze the materials' microcrystal structure and lattice imperfections, e.g. particle size, lattice fringe, grain size

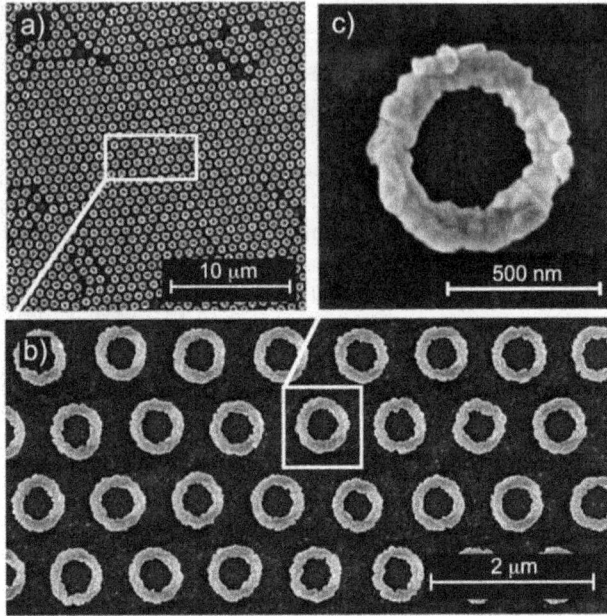

FIGURE 3.7 SEM Morphology of Au Nanorings.

Source: (Reprinted with permission from MDPI Publisher) [20]

distribution, and the defect in the grain structure and grain boundaries [22]. The presence of metallic clusters at the surface of materials for as-optimized sensors is investigated using TEM coupled with X-ray photoelectron spectroscopy (XPS).

3.3.4 RAMAN, PHOTOLUMINESCENCE, AND DIFFUSE-REFLECTANCE SPECTROSCOPY

A similar characterization concept in energetics interaction between molecules and materials is observed in Raman spectroscopy. Raman spectroscopy has been used to analyze the interaction between molecules and their polarization on the surface materials. Light interacts with a molecule through Raman scattering, polarizing the electron cloud and raising the molecule to a short-lived form known as a virtual state. The energy differential between the incoming and scattered photons is evidence of a molecule's shift between vibrational states [23]. In solid materials for the sensor, Raman spectroscopy was performed to characterize the materials' crystal structure. The observed peaks of Raman show the characteristic of the atomic structure of the materials, and the intensity of the Raman spectra reflects the quality of grain size and crystallinity of the materials. Bands in the Raman spectrum might be attributed to grain size effects and include the materials' impurities in the sensor application [17]. The origin of the sensitivity measurement in Raman spectroscopy for materials sensors might be because of surface plasmon resonance (SPR). SPR increases in local electromagnetic field strength, leading to the enhancing of the Raman signal from molecules at the surface material sensors [24].

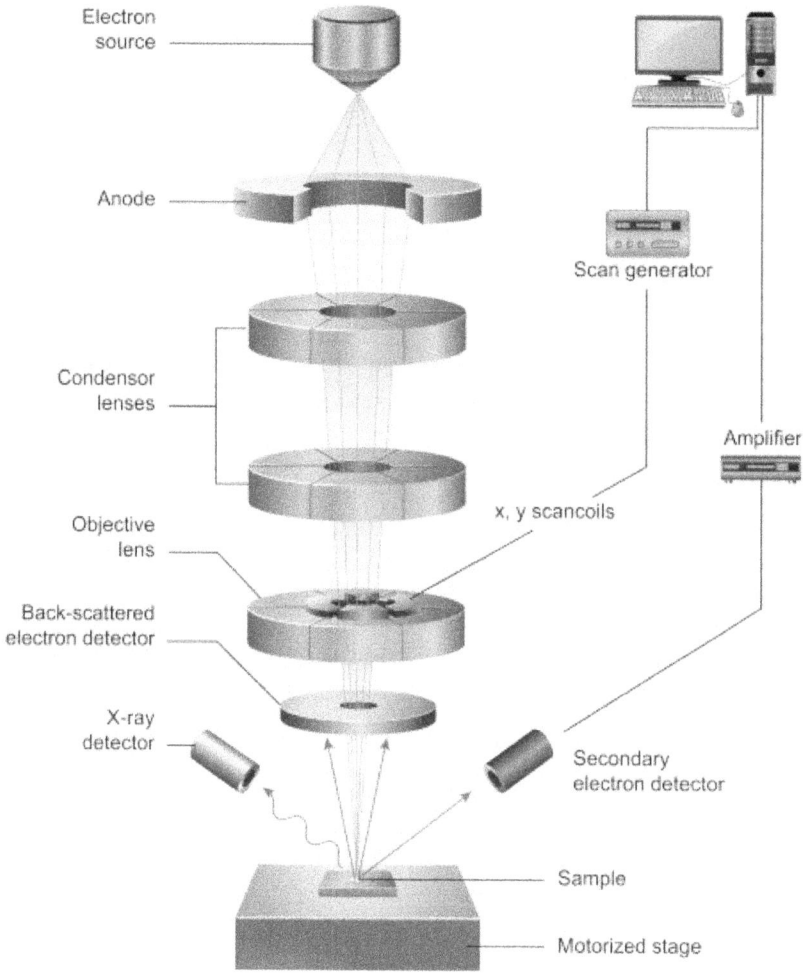

FIGURE 3.8 Schematic representation of SEM.

Source: (Created by authors)

In addition, photoluminescence (PL) spectroscopy can analyze the potential oxidation or reduction of samples on the surface. These oxidation-reduction processes caused the local charge transfer to induce electric positive or negative dipoles. When the molecular species are adsorbed onto the material sensor surfaces, the potential height may dramatically change the electronic structure into two categories: Lewis acids and bases. Lewis acids have a massive electron affinity, while Lewis bases have a slight affinity relative to the material's work function. Lewis acids attract electrons from the bulk into the surface electronic states, whereas Lewis bases release electrons back into the bulk from the surface states [25].

Tolouei et al. (2020) revealed the photoluminescence sensitivity of porous silicon-based sensors to CO_2 and propane gases [26]. Modified porous silicon has been used to detect CO_2 and propane directly. To suppress or reduce non-radiative recombination centers, a porous silicon surfaces layer was placed on porous silicon surfaces to enhance luminescence. The operation sensor effect is based on changes in the photoluminescence of the CH_x/porous silicon area caused by gaseous interactions. Carbon dioxide diminishes photoluminescence intensity, while propane has the reverse effect. Photoluminescence quenching leads to optical sensors.

Figure 3.9 shows the photoluminescence spectra of porous silicon as produced (a), coated with CH_x (b), and annealed at 200°C (c), 400°C (d), and 600°C (e). A large asymmetric PL band with an apparent maximum at 670 nm is presented in Figure 3.9. It was found that a 240 CH_x layer increased the band intensity. Moreover, the orange light is enough to be observed at 623 nm. Strong Coulomb interactions could result in photo-generated electron-hole pairs inside π-bonded grains in polymer-like carbon films [26]. These inside surface phenomena explain the signal's origin produced in PL material sensors.

Moreover, in a material-based semiconductor for sensor application, diffuse-reflectance spectroscopy (DRS) has been widely used to evaluate the bandgap energy of the material. Material sensors with a semiconductor structure are usually used to detect gas-phase molecules. The wide bandgap semiconductor could enhance sensing features like sensitivity and reaction time. When an electron donor (e.g. adsorbed hydrogen) or electron acceptor (e.g. adsorbed oxygen) is attached to the surface of a wide bandgap solid-gas sensor, the sensitivity principle is described. The electron donors or acceptors create surface states, followed by electron exchanges within the

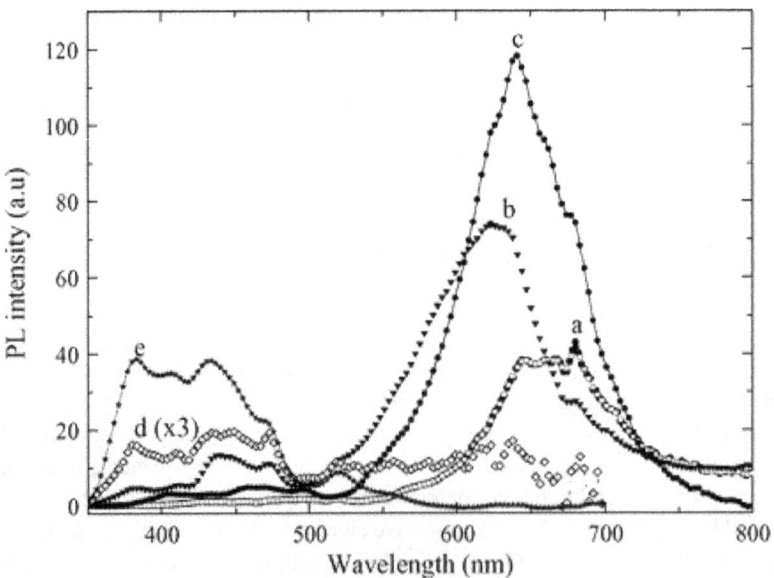

FIGURE 3.9 PL spectra of samples at room temperature (a), coated with CHx (b), and (c-e) annealed at 200°C, 400°C, an100°C (e).

Source: (Reprinted with Copyright permission from Elsevier B.V.) [26]

semiconductor's interior, resulting in a space charge layer near the surface [25]. The conductivity of the space charge area is controlled by altering the surface concentration of donors/acceptors.

3.4 ADVANCED CHARACTERIZATION

3.4.1 ELECTROCHEMICAL IMPEDANCE

Electrochemical impedance spectroscopy (EIS) is becoming increasingly used in biosensors and chemical sensors, not only as a characterization tool but also as a quantitative diagnostic and detection method. This characterization is necessary because most sensors are inherently non-uniform at the nanoscale, so the non-uniformity at the electrode interface makes the electrochemical characteristics of the sensor challenging to observe either by cyclic voltammetry, pulse voltammetry, or square wave voltammetry techniques. Electrochemical impedance is usually determined by applying an AC potential to an electrochemical cell, which measures the amount of current flowing in the cell system as shown in Figure 3.10a. The current signal generated from this mechanism can be measured as a sum of sinusoidal functions (Fourier series). Generally, the measured impedance uses a signal with a low excitation to produce a pseudo-linear cell response in a linear (or pseudo-linear) system.

Tolouei et al. (2020) developed microelectromechanical systems (MEMs) biochemical sensors for the detection of di(2-ethylhexyl) phthalate (DEPH) [26], an electrochemical impedance characterized the electrochemical properties. This EIS characterization is widely used because it has a high sensitivity, especially for electrode surfaces consisting of more than one layer, as in the case of biosensors. This impedance value is measured as a function of the frequency from 0.001–105 Hz with an AC voltage value from 5–500 mV in applied amplitude (rms) to the layer. The resulting output format is a Bode plot, consisting of a log-log plot of impedance (log Z) versus frequency (log f) as shown in Figure 3.10b. Information about relative permeability and barrier layers can be obtained in the low-frequency portion of the Bode plot. However, the interpretation of this system is considered complicated;

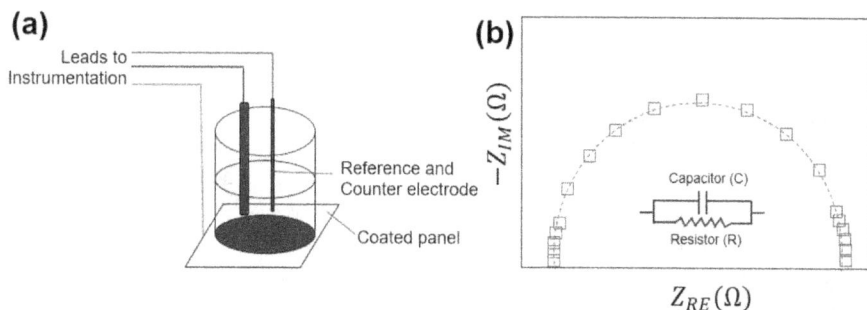

FIGURE 3.10 EIS schematic cell diagram (a) and the signal output of Bode plot form impedance sensor characteristics (b).

Source: (Reprinted with permission from The Korean Electrochemical Society. [27]

thus, the log Z value at 0.1 Hz is re-plotted to monitor the change in the layer's impedance as it interprets surface electrochemical properties.

3.4.2 VOLTAMMETRY

Voltammetry (derived from voltamperometry) techniques have attracted many researchers for developing sensors, including functional materials-based sensors [28]. This voltammetry techniques has been employed for clinical diagnostics [29], food securities [30], essential oils [31], drugs analysis [32], and environmental monitoring [33]. Voltammetry is a controlled potential method that is a part of electroanalytical methods which involve the application of a DC potential (volt) to an electrode to drive redox reactions, then the resulting current (ampere) flowing through the electrochemical cell is measured. Voltammetry data is provided in a current-voltage curve, which is usually known as a voltammogram. Furthermore, qualitative and quantitative analysis could be performed through voltammetry techniques.

Practically, various voltammetry techniques have been employed in the electrochemical sensors such as cyclic voltammetry (CV), linear sweep/scan voltammetry (LSV), alternating current voltammetry (ACV), square wave voltammetry (SWV), normal pulse voltammetry (NPV), differential pulse voltammetry (DPV), staircase voltammetry (SV), anodic stripping voltammetry (ASV), cathodic stripping voltammetry (CSV), adsorptive stripping voltammetry (AdSV), and square wave anodic stripping voltammetry (SWASV), as classified in Figure 3.11.

3.4.2.1 Cyclic Voltammetry (CV)

CV is the most widely employed for the first evaluation of electrochemical sensing. CV could be performed by applying forward potential on the working electrode from initial to a certain potential, then backward to the initial potential in the unstirred analyte solution. The CV is usually scanned from negative (less) to the positive (higher) potential, then returns to the initial. CV could provide qualitative information about electrochemical reaction, including redox potential of the electroactive species, effect of media (electrolyte/pH) toward redox reaction, phenomena on the electrode surface, thermodynamics and kinetics reaction, and catalysis process, as well as the preliminary study for the development of electrode modification using various functional materials. Furthermore, the type of electron transfer reversibility and the mass transfer of analyte from the CV could be determined [34].

Likasari et al. (2021) employed the CV technique to characterize the catalytic activity of the green synthesized NiO nanoparticles that been modified to the graphite paste (NiO/GPE) electrode for the detection of glucose [29]. The CV voltammogram showed the excellent electrocatalytic activity of the NiO NPs toward oxidation of glucose compared to the bare of GPE, as shown in Figure 3.12a. Furthermore, the CV evaluation using several scan rates showed the linear correlation of the square root of scan rate against the anodic peak current ($R^2 = 0.9971$), revealing the mass transfer on the NiO/GPE electrode was mainly controlled by diffusion (Figure 3.12b–c). Moreover, CV also could perform the quantitate sensing, by evaluating the increase of the redox currents against the increase of the analyte concentration. Figure 3.12d–e depicted a linear plot between the concentration of glucose and the anodic peak

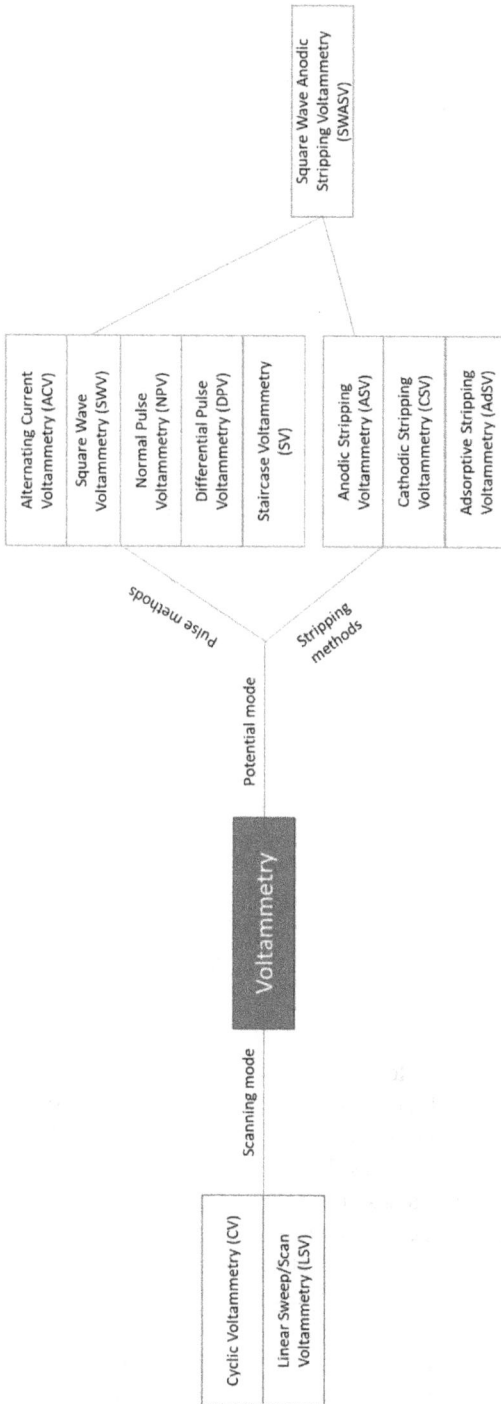

FIGURE 3.11 Classification of voltammetry-based potential and scanning mode, some of the emerging voltammetry techniques.

Source: (Created by authors)

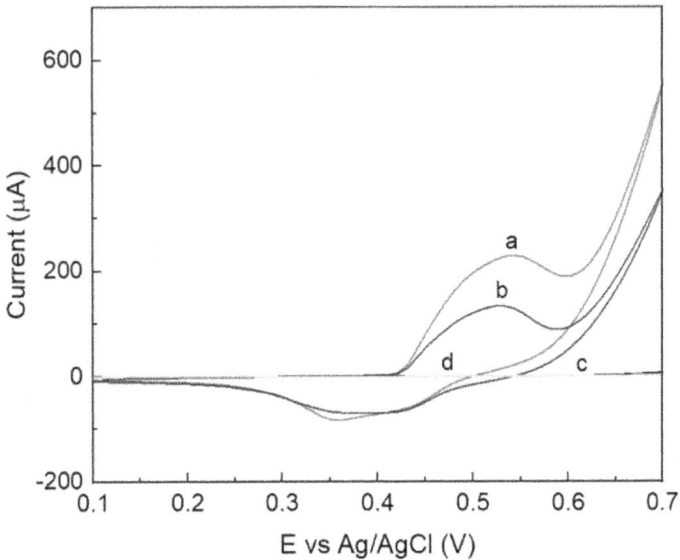

FIGURE 3.12   CVs voltammogram of the glucose on the NiO/GPE electrode.

Source: (Created by authors)

current. Additionally, the validation of analytical parameters, including detection limit, precise, and accuracy also could be performed from the CV graph.

Not only for small organic molecule sensing but the CV technique was also employed to evaluate the performance of tin (IV) oxide nanoparticles (SnO_2 NPs) modified to bare Au electrode for the detection of Hg^{2+} heavy metal. From the CV curve, the electrocatalytic of SnO_2 NPs could be studied. The feasibility of the CV as the technique for evaluating Hg^{2+} electrochemical sensing using SnO_2 NPs/Au electrode was shown by the very low detection limit down to 1.97 ppb [33].

3.4.2.2   Differential Pulse Voltammetry (DPV)

Among the pulse voltammetry techniques, different pulse voltammetry (DPV) was also massively employed in functional material-based electrochemical sensing. The DPV shows excellent performance for the analysis in the very low-level concentration of both organic and inorganic species; DPV could produce sensitive detection due to potential pulses superimposed to a staircase ramp applied on the working electrode at the end of a potential step that causes subtracting current measured. As consequence, the current is recorded twice (the first is just before the pulse is applied and the second is the end of the pulse), affecting the increase in the sensing sensitivity.

Fadillah and co-workers (2020) employed the DPV technique on the functionalized graphene oxide decorated tin (IV) oxide modified to GPE for the determination of small organic molecules of eugenol [31]. The modified electrode reported performed good stability, selectivity, and sensitivity with the detection limit down to

0.02 µM. Moreover, this developed electrode showed good recovery and fit with the other methods with a high confidence level. DPV has also been performed for the determination of heavy metals analytes. Deshmukh et al. reported the determination of Cu^{2+} on EDTA-modified PANI/SWNTs nanocomposite electrode using DPV technique, while Deshmukh et al. (2018) employed the DPV for the simultaneous detection of Pb^{2+}, Cu^{2+}, and Hg^{2+} on hydroxyapatite film modified to indium tin oxide electrode [35].

3.4.2.3 Square Wave Voltammetry (SWV)

Square wave voltammetry (SWV) is one of the voltammetry techniques that is powerful for analysis of any analyte, especially the biochemical system, as well as for studying the mechanism of electrode process and electro kinetics [36]. SWV is performed by applying the symmetrical square wave that is superimposed on a staircase potential to the working electrode, then the current is recorded twice (the first is at the end of the forward pulse and the second is at the end of reverse pulse), resulting in a net current. Interestingly, the net current is higher than both the forward and reverse currents, affecting the excellent sensitivity of the technique.

SWV technique has been used in broad sensing and biosensing applications, including the detection of biomarkers, heavy metals, and microorganism in food contaminants that employed functional material for improving the performance. For example, Ji et al. (2020) used the SWV technique for the detection of norepinephrine (an important catecholamine neurotransmitter) on the graphene screen-printed electrode [37]. From their experiment, a very low detection of 0.265 µM could be achieved. SWV technique could perform a low background current that will increase the sensitivity, even if in the biochemical system. On the other hand, SWV has been also employed for the simultaneous determination of folic, uric, and ascorbic acids in urine matrix using gold nanoparticles modified to carbon paste electrode (CPE) [38].

3.4.2.4 Stripping Voltammetry

Stripping voltammetry is well-known as a very powerful technique for the analysis of trace metals. The high sensitivity. The technique has two steps: the first is the electrolytic deposition of the metals ions, then the second is the stripping (measurement) step by the dissolution of the deposit. The high sensitivity performance of the technique is due to the preconcentration process in the deposition step before being analyzed [39]. The most widely used is anodic stripping voltammetry (ASV) instead of cathodic stripping voltammetry.

3.4.2.5 Anodic Stripping Voltammetry (ASV)

ASV is an electroanalytical technique that is performed by the electrodeposition using cathodic deposition to preconcentrate the analyte, then followed by an anodic stripping process to re-oxidize the metal and the current is measured [39]. The ASV technique provides the advantages of the possibility to simultaneously detect several metals because it has owned different reduction standard potential (E°). Ivandini et al. (2015) developed melamine detection by electrochemical immunochromatographic strip tests using gold nanoparticles (AuNPs) as the label and employed ASV for the measurement [40]. The AuNPs (Au^0) were released from conjugated antibodies using strong acid-producing Au^{3+}, then preconcentrated and stripped using ASV technique, resulting in a low detection of 0.069 µg/mL. On the other hand, Pan et al. fabricated the nanohydroxyapatite (NHAP)/ionophore/Nafion-modified to glassy carbon electrode and employed for the detection of Pb^{2+} using ASV technique [41]. Interestingly, the accumulation process of Pb^{2+} occurred before the electrodeposition step due to the presence of strong adsorption of NHAP, increasing the sensitivity. Furthermore, the lead ionophore species could significantly improve the selectivity towards Pb^{2+}.

3.4.3 POTENTIOMETRY

Potentiometry is a popular classical electroanalytical technique that is powerful for analysis of the concentration of the analytes by measuring the potential difference between two electrodes. The technique widely used is using ion-selective electrodes (ISEs) for the detection of pH, cations, and anions, which are useful for environmental monitoring, clinical diagnostics, food security, and quality analysis, as well as process control in industries. The techniques rapidly developed due to their advantages of being low cost, accurate, and rapid [39].

The electrode in the potentiometry measurement consists of an indicator (working) electrode and a reference electrode. The development in potentiometry is to develop the working ISEs that can selectively detect the presence of ionics species. The ion-selective electrodes are mainly membrane-based and consist of permselective ion-conducting materials. Thus, the emerging research ahead is to develop unique membrane materials. Some key points for the membrane materials development for potentiometry are that it is mechanically stable, insoluble in water, and performs a selective interaction/binding property (i.e. ion exchange) to the analyte target. The selective interaction/binding generating potential changed in the electrode that is corresponded to the analyte concentration. Besides the ISEs-based transduction, recently rapidly developed other types of potentiometry based on the solid contact electrodes (SCEs) include light addressable potentiometric sensors (LAPS), coated wire electrodes (CWEs), and ion-sensitive field-effect transistors (ISFETs) [42].

For example, Kaur et al. (2020) developed a nano-composite membrane of boron-doped graphene oxide-aluminum fumarate metal-organic framework (BGO/AlFu MOF) for the selective detection of bromide (Br^-) ions based on potentiometric technique [43]. The composite performed a selective detection to the Br^- ions with the detection limit to 71 nM due to the concerted effect including the porous and charges properties. On the other hand, Niemiec et al. fabricated graphene flakes decorated with dispersed gold nanoparticles (AuNPs) layer for the potassium (K^+) ions [44]. The addition of AuNPs could increase the hydrophobic properties that improved the mechanical properties. Moreover, the dispersed AuNPs also increased the surface area that beneficial to the increase in the electrical conductivity, producing improved sensing materials for K^+ detection.

3.5   CONCLUSION

In this chapter, surface analysis techniques and advanced characterization become an essential part of developing new materials for sensor and biosensor applications. Characterization of the surface of the material can be done by various techniques such as spectroscopy techniques, while advanced characterization can be carried out according to the application, such as by electrochemical impedance, amperometry, resistance, voltammetry, etc. The emphasis on this characterization technique is applied to the use of combined characterization techniques to study the sensor and biosensor materials' characteristics thoroughly. The combined application using the experimental concept can produce the characteristics of the functional material synergistically to produce a sensor and biosensor material that meets the characteristics according to the standard. Finally, each characterization technique can describe the properties of different materials. Therefore, careful selection of methods is a step that must be considered in developing materials for sensors and biosensors.

REFERENCES

[1] G. Fadillah, O.A. Saputra, T.A. Saleh, Trends in polymers functionalized nanostructures for analysis of environmental pollutants, *Trends in Environmental Analytical Chemistry*, **26** (2020) e00084.

[2] T.A. Saleh, G. Fadillah, Recent trends in the design of chemical sensors based on graphene: Metal oxide nanocomposites for the analysis of toxic species and biomolecules, *TrAC Trends in Analytical Chemistry*, **120** (2019) 115660.

[3] B. Ryplida, B.C. Lee, S.Y. Park, Conductive membrane sensor-based temperature and pressure responsive f-polymer dot hydrogels, *Composites Part B: Engineering*, **234** (2022) 109755.

[4] A. Bratek-Skicki, Towards a new class of stimuli-responsive polymer-based materials: Recent advances and challenges, *Applied Surface Science Advances*, **4** (2021) 100068.

[5] F. Rasch, V. Postica, F. Schütt, Y.K. Mishra, A.S. Nia, M.R. Lohe, X. Feng, R. Adelung, O. Lupan, Highly selective and ultra-low power consumption metal oxide based hydrogen gas sensor employing graphene oxide as molecular sieve, *Sensors and Actuators B: Chemical*, **320** (2020) 128363.

[6] B.G. Ghule, N.M. Shinde, S.D. Raut, S.F. Shaikh, A.M. Al-Enizi, K.H. Kim, R.S. Mane, Porous metal-graphene oxide nanocomposite sensors with high ammonia detectability, *Journal of Colloid and Interface Science*, **589** (2021) 401–410.

[7] H. Jiang, D. Jiang, X. Liu, J. Yang, A self-driven PET chip-based imprinted electrochemical sensor for the fast detection of Salmonella, *Sensors and Actuators B: Chemical*, **349** (2021) 130785.

[8] F.D.L. Meza López, S. Khan, G. Picasso, M.D.P.T. Sotomayor, A novel highly sensitive imprinted polymer-based optical sensor for the detection of Pb(II) in water samples, *Environmental Nanotechnology, Monitoring & Management*, **16** (2021) 100497.

[9] L. Fang, M. Jia, H. Zhao, L. Kang, L. Shi, L. Zhou, W. Kong, Molecularly imprinted polymer-based optical sensors for pesticides in foods: Recent advances and future trends, *Trends in Food Science & Technology*, **116** (2021) 387–404.

[10] D. Li, X. Qiu, H. Guo, D. Duan, W. Zhang, J. Wang, J. Ma, Y. Ding, Z. Zhang, A simple strategy for the detection of Pb(II) and Cu(II) by an electrochemical sensor

based on Zn/Ni-ZIF-8/XC-72/Nafion hybrid materials, *Environmental Research*, **202** (2021) 111605.

[11] K.S. Pasupuleti, M. Reddeppa, D.-J. Nam, N.-H. Bak, K.R. Peta, H.D. Cho, S.-G. Kim, M.-D. Kim, Boosting of NO2 gas sensing performances using GO-PEDOT:PSS nano-composite chemical interface coated on langasite-based surface acoustic wave sensor, *Sensors and Actuators B: Chemical*, **344** (2021) 130267.

[12] R. Das, M. Ali, S.B. Abd Hamid, Current applications of x-ray powder diffraction: A review, *Reviews on Advanced Materials Science*, **38** (2014) 95–109.

[13] A.N. Popova, Crystallographic analysis of graphite by X-Ray diffraction, *Coke and Chemistry*, **60** (2017) 361–365.

[14] D.M. Hausmann, R.G. Gordon, Surface morphology and crystallinity control in the atomic layer deposition (ALD) of hafnium and zirconium oxide thin films, *Journal of Crystal Growth*, **249** (2003) 251–261.

[15] M.L. García-Betancourt, C. Magaña-Zavala, A. Crespo-Sosa, Structural and optical properties correlated with the morphology of gold nanoparticles embedded in synthetic sapphire: A microscopy study, *J Microsc Ultrastruct*, **6** (2018) 72–82.

[16] M. Bouroushian, Characterization of thin films by low incidence X-ray diffraction, *Crystal Structure Theory and Applications*, **01** (2012) 35–39.

[17] J. Kappler, N. Bârsan, U. Weimar, A. Dièguez, J.L. Alay, A. Romano-Rodriguez, J.R. Morante, W. Göpel, Correlation between XPS, raman and TEM measurements and the gas sensitivity of Pt and Pd doped SnO2 based gas sensors, *Fresenius' Journal of Analytical Chemistry*, **361** (1998) 110–114.

[18] R. Toyoshima, T. Tanaka, T. Kato, K. Uchida, H. Kondoh, In situ AP-XPS analysis of a Pt thin-film sensor for highly sensitive H2 detection, *Chemical Communications*, **56** (2020) 10147–10150.

[19] J.F. Watts, J. Wolstenholme, Comparison of XPS and AES with other analytical techniques, in: An Introduction to Surface Analysis by XPS and AES, 2019, pp. 223–227.

[20] K.J. Stine, Biosensor applications of electrodeposited nanostructures, *Applied Sciences*, **9** (2019) 797.

[21] W. Kuo, M. Briceno, D. Ozkaya, Characterisation of catalysts using secondary and backscattered electron in-lens detectors, *Platinum Metals Review*, **58** (2014) 106–110.

[22] M. Predikaka, A. Bähr, C. Koffmane, J. Ninković, E. Prinker, R. Richter, J. Treis, A. Wassatsch, EDET DH80k: Characterization of a DePFET based sensor for TEM direct electron imaging, *Nuclear Instruments and Methods in Physics Research Section A: Accelerators, Spectrometers, Detectors and Associated Equipment*, **958** (2020) 162544.

[23] C.S.S.R. Kumar, Raman Spectroscopy for Nanomaterials Characterization, Springer Berlin Heidelberg, Berlin, Heidelberg, 2012.

[24] P.-M. Allemand, K.C. Khemani, A. Koch, F. Wudl, K. Holczer, S. Donovan, G. Grüner, J.D. Thompson, Organic molecular soft ferromagnetism in a fullerene C_{60}, *Science*, **253** (1991) 301–302.

[25] D. Degler, N. Barz, U. Dettinger, H. Peisert, T. Chassé, U. Weimar, N. Barsan, Extending the toolbox for gas sensor research: Operando UV/vis diffuse reflectance spectroscopy on SnO2-based gas sensors, *Sensors and Actuators B: Chemical*, **224** (2016) 256–259.

[26] N.E. Tolouei, S. Ghamari, M. Shavezipur, Development of circuit models for electrochemical impedance spectroscopy (EIS) responses of interdigitated MEMS biochemical sensors, *Journal of Electroanalytical Chemistry*, **878** (2020) 114598.

[27] W. Choi, H.-C. Shin, J.M. Kim, J.-Y. Choi, W.-S. Yoon, Modeling and applications of Electrochemical Impedance Spectroscopy (EIS) for lithium-ion batteries, *J. Electrochem. Sci. Technol*, **11** (2020) 1–13.

[28] F. Scholz, Voltammetric techniques of analysis: The essentials, *ChemTexts*, **1** (2015) 17.

[29] I.D. Likasari, R.W. Astuti, A. Yahya, N. Isnaini, G. Purwiandono, H. Hidayat, W.P. Wicaksono, I. Fatimah, NiO nanoparticles synthesized by using Tagetes erecta L leaf extract and their activities for photocatalysis, electrochemical sensing, and antibacterial features, *Chemical Physics Letters*, **780** (2021) 138914.

[30] G. Fadillah, S. Triana, U. Chasanah, T.A. Saleh, Titania-nanorods modified carbon paste electrode for the sensitive voltammetric determination of BPA in exposed bottled water, *Sensing and Bio-Sensing Research*, **30** (2020) 100391.

[31] G. Fadillah, W.P. Wicaksono, I. Fatimah, T.A. Saleh, A sensitive electrochemical sensor based on functionalized graphene oxide/SnO2 for the determination of eugenol, *Microchemical Journal*, **159** (2020) 105353.

[32] S. Deepa, B.E. Kumara Swamy, K. Vasantakumar Pai, Voltammetric detection of anti-cancer drug Doxorubicin at pencil graphite electrode: A voltammetric study, *Sensors International*, **1** (2020) 100033.

[33] W.P. Wicaksono, I. Sahroni, A.K. Saba, R. Rahman, I. Fatimah, Biofabricated SnO2 nanoparticles using Red Spinach (Amaranthus tricolor L.) extract and the study on photocatalytic and electrochemical sensing activity, *Materials Research Express*, **7** (2020) 075009.

[34] N. Elgrishi, K. Rountree, B. McCarthy, E. Rountree, T. Eisenhart, J. Dempsey, A practical beginner's guide to cyclic voltammetry, *Journal of Chemical Education*, **95** (2017).

[35] M.A. Deshmukh, H.K. Patil, G.A. Bodkhe, M. Yasuzawa, P. Koinkar, A. Ramanaviciene, M.D. Shirsat, A. Ramanavicius, EDTA-modified PANI/SWNTs nanocomposite for differential pulse voltammetry based determination of Cu(II) ions, *Sensors and Actuators B: Chemical*, **260** (2018) 331–338.

[36] V. Mirceski, R. Gulaboski, M. Lovric, I. Bogeski, R. Kappl, M. Hoth, Square-wave voltammetry: A review on the recent progress, *Electroanalysis*, **25** (2013) 2411–2422.

[37] D. Ji, Z. Shi, Z. Liu, S.S. Low, J. Zhu, T. Zhang, Z. Chen, X. Yu, Y. Lu, D. Lu, Q. Liu, Smartphone-based square wave voltammetry system with screen-printed graphene electrodes for norepinephrine detection, *Smart Materials in Medicine*, **1** (2020) 1–9.

[38] M. Arvand, A. Pourhabib, M. Giahi, Square wave voltammetric quantification of folic acid, uric acid and ascorbic acid in biological matrix, *Journal of Pharmaceutical Analysis*, **7** (2017) 110–117.

[39] N.S. Lawrence, Analytical electrochemistry, second edition. By J. Wang, John Wiley & Sons: Chichester, England. £53.95. 207 pp. ISBN 0471–282272–3, *The Chemical Educator*, **7** (2002) 180–181.

[40] T.A. Ivandini, W.P. Wicaksono, E. Saepudin, B. Rismetov, Y. Einaga, Anodic stripping voltammetry of gold nanoparticles at boron-doped diamond electrodes and its application in immunochromatographic strip tests, *Talanta*, **134** (2015) 136–143.

[41] D. Pan, Y. Wang, Z. Chen, T. Lou, W. Qin, Nanomaterial/ionophore-based electrode for anodic stripping voltammetric determination of lead: An electrochemical sensing platform toward heavy metals, *Analytical Chemistry*, **81** (2009) 5088–5094.

[42] A. Bratov, N. Abramova, A. Ipatov, Recent trends in potentiometric sensor arrays: A review, *Analytica Chimica Acta*, **678** (2010) 149–159.

[43] N. Kaur, J. Kaur, R. Badru, S. Kaushal, P.P. Singh, BGO/AlFu MOF core shell nanocomposite based bromide ion-selective electrode, *Journal of Environmental Chemical Engineering*, **8** (2020) 104375.

[44] B. Niemiec, N. Lenar, R. Piech, K. Skupień, B. Paczosa-Bator, Graphene flakes decorated with dispersed gold nanoparticles as nanomaterial layer for ISEs, *Membranes*, **11** (2021) 548.

4 Conducting Polymer Nanocomposites as Sensors

Shumaila, Sunny Khan, M. Zulfequar,
Preeti Singh, Praveen Yadav,
and Om Prakash Yadav

CONTENTS

4.1 INTRODUCTION

Most polymers are considered as insulators since they consist of only covalent bonds without free movable electrons or ions. The advent of intrinsically conducting polymers (CPs) paved the way for understanding polymeric material's chemical composition and electrical characteristics [1]. Conjugated chains with alternating single and double bonds characterize this family of materials. The existence of π-electrons in their conjugated backbone give rise to delocalization of charges into a conduction band, therefore resulting in the conductivity of materials [2]. Furthermore, the nature of the inherent quasi-one-dimensionality of CPs, as well as the level of intra- and interchain delocalization of electrons, influence their physiochemical properties. Alan Heeger, Alan MacDiarmid, and Hideki Shirakawai discovered first CP i.e.

DOI: 10.1201/9781003263852-4

Polyacetylene (PAs) in 1977. They were awarded the Nobel Prize in chemistry for this work in 2000 [3]. Other examples of CPs include polyaniline (PAn), polypyrrole (PPy), polythiophene (PTh), and their derivatives (Figure 4.1) [4].

CPs generally show conduction ranging from 10^{-10} to 10^{-5} S/cm, which can be further increased through doping using a variety of materials. The process of doping combines the conducting behavior of dopants with many advantages from polymeric materials. By altering the level of doping up to a particular percentage of weight, the conductivity of the CPs can be controlled from insulating to highly conducting. As a result, researchers have been drawn to CPs to investigate their potential uses in domains such as EMI shielding, energy storage electrodes, sensors, and flexible electronics. CPs, on the other hand, have a few limitations, such as poor solubility, biocompatibility, and mechanical qualities, which can be solved by chemical modification or compounding with other polymers or nanofillers, resulting in nanocomposite forms.

Polymer nanocomposites (PNCs) are composites that have at least one component of the nanometer scale and a polymer as the major component, or PNCs can also be defined as combinations of a polymer and one or more inorganic/organic nanoparticles. This combination takes place in such a manner so that novel properties of the synthesized nanomaterial can be taken together with the current characteristics of polymer to give new characteristics [5]. Introducing a nanocomponent to a conducting polymer matrix—such as graphene, nanotubes, metal alloys, salts, conventional polymeric materials (blending), and biological agents—is among the best ways to overcome the drawbacks of CPs because the nanoconstituent can change the affinity of the composites, enhance the transport properties of CPs, or even act as a catalyst. In some cases, nanostructures of polymers can also be taken as a constituent of PNCs

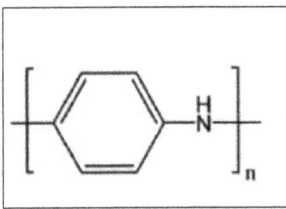

Polyaniline (PAni)

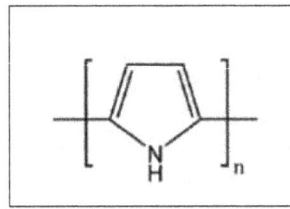

Polypyrrole(PPy)

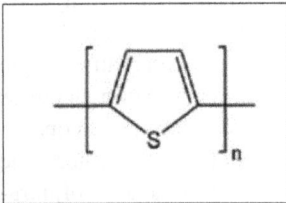

Polythiophene (PTh)

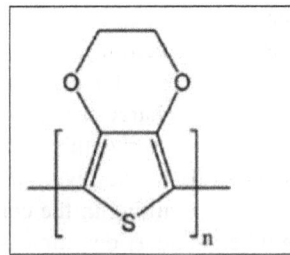

Poly(3,4-ethylenedioxythiophene) (PEDOT)

FIGURE 4.1 Chemical structures of PAni, PPy, PTh and PEDOT.

along with the other organic/inorganic filler which may or may not have a size in nanometer range. The reaction time of the sensor could be substantially faster than standard CPs due to its porous structure and large specific surface area of the nanostructured conducting polymer [6]. In general, nanocomposites have one or perhaps more discontinuous or fragmented phases of scattered particles in one dispersion medium as their fundamental feature. The discontinuous phase is known as "reinforcement" and discontinuous phase is named as "matrix". Discontinuous phase is usually stiff and strong as compared to the continuous phase [7]. Along with other properties, morphology of the constituent components also plays an significant part in defining the properties of resultant nanocomposites [8]. A vast variety of novel nanocomposites with a synergistic activity can be created thanks to the interface between electron donor and acceptor, with probable applications in sensors. PNCs can be simply divided into the following two components.

4.1.1 POLYMER MATRIX

The polymer matrix is major component of composite. This is the continuous phase which holds the reinforcing agent in its place and persuades most of the properties of the PNCs. The molecular structure, kind, and mode of positioning of the electron pair donating atoms or groups in the polymer matrix all play a role in charge flow inside PNCs. The percolation behaviors of PNCs are also influenced by the molecular mass and modulus of the parent polymeric matrix.

4.1.2 FILLERS

fillers play very important roles in the chemistry of PNCs in spite of being a minor component of them. The addition of minute amounts of fillers to the polymer can improve a variety of qualities such as compressive, resistance, temperature, heat resistance, and electrical characteristics. Miscibility/compatibility and morphology, interfacial properties, and performance of the PNCs can also be altered by fillers depending on their localization, their connections with polymer components, and the method of accommodation of these additives within polymer matrix. For the production of PNCs with low percolation thresholds, high aspect-ratio conductive fillers (such as carbon nanotubes [CNTs], carbon nanofiber, nanowires, metal nanoflakes, and graphene) are advised. Nanofillers are divided into three categories based on their dimensions: 1D (one-dimensional), 2D (two-dimensional), and 3D (three-dimensional) fillers.

4.2 SYNTHESIS OF HIGH-PERFORMANCE CONDUCTING POLYMER NANOCOMPOSITES (PNCS)

Lowering the percolation threshold and increasing the conductivity are two major goals for producing high-performance PNCs. The conductive fillers, polymer matrix, and production process all have an impact on these qualities. In order to prepare high-performance PNCs, conductive fillers need to be added as little as possible. Synthesis techniques of conducting polymer nanocomposites (PNCs) can be classified mainly into two categories: chemical and electrochemical synthesis.

4.2.1 CHEMICAL SYNTHESIS OF PNCs

4.2.1.1 Chemical In-Situ Synthesis

The in situ polymerization method is one of the most used methods for synthesis of polyaniline (PAni) nanocomposites. Because of the strong contacts between the polymer matrix and the nanofiller, this technique allows for a more intimate mixing of the two components [9]. In this method, the monomer and filler are added to a common solvent, followed by a polymerization process. To generate the polymer nanocomposite, a nanofiller precursor is sometimes added to the monomer solution, followed by heat or electrochemical treatment. The morphologies of the nanostructures generated in this approach are homogenous inside the polymer matrix [10].

4.2.1.2 Interfacial Synthesis

Interfacial polymerization is usually adopted facile to obtain bulk quantities of polymer nanostructures [11]. A slow oxidation of monomer occurs at the interface of two immiscible solvents in this technique of synthesis, whereby the oxidizing agent is primarily disseminated in aqueous medium and the relevant monomer is deposited in organic medium.

4.2.1.3 Solution/Dispersion Mixing

This is one of the most straightforward ways to make polymer composites. In this method, a polymer is dissolved in a suitable solvent, and then nanofiller is added to the polymer solution while it is constantly stirred. The main benefit of the solvent casting method is its simplicity, as it does not require any special equipment. Polymer structure, quantity, solvent type, temperature, and stirring frequency are a few of the many elements that influence the final result.

4.2.1.4 Emulsion Polymerization

An emulsion system is generally made using water, oil, and surfactant, with a superior surfactant being utilized to stabilize the emulsion system. The surfactant's function in an emulsion is to improve the stability of its droplets (oil or water), and the amount affects the size of the droplets, allowing for the selective synthesis of either nanograins or nanofibers.

4.2.2 ELECTROCHEMICAL SYNTHESIS

A three-electrode cell with a working electrode, a counter electrode, and a reference electrode in an electrolyte solution is used in a common electrochemical approach for the manufacture of PNCs. Electrochemical synthesis of nanocomposites has several advantages over chemical synthesis, including the elimination of oxidizing agents; the polymer obtained is expected to be in a relatively more pure form (as no additional chemicals such as surfactant or oxidant are used); the electrochemically synthesized polymer has better adherence than chemically synthesized polymer; and the thickness and morphology of the polymer film can be easily controlled [11, 12].

4.3   SENSOR APPLICATIONS OF PNCS

Sensor technology is mostly based on different parameters like sensitivity, selectivity, and response time, which are crucial for the functioning of a sensor [13]. The two primary components of a sensor are the receptor, which is responsible for the surface reaction between the host material and the material to be sensed, and the transducer, which is responsible for translating the surface reaction into changes in the electrical resistance of the sensor. (Figure 4.2) [14]. Conducting polymers have also been employed in a variety of sensors as a signal transducer. In most circumstances, a response time of few seconds is sufficient for human recognition to meet the requirements more easily than other metrics. Owing to their chemical and structural diversity, CPs outreach the inorganic counterparts in achieving good sensitivity and selectivity. Redox reactions, volume and weight changes, chain conformational changes, screening, and charge transfer are all used to detect CPs. PNCs are the most attractive choices for sensor applications due to their strengths, low-temperature synthesis, ease of processing, large surface-to-volume area, versatility, and cost effectiveness. On the basis of different transduction modes, PNCs-based sensors can be classified into the following five main classes [15].

1. Conductometric mode
2. Potentiometric mode
3. Amperometric mode
4. Colorimetric mode
5. Gravimetric mode

The change in electrical conductivity of the PNCs in respect to analyte interaction is monitored in conductometric mode. The conductivity is usually calculated as a function of the concentration of the analyte. A fixed potential in solution can also be used to monitor it. When no current is flowing, the potentiometric mode measures the change in the system's chemical potential, which is generated by the analyte. Changes in the system's chemical potential are caused by shifts in the anion balance inside the polymer film caused by doping/de-doping or redox activities. The current generated by the redox reaction of an analyte at a sensing (working) electrode is measured in amperometric mode. Faraday's law and a dynamic reaction govern the current, resulting in steady-state conditions in the system. Sensors in the colorimetric mode measure changes in optical absorption properties that are influenced

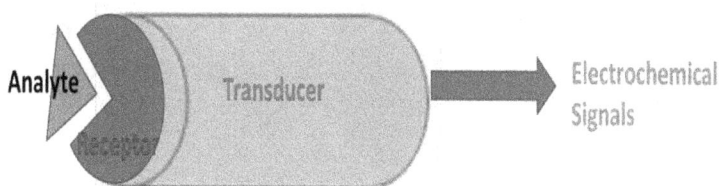

Analyte   Transducer   Electrochemical Signals

Receptor

FIGURE 4.2   Schematic representation of a sensor.

by the local electrical structure. Because conducting polymers' bandgaps are sensitive to analyte-induced changes, this sort of sensor is a potential option. The weight change of PNCs as a result of analyte–polymer interaction is the most important factor in gravimetric mode. A quartz crystal microbalance can easily detect even minute weight changes in polymer. Various chemical and biological items have been detected using sensors based on conducting polymer nanocomposites that follow these transduction principles.

4.3.1 PNCs in Biosensors

Biosensors are analytical devices that connect biological matter such as monosaccharides, cells, and nucleic acids to a detector/transducer, causing biomolecules to respond, which the detector/transducer then turns into a computable signal (Figure 4.3). Biosensors are made up of three basic parts: a biorecognition element, an immobilization surface with PNCs, and a detector/transducer unit [16]. PNCs are the most adaptable materials for biosensing since they may be used as immobilization matrices, receptors, and transducers (redox systems for the transmission of electrical charge) in a biosensor. Recently, nanostructure conducting polymer-based sensors have come up in limelight due to the capability to control their dimensions and structures, and henceforth the properties of final nanocomposites.

Among all the CPs, PAni has shown utmost importance because of its amazing characteristics, such as ease of synthesis, control over shape and size, and low cost of monomer as compared to other monomers of the class [17]. PAni also demonstrated two redox couples that facilitate charge transfer between the polymer and the enzyme. As a result, PAni works as a self-contained electron transfer mediator in the system, and no extra diffusional mediators are required in the biosensor. Subsequently, the biosensor can achieve excellent long-term stability because the mediator is localized to the sensor's surface, which prevents the mediator from leaching into the media [18]. Many different types of biosensors using conducting polymer nanocomposites

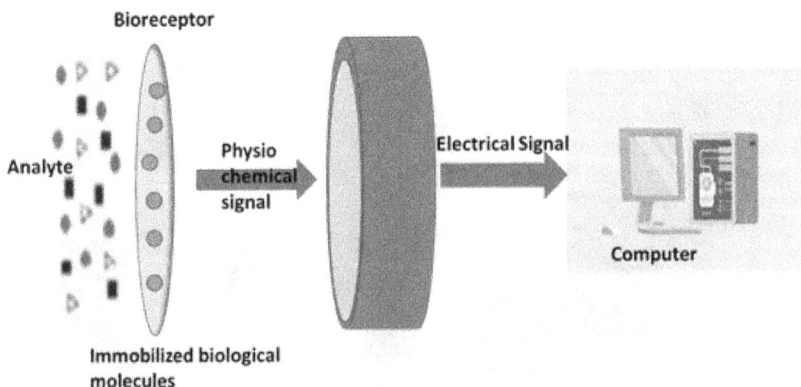

FIGURE 4.3 Schematic representation of a biosensor.

have been reported to detect glucose [19–21], lactation [22], urea [23], DNA [24, 25] Carcinoembryonic antigens [26, 27], cholesterol [28–30], tuberculosis [31, 32], etc.

4.3.2 PNCs in Gas Sensors

PNCs have been utilized to develop gas/vapor sensors in various transduction modes during the last few years. These PNCs-based gas sensors are more sensitive and have a shorter temporal response at ambient temperature than currently available metal oxide (MO)-based gas sensors, which control both chemical and physical properties by applying different substituents. The gas-sensing response of PNCs-based sensors is caused by physical absorption of target analyte molecules onto sensing films, as well as the electron capture/donation process of the polymer matrix embedded with nanomaterials, which is aided by junction effects at the conducting polymer-inorganic interfaces. Nonetheless, the majority of MO-based gas sensors operate at high temperatures (200–500°C) and require a lot of electricity to function. As a result, great work has gone into developing gas sensors that consume little power and operate at low temperatures. Creating a p–n heterojunction between p-type polymers and n-type nanoscale semiconducting metal oxides looks to be a potential method for increasing their efficiency at ambient temperatures. Many studies on gas sensors based on the combination of conducting polymers and MO nanostructures have been published, with impressive results [33, 34].

Carbon nanotubes (CNTs) have also been included into CPs to increase gas-sensing properties including sensitivity and selectivity, lower detection limits, and expansion of detection capacity of a growing number of gases at ambient temperature. Furthermore, by embedding CNTs in a polymer matrix, the dispersibility problem of CNTs is reduced, and more flexibility is offered for modifying CNT

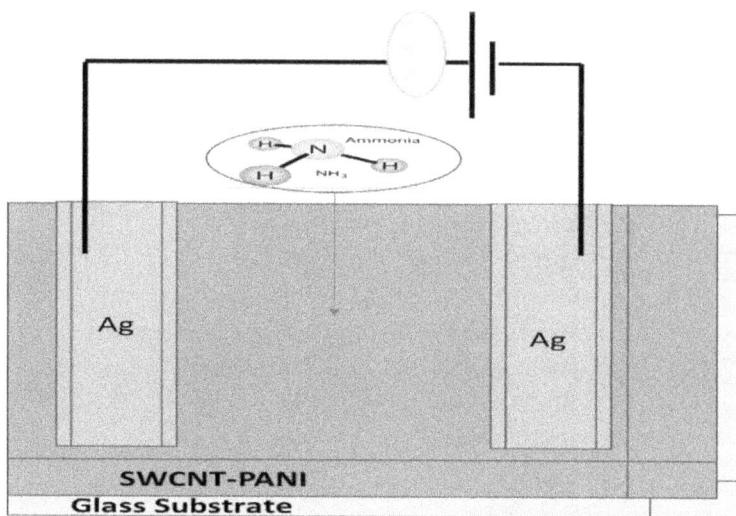

FIGURE 4.4 Schematic diagram of PAni/CNTs based gas (ammonia) sensor.

electronic characteristics, making them perfect candidates for gas-sensing applications [35]. To increase the sensing capability of graphene-based binary nanocomposites, graphene-based ternary nanocomposites have also been created. In this study, a noble metal-conducting polymer or a metal oxide-conducting polymer was hybridized with graphene to combine their advantages in the production of a suitable sensor [36]. PNCs have been combined with different fillers including metal oxides, nanostructures of metals, CNTs, and graphene and multicomponent systems in order to make a number of gas sensors (Figure 4.4). A summary of different conducting polymer nanocomposite-based sensors is presented in Table 4.1.

4.4 CONCLUSION AND FUTURE PROSPECTS

Applications of conducting polymer nanocomposites as sensors have been discussed in this chapter. Mainly fabrication methods, working types, importance, and applications of the nanocomposites of CPs have been presented and evaluated. Both the advantages and drawbacks of PNCs as gas sensors, as well as biosensors, are discussed. As compared to their bulk counterparts, CPs in their nanostructured forms are more efficient in sensing because of their high surface area/high surface-to-volume ratio for the diffusion of analyte into the polymer matrix, and vice versa. Apart from this the morphology and thickness also play a crucial role in the sensing

TABLE 4.1
A Summary of Different Conducting Polymer Nanocomposite-Based Sensors

Contucting Polymer	Filler	Analyte	Reference
PAni	TiO_2, SnO_2	CO	[33]
PPy	WO_3	NO_2	[34]
PAni	SWCNTs	NH_3	[35]
PAni	Graphene/TiO_2	NH_3	[36]
PPy	SWCNTs	Glucose	[19]
PAni	NiO2/Graphene oxide	Glucose	[20]
PAni	Pt	Glucose	[21]
PPy	MWCNTs/Lactate oxidase	Lactate	[22]
PPy	MWCNTs/Urease	Urea	[23]
PAni	Gold	DNA	[24, 25]
PAni	Potassium ferricyanide	Carcinoembryonic antigen	[26]
PAni	Poly (carboxybetaine)	Carcinoembryonic antigen	[27]
PAni	MWCNTs/Pt	Cholesterol	[28]
PAni	MWCNTs/Starch	Cholesterol	[29]
PAni	Graphene	Cholesterol	[30]
PAni	Graphene/Gold	Tuberculosis	[31]
PAni	CNTs	Tuberculosis	[32]

because of inter-domain spacing, and further reducing the analyte-polymer interactions. The nanocomposites of polyaniline and polypyrrole have come out to be most efficient for use as gas/biosensor. The main challenge here is to choose a suitable nanoparticle/polymer for the fabrication of a compact PNCs sensor. In addition, the uniform mixing of polymer and the filler is also challenging to get the best results as a sensor, especially with cost-effective and fast methods. Some of these sensors have shown few minor problems, such as difficult mass production for commercialization and also relatively low repeatability of the process compared to other commercialized methods. More combinations of these polymers are needed to be explored for the detection of oxidizing and reducing gases. In the future, it is expected to use more novel nanoparticles/nanostructures to be combined with CPs for application in sensors to enhance the sensitivity and repeatability.

REFERENCES

[1] Shirakawa, H. (2001). The Discovery of Polyacetylene Film: The Dawning of an Era of Conducting Polymers (Nobel Lecture). *Angew Chem Int Ed*, 40, 2574.
[2] Kaur, G., Adhikari, R., Cass, P., Bown, M., & Gunatillake, P. (2015). Electrically conductive polymers and composites for biomedical applications. *RSC Advances*, 5, 37553.
[3] Huang, J., & Kaner, R. B. (2006). Conjugated Polymers: Theory, Synthesis, Properties, and Characterization, 3rd ed.; Skotheim, T. A., & Reynolds, J., Eds.; CRC Press: Boca Raton, FL.
[4] Zare, E. N., Lakouraj, M. M., Moghadam, P. N., & Azimi, R. (2013). Novel Polyfuran/Functionalized Multiwalled Carbon Nanotubes Composites with Improved Conductivity: Chemical Synthesis, Characterization, and Antioxidant Activity. *Polym Compos.*, 34, 732–739.
[5] Gangopadhyay, R., & De, A. (2000). Conducting Polymer Nanocomposites: A Brief Overview. *Chem. Mater.*, 12, 608.
[6] Huang, J. X., Virji, S., Weiller, B. H., & Kaner, R. B. (2003). Polyaniline Nanofibers: Facile Synthesis and Chemical Sensors. *J. Am. Chem. Soc.*, 125, 314.
[7] Weisenberger, M. C., Grulke, E. A., Jacques, D., Ramtell, T., & Andrews, R. (2003). Enhanced Mechanical Properties of Polyacrylonitrile: Multiwall Carbon Nanotube Composite Fibers. *J. Nanosci. Nanotechnol.*, 3(6), 535.
[8] Dalton, A. B., Coolins, S., Muñoz, E., Razal, J. M., Ebron, V. H., & Ferraris, J. P. (2003). Super-Tough Carbon-Nanotube Fibres. *Nature*, 423, 703.
[9] Zare, E. N., & Lakouraj, M. M. (2014). Biodegradable Polyaniline/Dextrin Conductive Nanocomposites: Synthesis, Characterization, and Study of Antioxidant Activity and Sorption of Heavy Metal Ions. *Iran. Polym. J.*, 23, 257–266.
[10] Zhan, C., Yu, G., Lu, Y., Wang, L., Wujcik, E., & Wei, S. (2017). Conductive Polymer Nanocomposites: A Critical Review of Modern Advanced Devices. *J. Mater. Chem. C*, 5(7), 1569–1585.
[11] Shao, D., Hu, J., Chen, C., Sheng, G., Ren, X., & Wang, X. (2010). Polyaniline Multiwalled Carbon Nanotube Magnetic Composite Prepared by Plasma-Induced Graft Technique and Its Application for Removal of Aniline and Phenol. *J. Phys. Chem. C*, 114, 21524.
[12] Ameen, S., Ansari, S. G., Song, M., Kim, Y. S., & Shin, H. S. (2009). Fabrication of Polyaniline/TiO_2 Heterojunction Structure Using Plasma Enhanced Polymerization Technique. *Superlattices Microstruct.*, 46, 745.
[13] Rahman, M. A., Kumar, P., Park, D. S., & Shim, Y. B. (2008). Electrochemical Sensors Based on Organic Conjugated Polymers. *Sensors*, 8, 118–141.

[14] Cabot, A., Vila, A., & Morante, J. R. (2002). *Sens Actuators B*, 84, 12.

[15] Stetter, J. R., & Li, J. (2008). Amperometric Gas Sensors: A Review. *Chem. Rev.*, 108, 352–366.

[16] Xia, L., Wei, Z., & Wan, M. (2010). Conducting Polymer Nanostructures and Their Application in Biosensors. *J. Colloid Interface Sci.*, 341, 1–11.

[17] Xia, L., Wei, Z., & Wan, M. (2010). Conducting Polymer Nanostructures and Their Application in Biosensors. *J. Colloid Interface Sci.*, 341, 1–11.

[18] Dhand, C., Das, M., Datta, M., & Malhotra, B. D. (2011). Recent Advances in Polyaniline Based Biosensors. *Biosens. Bioelectron.*, 26(6), 2811–2821.

[19] Callegari, A., Cosnier, S., Marcaccio, M., Paolucci, D., Paolucci, F., Georgakilas, V., Tagmatarchis, N., Vazquez, E., & Prato, M. (2004). Functionalised Single Wall Carbon Nanotubes/Polypyrrole Composites for the Preparation of Amperometric Glucose Biosensors. *J. Mater. Chem.*, 14, 807.

[20] Zhuang, X., Tian, C., Luan, F., Wu, X., & Chen, L. (2016). One-Step Electrochemical Fabrication of a Nickel Oxide Nanoparticle/Polyaniline Nanowire/Graphene Oxide Hybrid on a Glassy Carbon Electrode for Use as a Non-Enzymatic Glucose Biosensor. *RSC Adv.*, 6(95), 92541–92546.

[21] Zhai, D., Liu, B., Shi, Y., Pan, L., Wang, Y., Li, W., Zhang, R., & Yu, G. (2013). Highly Sensitive Glucose Sensor Based on Pt Nanoparticle/Polyaniline Hydrogel Heterostructures. *ACS Nano*, 7(4), 3540–3546.

[22] Meshram, B. H., Kondawar, S. B., Mahajan, A. P., Mahore, R. P., & Burghate, D. K. (2014). Conducting Polymer Nanocomposites for Sensor Applications. *J Chin Adv Mater Soc*, 2, 223.

[23] Guo, M., Chen, J., Li, J., Tao, B., & Yao, S. (2005). Fabrication of Polyaniline/Carbon Nanotube Composite Modified Electrode and Its Electrocatalytic Property to the Reduction of Nitrite. *Anal. Chim. Acta*, 532(1), 71–77.

[24] Shoaie, N., Forouzandeh, M., & Omidfar, K. (2018). Highly Sensitive Electrochemical Biosensor Based on Polyaniline and Gold Nanoparticles for DNA Detection. *IEEE Sens. J.*, 18(5), 1835–1843.

[25] Norouzi, P., Eshraghi, M. A., & Ebrahimi, M. (2019). DNA Biosensor for Determination of 5-Fluorouracil Based on Gold Electrode Modified with Au and Polyaniline Nanoparticles and FFT Square Wave Voltammetry. *J. Appl. Chem. Res.*, 13(1), 24–35.

[26] He, S., Wang, Q., Yu, Y., Shi, Q., Zhang, L., & Chen, Z. (2015). One-Step Synthesis of Potassium Ferricyanide-Doped Polyaniline Nanoparticles for Label-Free Immunosensor. *Biosens. Bioelectron.*, 68, 462–467.

[27] Wang, J., & Hui, N. (2019). Zwitterionic Poly (Carboxybetaine) Functionalized Conducting Polymer Polyaniline Nanowires for the Electrochemical Detection of Carcinoembryonic Antigen in Undiluted Blood Serum. *Bioelectrochemistry*, 125, 90–96.

[28] Xu, Z., Cheng, X., Tan, J., & Gan, X. (2016). Fabrication of Multiwalled Carbon Nanotube-Polyaniline/Platinum Nanocomposite Films toward Improved Performance for a Cholesterol Amperometric Biosensor. *Biotechnol. Appl. Biochem.*, 63(6), 757–764.

[29] Gautam, V., Singh, K. P., & Yadav, V. L. (2018). Polyaniline/MWCNTs/Starch Modified Carbon Paste Electrode for Non-Enzymatic Detection of Cholesterol: Application to Real Sample (Cow Milk). *Anal. Bioanal. Chem.*, 410(8), 2173–2181.

[30] Lakshmi, G., Sharma, A., Solanki, P. R., & Avasthi, D. K. (2016). Mesoporous Polyaniline Nanofiber Decorated Graphene MicroFlowers for Enzyme-Less Cholesterol Biosensors. *Nanotechnology*, 27(34), 345101–345111.

[31] Mohd Azmi, U., Yusof, N., Kusnin, N., Abdullah, J., Suraiya, S., Ong, P., Ahmad Raston, N., Abd Rahman, S., & Mohamad Fathil, M. (2018). Sandwich Electrochemical

Immunosensor for Early Detection of Tuberculosis Based on Graphene/Polyaniline-Modified ScreenPrinted Gold Electrode. *Sensors*, 18(11), 3926–3940.

[32] Chen, Y., Guo, S., Zhao, M., Zhang, P., Xin, Z., Tao, J., & Bai, L. (2018). Amperometric DNA Biosensor for Mycobacterium Tuberculosis Detection Using Flower-Like CarbonNanotubes-Polyaniline Nanohybrid and Enzyme-Assisted Signal Amplification Strategy. *Biosens. Bioelectron.*, 119, 215–220.

[33] Ram, M., Yavuz, O., Lahsangah, V., & Aldissi, M. (2005). CO Gas Sensing from Ultrathin Nano-Composite Conducting Polymer Film. *Sens Actuators B*, 106, 750.

[34] Mane, A. T., Navale, S. T., & Patil, V. B. (2015). Room Temperature NO_2 Gas Sensing Properties of DBSA Doped PPy–WO_3 Hybrid Nanocomposite Sensor. *Org Electron*, 19, 15.

[35] Ansari, N., Lone, M. Y., Shumaila, Ali, J., Zulfequar, M., Husain, M., Islam, S. S., & Husain, S. (2020). Trace Level Toxic Ammonia Gas Sensing of Single-Walled Carbon Nanotubes Wrapped Polyaniline Nanofibers. *J. Appl. Phys.*, 127, 044902, https://doi.org/10.1063/1.5113847

[36] Tian, J., Yang, G., Jiang, D., Su, F., & Zhang, Z. (2016). A Hybrid Material Consisting of Bulk-Reduced TiO_2, Graphene Oxide and Polyaniline for Resistance Based Sensing of Gaseous Ammonia at Room Temperature. *Microchim Acta*, 183, 2871–2878, https://doi.org/10.1007/s00604-016-1912-6

5 Carbon-Based Functional Nanomaterials for the Detection of Volatile Organic Compounds

Bhasker Pratap Choudhary and N. B. Singh

CONTENTS

DOI: 10.1201/9781003263852-5

5.1 INTRODUCTION

Nanomaterials (NMs) are the major component of nanoscience and nanotechnology. It is an interdisciplinary and broad area of research and is growing worldwide with great speed. It has already started revolutionizing the materials for use in different sectors. It has also significant commercial impact, which will increase considerably in coming days. The word nano comes from Greek word "nanos", meaning "dwarf". It is used to describe one-billionth of something, and is represented by 10^{-9}. NMs are normally defined as materials having size less than "100 nm at least in one dimension" (Zhu Today et al., 2021). The conceptual foundation of nanotechnology was given by Richard Feynman, in his lecture entitled "There's Plenty of Room at the Bottom" in 1959 (Feynman, 1960). It was found by the researchers that those physicochemical properties of NMs depend on size, and because of this, NMs attracted considerable interest of scientists and researchers. The basic differences between NMs and bulk materials are given in Figure 5.1.

Materials and NMs are the backbone of modern society. Out of different materials, carbon-based materials play an important role in civilization. Carbon, the most important element in the periodic table, has special importance (Hirsch, 2010; Speranza, 2021). Carbon has a wide spectrum of compounds and allotropic forms. Carbon at the nanoscale (e.g., nanotubes, fullerenes, nanofibers, nanocones, nanodiamonds, graphene, graphene nanoribbons, etc.) with relatively large surface areas

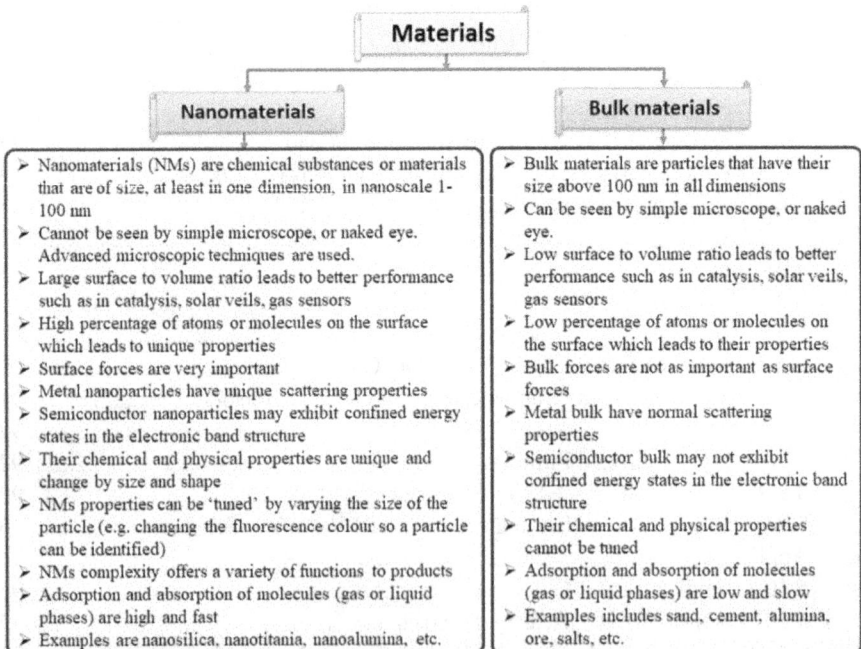

Materials

Nanomaterials

Bulk materials

Nanomaterials	Bulk materials
➢ Nanomaterials (NMs) are chemical substances or materials that are of size, at least in one dimension, in nanoscale 1-100 nm	➢ Bulk materials are particles that have their size above 100 nm in all dimensions
➢ Cannot be seen by simple microscope, or naked eye. Advanced microscopic techniques are used.	➢ Can be seen by simple microscope, or naked eye.
➢ Large surface to volume ratio leads to better performance such as in catalysis, solar veils, gas sensors	➢ Low surface to volume ratio leads to better performance such as in catalysis, solar veils, gas sensors
➢ High percentage of atoms or molecules on the surface which leads to unique properties	➢ Low percentage of atoms or molecules on the surface which leads to their properties
➢ Surface forces are very important	➢ Bulk forces are not as important as surface forces
➢ Metal nanoparticles have unique scattering properties	➢ Metal bulk have normal scattering properties
➢ Semiconductor nanoparticles may exhibit confined energy states in the electronic band structure	➢ Semiconductor bulk may not exhibit confined energy states in the electronic band structure
➢ Their chemical and physical properties are unique and change by size and shape	➢ Their chemical and physical properties cannot be tuned
➢ NMs properties can be 'tuned' by varying the size of the particle (e.g. changing the fluorescence colour so a particle can be identified)	➢ Adsorption and absorption of molecules (gas or liquid phases) are low and slow
➢ NMs complexity offers a variety of functions to products	➢ Examples includes sand, cement, alumina, ore, salts, etc.
➢ Adsorption and absorption of molecules (gas or liquid phases) are high and fast	
➢ Examples are nanosilica, nanotitania, nanoalumina, etc.	

FIGURE 5.1 Differences between NMs and bulk materials.

Source: (Saleh, 2020)

exhibit novel properties and used in different industrial sectors. Functionalization of carbon NMs affects different properties such as their compatibility to biocompatibility and toxicity towards the environment and living organisms. This makes functionalized NMs a material with huge scope. One of the important applications of carbon NMs is their use as sensors. In this chapter, different aspects of carbon NMs with special reference to functional carbon NMs for the detection of VOCs; a family of hydrocarbon compounds have been discussed.

5.2 CLASSIFICATION OF NMS

There are number of criteria to divide NMs, as represented in Figure 5.2 (Saleh, 2020).

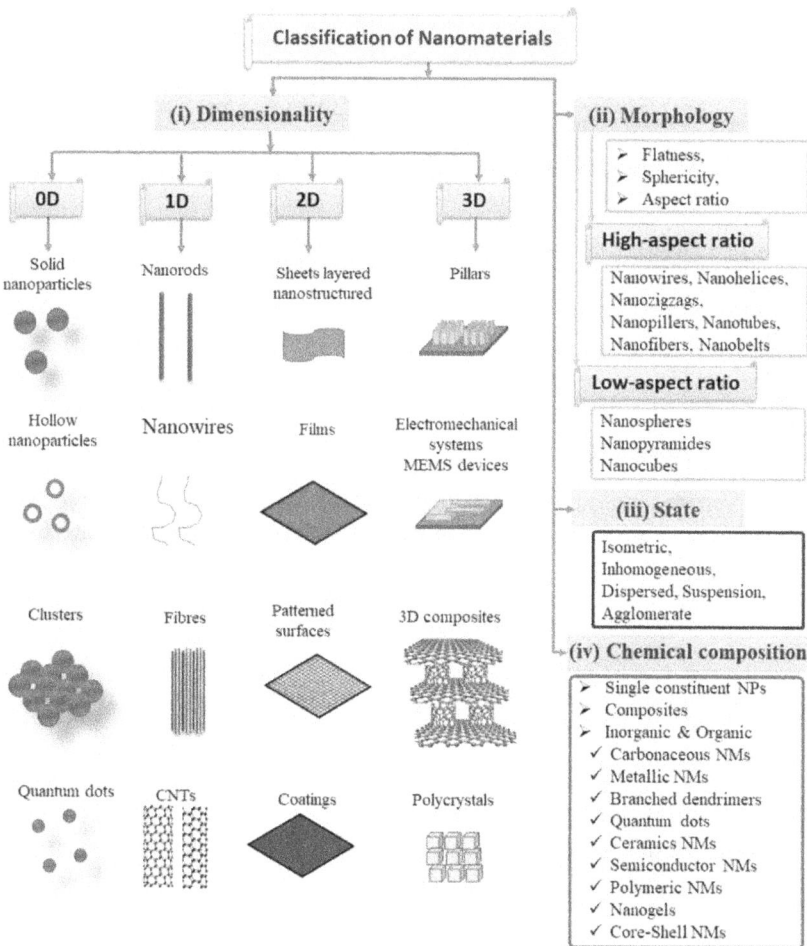

FIGURE 5.2 Classification of NMs.

Source: (Saleh, 2020)

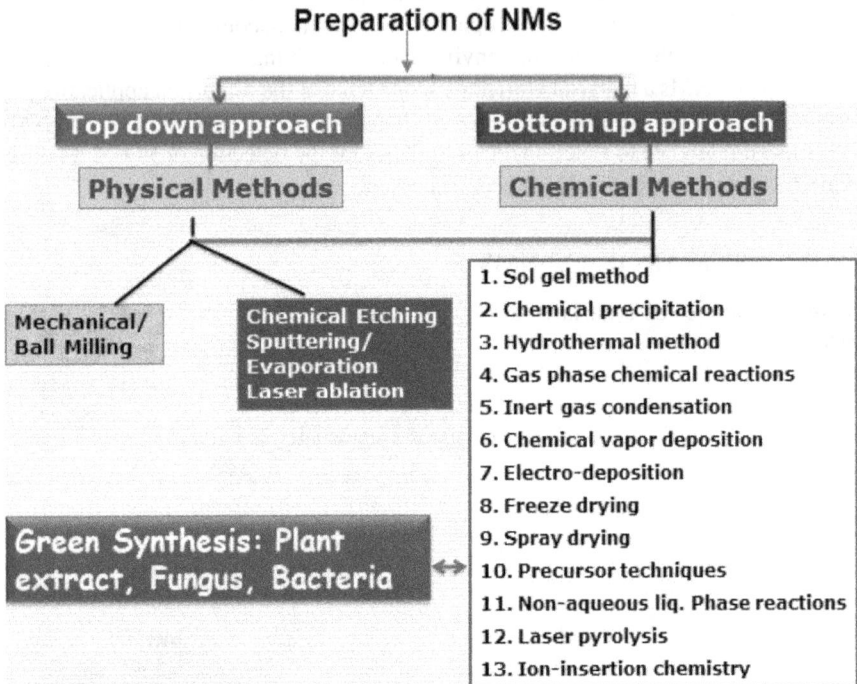

FIGURE 5.3   Preparation of NMs.

Source: author's creation

5.3 PREPARATION OF NMS

NMs are prepared in different ways. Basically, there are two approaches: top-down and bottom-up approaches. In the top-down approach, physical methods such as mechanical/ball milling, sputtering, laser ablation, etc., are used. In chemical methods, there are two approaches: chemical routes and green routes. Different methods are given in Figure 5.3.

5.4 FUNCTIONALIZATION OF NMS

Despite the fact that NMs are highly reactive and possess tunable, size-dependent properties, they have number of limitations which can be tackled by functionalization. Functionalization improves the characteristics of the NMs. Surface modification of NPs affects a number of characteristics such as surface chemistry, size, charge, hydrophilicity and hydrophobicity, etc., and thus could change the solubility, biodistribution, biocompatibility, and clearance of developed nanoprobes. They have also good physical properties and anti-agglomeration, anti-corrosion and non-invasive characteristics. Functionalized NMs have applications in various fields such as biomedicine, bioenergy, and biosensors—and as a result have drawn lot of attention from scientists.

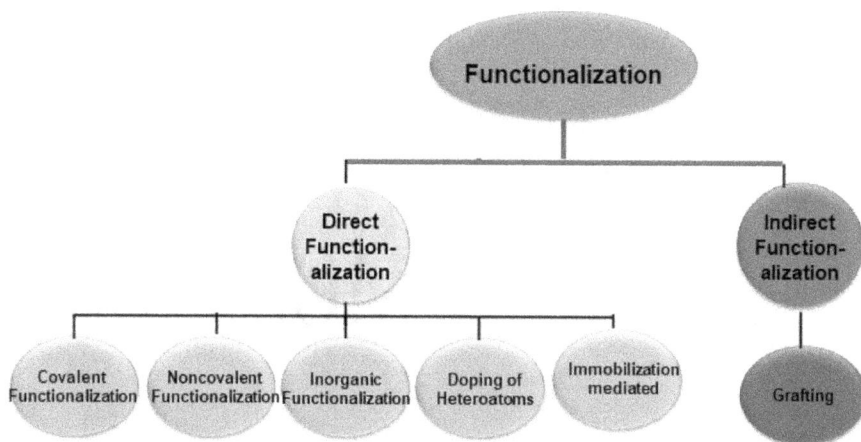

FIGURE 5.4   Types of functionalization.

Basically, there are two methods of functionalization: direct and indirect functionalizations, whereby a number of groups can be attached at the surface of NMs (Figure 5.4) (Darwish and Mohammadi, 2018). In direct functionalization method, species to be attached (ligand) should contain at least two functional groups: one for binding with the nanoparticle surface (complexing agent/chelating group) and another as active functional group (surface modifying group/functional group). The basic advantage of the method is that it is a one-step process and a variety of bifunctional moieties are available for functionalization. In post-synthesis functionalization, the ligand is grafted on the surface of the nanoparticle after the synthesis. Like the previous method, here also a bifunctional ligand is used. In the first step, the binding group can attach itself on the nanoparticle surface, while in the second step, the active functional groups can be modified (Subbiah et al., 2010).

5.5 CARBON NANOMATERIALS (CNMS) AND THEIR CLASSIFICATION

Different type of Carbon nanoparticles (CNPs) was discovered during different times. Fullerenes were discovered by Kroto et al. (1985). Carbon nanotubes (CNTs) (Iijima, 1991), carbon nanodiamonds, carbon nanofibers, carbon nano-onions, and other forms were discovered from time to time (Kharisov and Kharissova, 2019; Mostofizadeh et al., 2011; Tiwari et al., 2012). Carbon materials are generally classified based on their dimensionality (D) from zero to three dimensions: 0D, 1D, 2D, and 3D. Carbon nanostructures may be considered as number of NMs (Figure 5.5) (Abd-Elsalam, 2020).

5.6 SYNTHESIS OF CNMS

A number of techniques have been developed for preparation of CNMs (Iijima, 1991; Li et al., 1996; Yudasaka et al., 1997). Some of the techniques are given in Figure 5.6 (Saleh and Gupta, 2016).

(A)

(B)

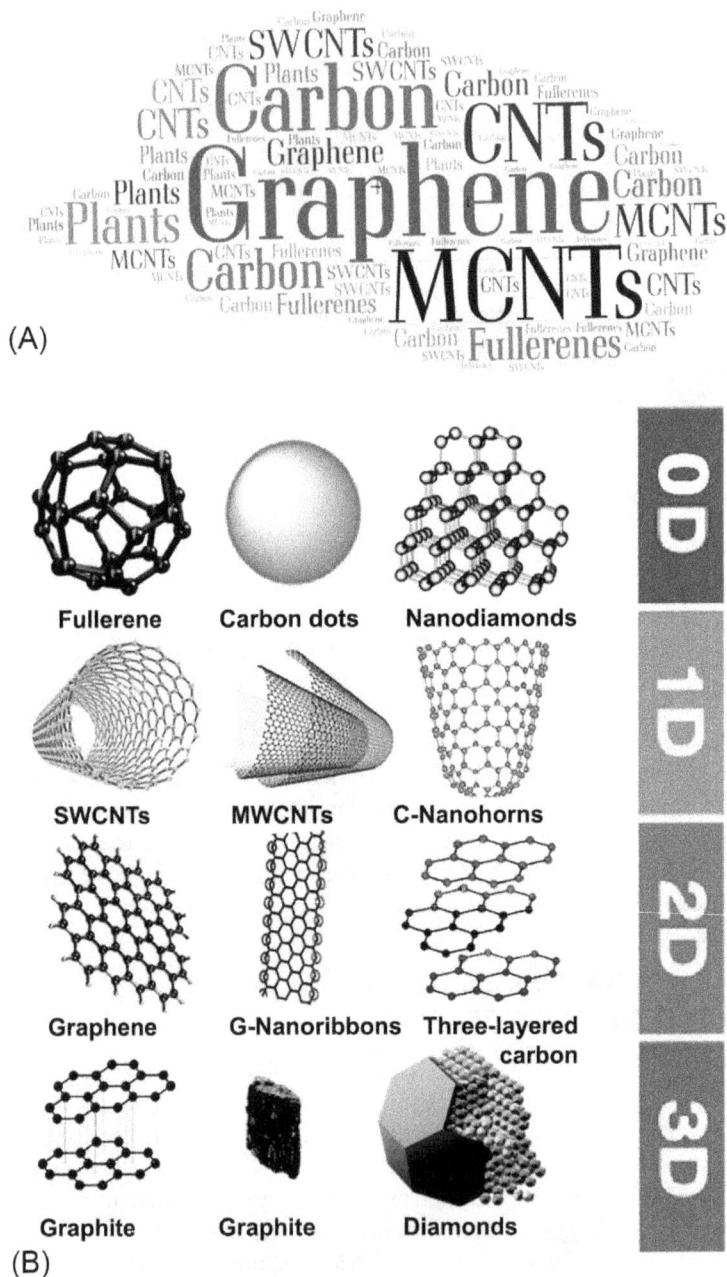

FIGURE 5.5   Different types of carbon NMs.

Source: (Abd-Elsalam, 2020)

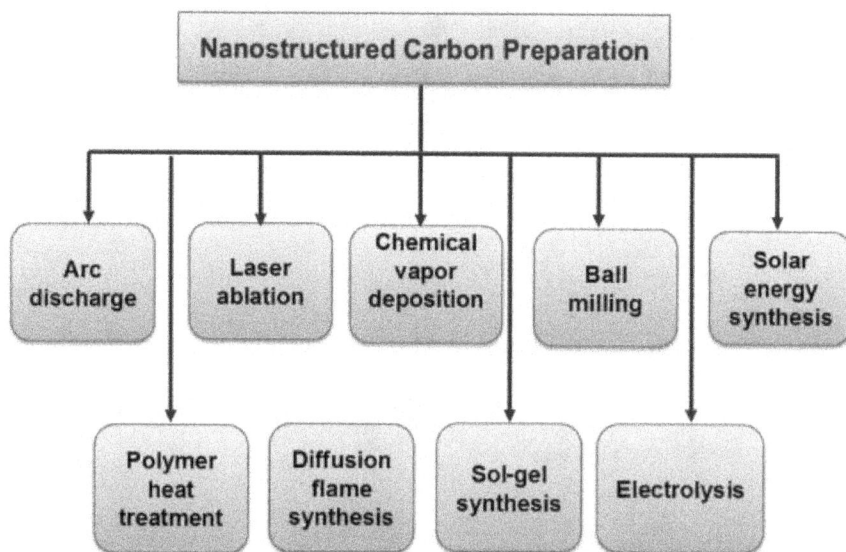

FIGURE 5.6   Different methods of preparation of CNMs.

5.7 FUNCTIONALIZATION OF CNMS

Special structures of CNMs force them to interact with compounds through cova-lent and non-covalent bonds. Non-covalent interactions are generally van der Waals forces, hydrogen bonding, π-π stacking, hydrophobic interactions, etc. In this sec-tion, functionalizations of CNMs (CNT, graphene, fullerene, CND, etc.) have been discussed.

5.7.1 FUNCTIONALIZATION OF CARBON NANOTUBE (CNT)

CNT—having a diameter of the order of 1 nm and length ranging from a few nano-meters to microns—is the most important form of carbon. CNT is configurationally equivalent to a 2D graphene sheet rolled into a tube. Carbon nanotubes are prepared by number of methods, as given in Figure 5.7 (Purohit et al., 2014)

As a result of oxidation, carboxyl groups are added up on the surface of CNT, which are useful for further modifications. This facilitates covalent coupling of molecules through the creation of amide and ester bonds (Figure 5.8). Due to this, functional moieties of wide ranges are created. van der Waals interactions between CNTs are reduced in the presence of carboxyl groups, increasing the separation of nanotubes. Suitable group attachments increase the solubility in organic or aqueous solvents, with a possibility of further modifications (Fernando et al., 2004; Hamon et al., 1999; Chen et al., 2001; Balasubramanian and Burghard, 2005).

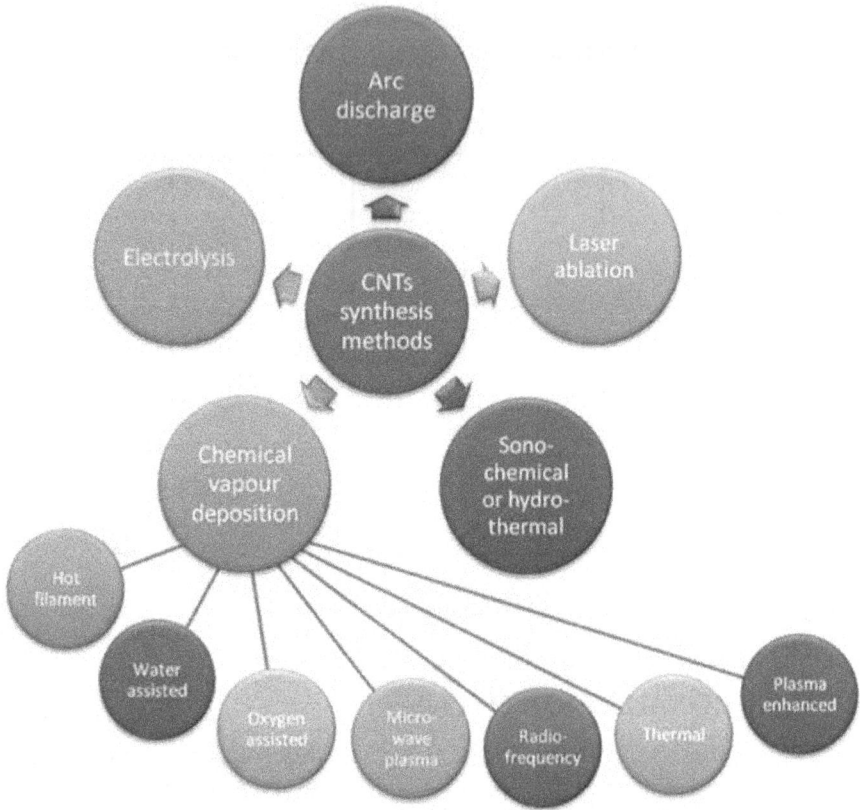

FIGURE 5.7   Synthesis methods of CNTs.

Source: (Purohit et al., 2014)

A number of addition reactions for functionalization are known, and the most important are shown in Figure 5.9.

Nucleophilic substitution reactions can occur on fluorinated CNTs (Figure 5.10).

5.7.2 FUNCTIONALIZATION OF GRAPHENE (G)

Graphene is considered to be the most important achievements in the field of science and technology since 2004. Graphene with π-rich electronic structure has two-dimensional honeycomb lattice of sp^2 hybridized carbon atoms. Graphene is easily converted to graphene oxide (GO) and reduced graphene oxide (rGO). G is hydrophobic, whereas GO is hydrophilic in nature. Because of this, GO is easily dispersible in water. GO contains both aliphatic (sp^3) and aromatic (sp^2) domains, and due to this, surface interactions are increased (Georgakilas et al., 2016). Forms of graphene are shown in Figure 5.11 (Tiwari et al., 2020).

Graphene can be functionalized both by covalent and non-covalent interactions. The surface interactions very much depend on the structure of graphene. Defects and

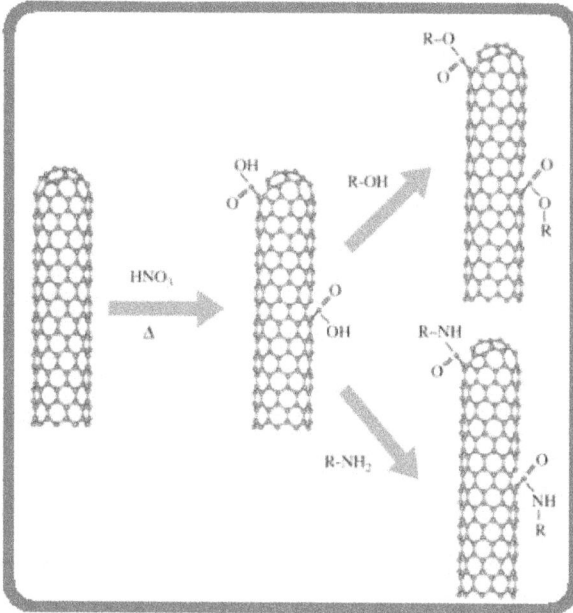

FIGURE 5.8 Thermal oxidation of CNT followed by esterification or amidization of the carboxyl groups.

Source: (Balasubramanian and Burghard, 2005)

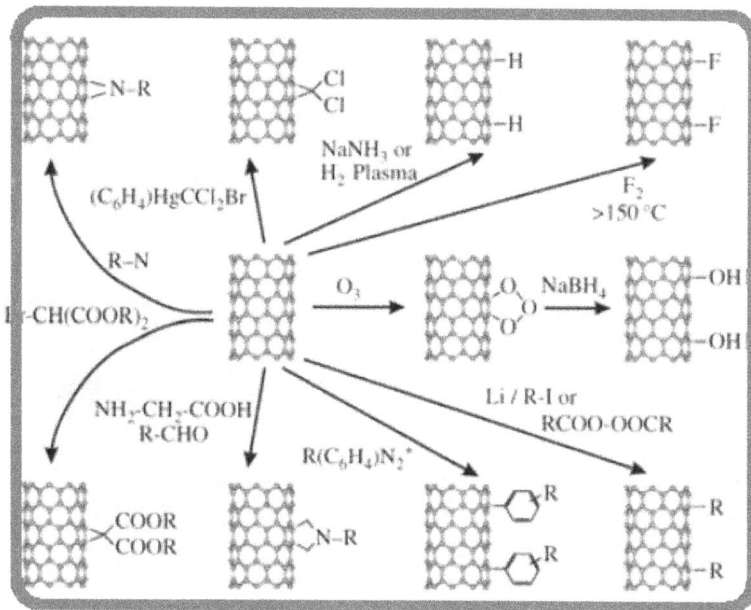

FIGURE 5.9 Possible addition reactions on CNT for functionalization.

Source: (Balasubramanian and Burghard, 2005)

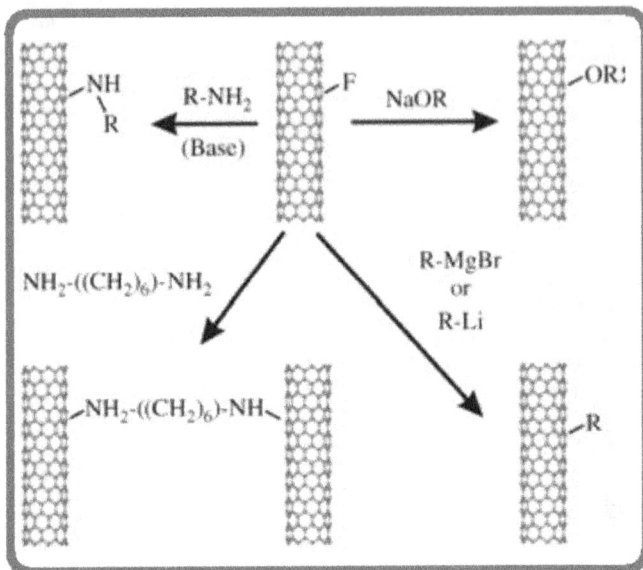

FIGURE 5.10   Nucleophilic substitution reactions can occur on fluorinated CNTs.

Source: (Balasubramanian and Burghard, 2005)

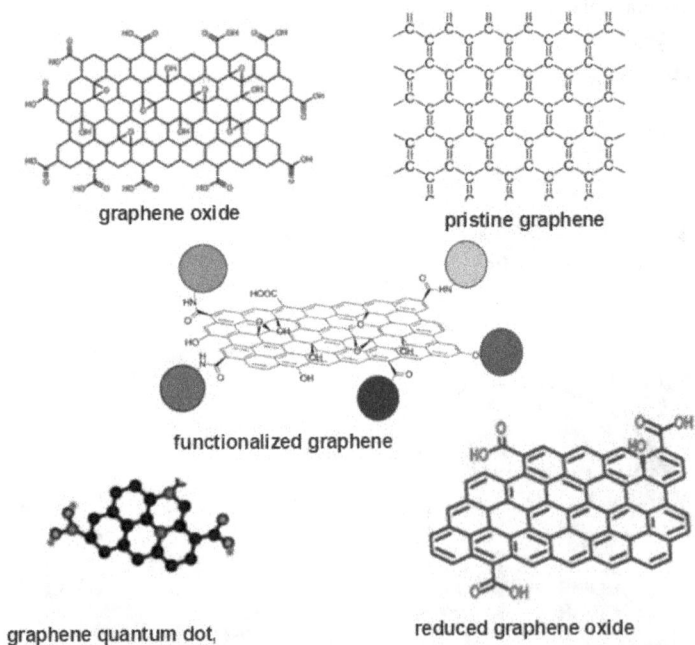

graphene oxide

pristine graphene

functionalized graphene

graphene quantum dot,

reduced graphene oxide

FIGURE 5.11   Different forms of graphene.

Source: (Tiwari et al., 2020)

functional groups make substantial differences in the surface chemistry of G, GO, and rGO (Figure 5.12) (Zhan et al., 2022).

5.7.3 FUNCTIONALIZATION OF C60

Fullerenes are molecular carbon allotropes and exhibit number of phenomena because of their π-electron nature which can be manipulated easily by chemical methods. Reactions such as nucleophilic additions, cycloaddition, hydroxylation, halogenations, free radical additions, and metal transition complexations have been reported for C60. Out of different processes, reactions involving the opening of the fullerene cage is most important. Fullerene can be functionalized by nucleophilic and cycloaddition reactions (Figure 5.13) (Guldi and Martin, 2011).

5.7.4 FUNCTIONALIZATION OF CARBON NANODIAMOND

Nanodiamonds (NDs) are nanosize carbon allotropes with variety of surface chemistry. NDs are obtained by detonation and followed by purification, surface modification, and surface functionalization. Two steps are involved in surface functionalization of

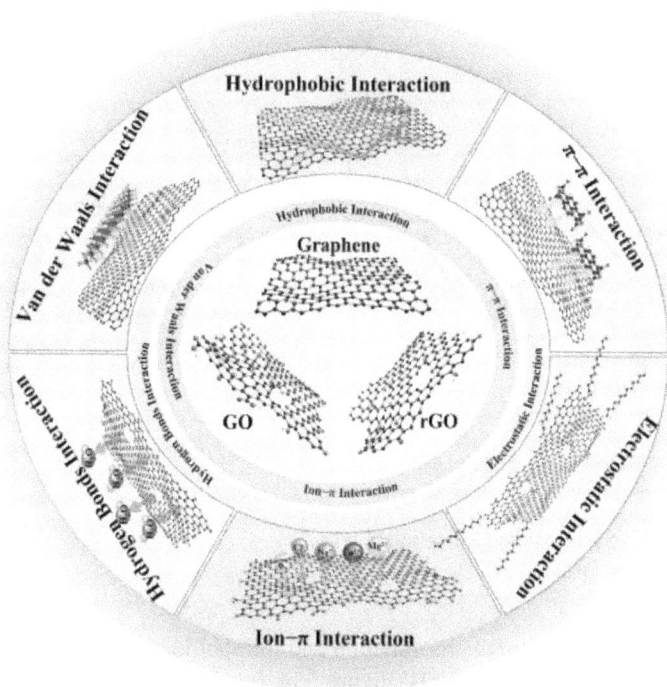

FIGURE 5.12 Different type of interactions at the surface of G, GO and rGO for functionalization.

Source: (Zhan et al., 2022)

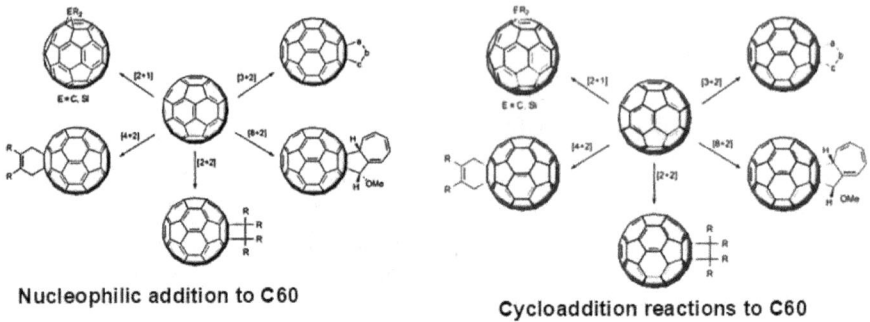

Nucleophilic addition to C60   **Cycloaddition reactions to C60**

FIGURE 5.13 Functionalization by nucleophilic and cycloaddition reaction on the surface of C60.

Source: (Guldi and Martin, 2011)

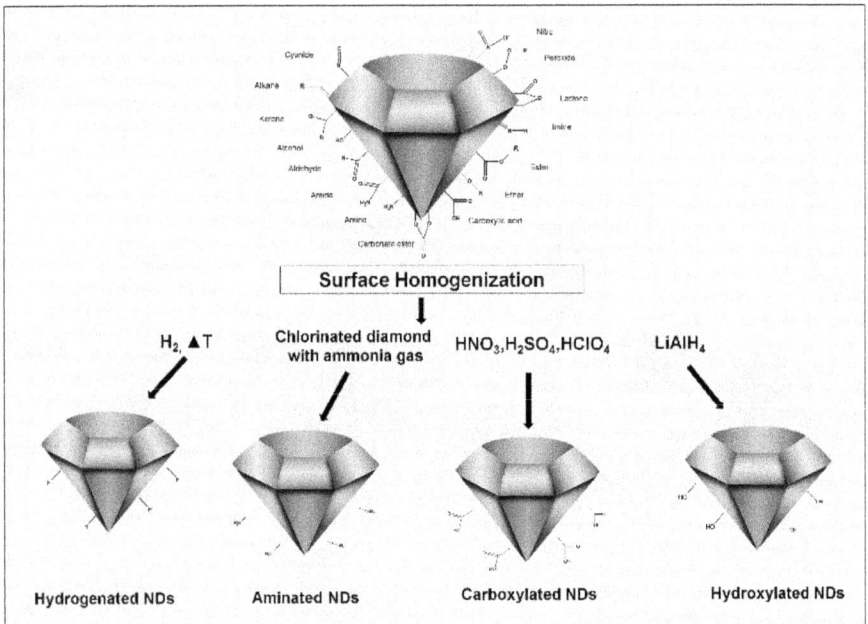

FIGURE 5.14 Surface homogenization of NDs.

Source: (Jariwala et al., 2020)

NDs. Steps are initial surface termination and immobilization of functional groups onto previously homogenized NDs (Figure 5.14, Figure 5.15) (Jariwala et al., 2020).

5.7.5 FUNCTIONALIZATION OF CARBON NANO-ONIONS (CNOs)

Discovery of carbon nano-onions (CNOs), a multi-shell fullerene, was made in 1992 (Ugarte, 1992). These are structured by concentric shells of carbon atoms. Different

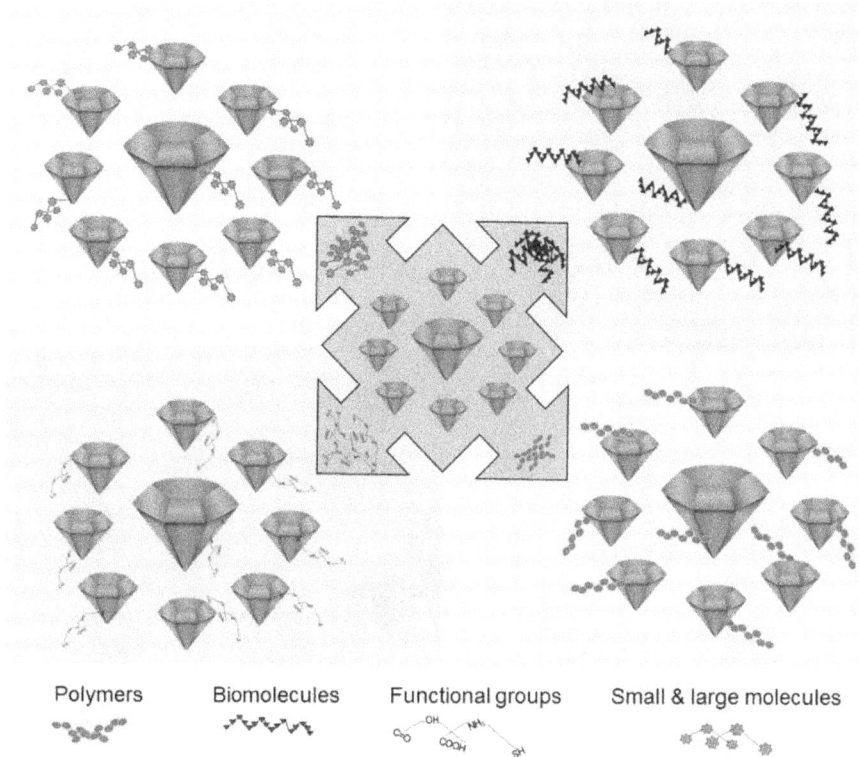

FIGURE 5.15 Surface functionalization of NDs.

Source: (Jariwala et al., 2020)

methods have been developed for its preparation, and properties have been studied. Functionalization of CNOs by chemical method has been investigated, and number of pathways were found to be useful for the introduction of different type of functional groups (Figure 5.16) (Bartelmess and Giordani, 2014; Mykhailiv et al., 2017).

5.8 SOURCES OF VOCS AND THEIR CLASSIFICATIONS

Painting the house, cooking, making a fire, cutting the grass, pesticides, breathing, etc., emit compounds of organic origin. Examples are alkanes, alkenes, ethers, carbonyls, alcohols, esters, aromatics, and amides. Human activities in general release organic compounds in the environment (Koppmann, 2007). Industrialization releases toxic gases and VOCs (Hailin et al., 2013). VOCs and their impacts are major threats to the environment. Indoor VOC concentrations are found much higher as compared to outdoors (Tiwary, 2009; Bartzis et al., 2015; Leidinger et al., 2014). A number of techniques have been developed to monitor VOCs (Teixeira et al., 2004; Vesely et al., 2003; Sabourin et al., 1988; Zeng et al., 2009). Automobiles are the major contributors of VOCs. Significant emissions of NO_x, diesel exhaust particles, and VOCs occur from Diesel fuel combustions. The following types of VOCs are found (Kumar et al., 2022).

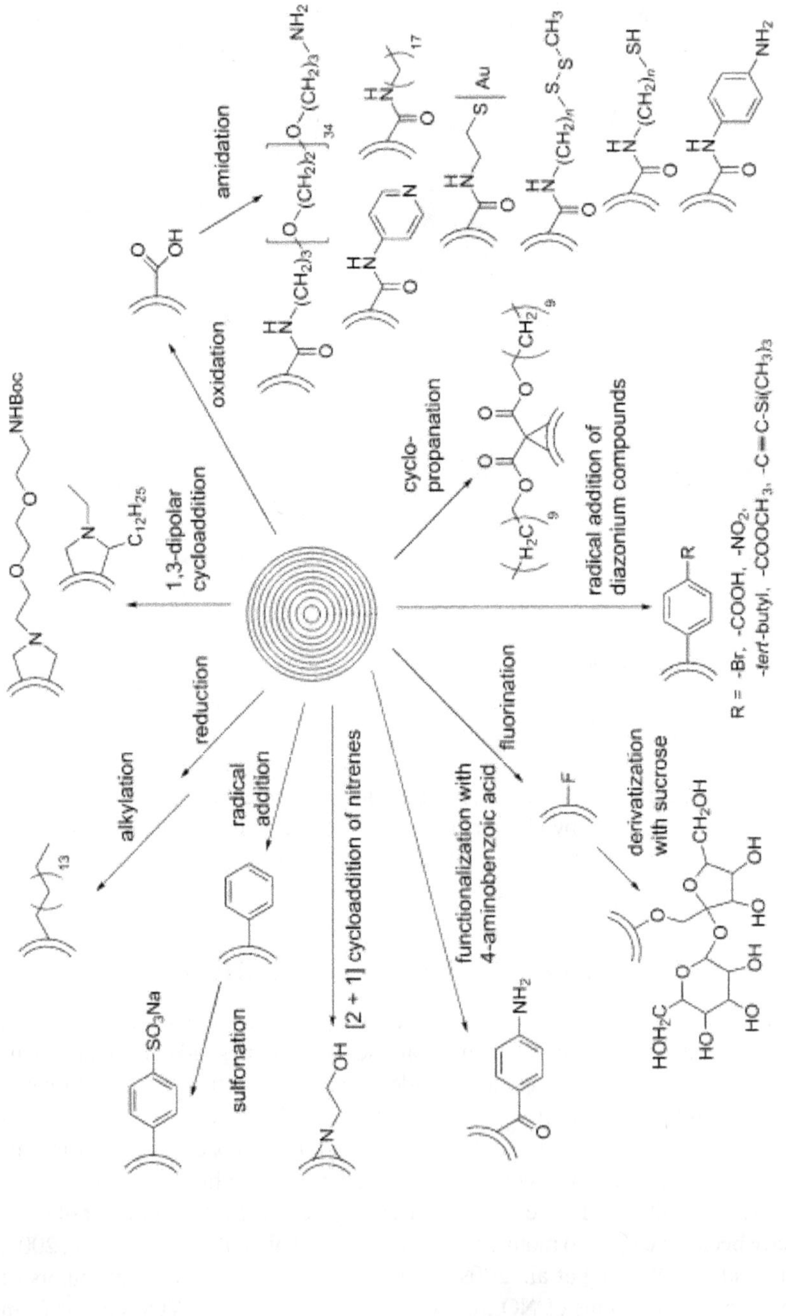

FIGURE 5.16 Covalent functionalization pathways for CNOs.

Source: (Mykhailiv et al., 2017)

TABLE 5.1
Sources of Major VOCs

Classification	Sources	Representatives
Aldehydes	Cosmetics and plastic adhesives Construction materials Biomass burning Fabrics decomposition Bio-waste VOCs degradation	Formaldehyde Acetaldehyde
Aromatic compounds	Adhesives Petroleum decomposition Liquid fuels combustion	Benzene Toluene Ethylbenzene
Alcohols	Preservative Antiseptics Cosmetics	Methanol Ethyl alcohol Isopropyl alcohol
Alkenes	Petrochemicals Varnishes Resins, Adhesives Pharmaceutical and perfumes	Propylene Ethylene
Ketones	Paint thinners, varnishes, adhesives, window cleaners	Acetone Ethyl butyl ketone
Halogenated VOCs	Paints, Chemical extractant Polymer syntheses Adhesives Water purification systems	CCl_4, C_6H_5Cl Trichloroethylene Dichloromethane
Polycyclic aromatic hydrocarbons	Combustion of coal, oil, organic matter, and biofuels	Phenanthrene Pyrene

Source: (Zhang et al., 2017)

1. *Aliphatic hydrocarbons:* the most important is hexane.
2. *Aromatic hydrocarbons:* benzene, toluene, and ethylbenzene.
3. *Halocarbons:* chlorinated VOCs (Cl-VOCs).
4. *Oxygenated organic compounds:* acid compounds, ketones, alcohols, aldehydes, ethers, phenols, esters, and formaldehyde.
5. *S, N-containing compounds:* normally obtained from leather industries and fossil fuel combustion.

VOCs are a big family of carbon-based chemicals with common members of more than 300 types. Table 5.1 gives list of major sources of VOCs (Zhang et al., 2017).

5.9 VOCS SENSING

CNMs is a vast class of materials with different dimensions (0D, 1D, and 2D). These exhibit unique physico-chemical properties. Some of the properties which make

CNMs very attractive for sensing are their charge transfer properties, electronic conduction, large surface area, and functionalization (Barros et al., 2021). These materials are used as sensors in different sectors (Figure 5.17) (Dariyal et al., 2021). One of the important areas where these materials are used as sensors is volatile organic compounds (VOCs), a group of carbon-based chemicals which easily evaporate at room temperature. Common VOCs include benzene, ethylene glycol, acetone, methylene chloride, formaldehyde, perchloroethylene, toluene, and xylene. Many household materials, such as paints and cleaning products, generate VOCs. These VOCs are very harmful to health and the environment. A number of techniques are being used to monitor VOCs, but they are costly. However, sensing technique is easy and cost effective. Therefore, lot of work is now being done in this area (Gakhar et al., 2022).

In recent years, a number of CNMs composites with metal oxides (MOs) have also been used as VOCs sensors (Figure 5.18) (Pargoletti and Cappelletti, 2020).

In general, CNMs-based sensors for VOCs sensing include capture, interaction, and diffusion of VOCs gas molecules, and as a result, different types of VOCs sensors have been designed. Some of the sensors are micro-gravimetric sensors, optical sensors, and resistive sensors (Yin et al., 2021). In the following section, different aspects of VOCs are discussed.

5.9.1 VOCs Sensors

In recent years, significant interest has been shown to explore the utilization of VOC gas sensors in diagnosis of non-invasive diseases and environmental gas monitoring

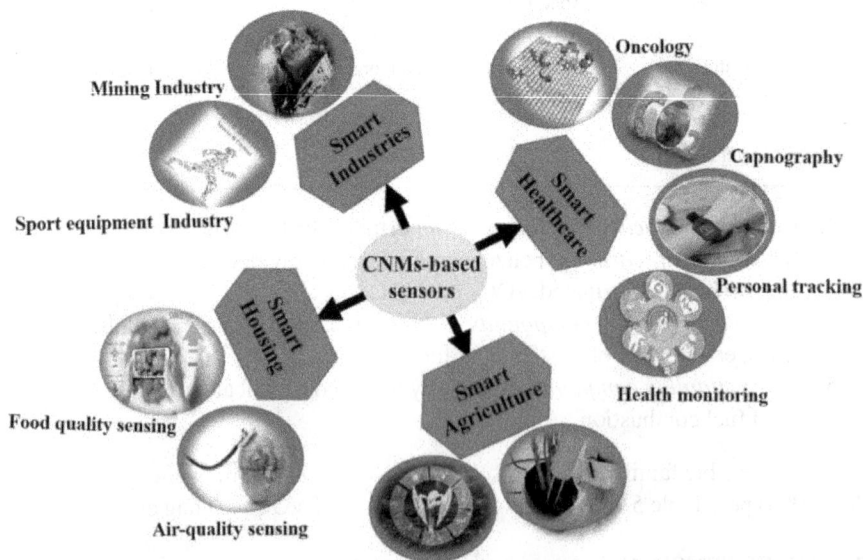

FIGURE 5.17 CNMs-based sensors for applications in various fields.

Source: (Dariyal et al., 2021)

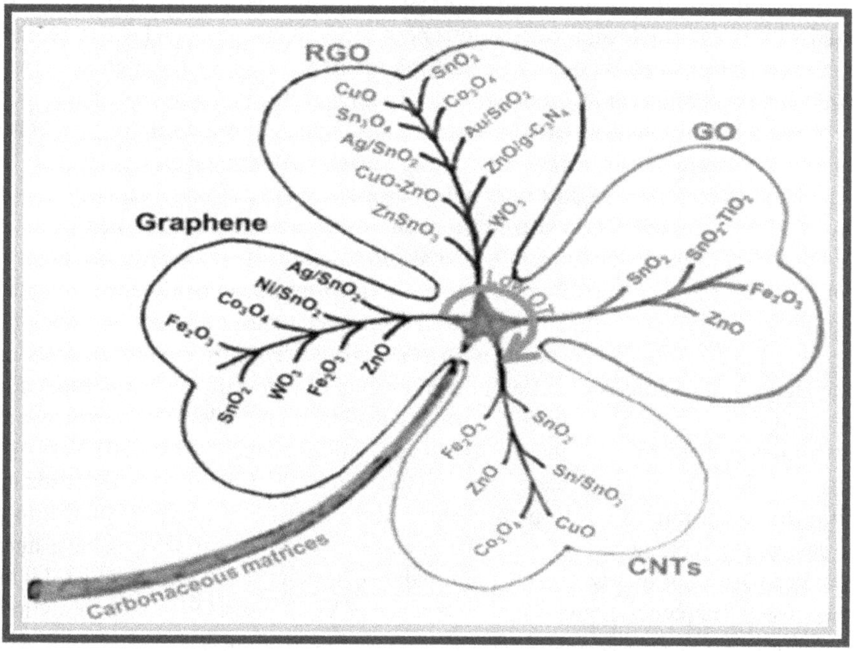

FIGURE 5.18 CNMs-MOs composites as VOCs sensors.

Source: (Pargoletti and Cappelletti, 2020)

(Panda et al., 2021; Andre et al., 2020; Ruzsanyi et al., 2021). Also, creation of gas-efficient VOC sensors has been proposed by researchers and scientists (Stewart et al., 2020; Das and Mohar, 2020). Proper knowledge of sensing mechanism system can offer new insights and information to develop a layout of efficient sensory devices to target the hazardous gases. Similarly, streamlined dimensioning and separation of the gas sensor system is critical to ensure the visual layout of a gas-efficient VOC, and should provide reliable dimensioning results for a viable experimental system. Therefore, an attempt is made to present the working principles of the VOC gas sensor and the practical arrangement used to measure the properties accordingly.

5.9.2 VOC SENSOR PRINCIPLE

There are three routes that have been reported to detect VOCs in gas sensors: (i) VOC gas molecules trapped; (ii) VOC gas molecules interacting with the active central sensor; and (iii) VOC gas molecules dispersed. A variety of VOC gas sensors—such as micro-gravimetric sensors, optical sensors, and resistance sensors—have been developed (Hussain et al., 2020; Chen et al., 2020; Elakia et al., 2020). Occurrence of optical sensory function depends on differences in spectrophotometry or colorimetry of the nerves caused by the intermolecular interaction of the VOCs with the active sensory center.

Micro-gravimetric sensors can eliminate VOC gas detection by various sensor loads due to intermolecular interactions between VOC analysts and active sensors. An excellent example is quartz crystal microbalance sensors for high sensitivity against vapors or gas at low temperatures. In the quartz crystal microbalance VOC process, a layer of carbon-based nanomaterials is coated over a quartz crystal resonator to form a sensor layer. Alternating current (AC) voltage is applied to the resonator electrodes. A quartz crystal microbalance gas sensor is modified with hydrophobic amino-functionalized graphene oxide and nanocomposites to detect aldehydes at room temperature (Chen et al., 2020). Benefits from hydrogen bonding interactions between active amino-functionalized graphene oxide and aldehydes, amino-functionalized graphene oxide sensors, and quartz crystal microbalance sensors have responded to changes in selected aldehyde frequency. In addition, the frequency responses of different components of aldehyde under the same concentration have shown a positive correlation with their molecular weight, which is consistent with the fact that quartz crystal microbalance sensors are very sensitive tools. On the other hand, gravimetric sensor is a multi-acoustic resonator film that incorporates a thin piezoelectric film instead of a quartz crystal film as a gas sensor (Guo et al., 2020a). The acoustic resonator film sensors have much smaller base than quartz crystal microbalance sensors, leading to a much higher response to frequency changes in the presence of VOC gases. Other small gravimetric options are available, such as surface acoustic wave sensors, micro-cantilevers, etc. However, these types of micro-gravimetric sensors suffer from a number of challenges, including low VOC concentration and non-linear response to instability. A few small gravimetric sensors with carbon-based nanomaterials were made for VOC analysis (McGinn et al., 2020; Guo et al., 2020a; Likhite et al., 2020; Luo et al., 2020; Kus et al., 2021).

Carbon-based nanomaterials have been extensively analyzed for VOCs due to their high sensitivity to critical response coupled with wide vision, rapid response speed, and excellent stability (Elakia et al., 2020; Hussain et al., 2020). Additionally, such resistance studies allow real-time monitoring and evaluation of VOC performance when using natural gas surveillance and the diagnosis of non-invasive diseases. A variety of CNMs have been used to create an integrated gas sensor of organic resistive volatile (Pargoletti and Cappelletti, 2020). Basically, the resistance sensor is based on distinguishing intermolecular resistance interactions between the VOC analyzer and the active sensor. The sensory response can be calculated with the help of Eq. 1 (Yin et al., 2021).

$$S = [(Rg - Ra) / Ra] \times 100\% \qquad (1)$$

Where S is the opposite sensory nerve response; Ra and Rg are the values of sensor in the air and the target VOC gas, respectively

In fact, the resistance sensor is often the preferred method for developing VOC gas sensors that work better for obvious advantages which include high sensitivity, ease of use, and excellent stability.

5.9.3 Gas Sensing Characterization

Unlike optical and micro-gravimetric sensors, gas sensor functions require the measurement of Ra and Rg sensor values in air and target gas. For accurate Ra and Rg values, the test set design serves as the viewing process. There are two basic methods of gas distribution: (i) powerful gas distribution system; and (ii) stable gas distribution system. The targeted VOC gas concentration in the test room is controlled by the injected volume of the liquid VOCs, and the corresponding information is obtained from Eq. 2 (Yin et al., 2021).

$$C = [(22.4 \times \phi \times \rho \times V_1) / M \times V_2] \times 1000 \qquad (2)$$

Where C is the concentration (ppm) of the target gas in the static test chamber, ϕ and M denote the volume fraction and molecular weight (g/mol) of the target gas, ρ is the density (g/mL) of the liquid VOCs, V_1 (mL) and V_2 (L) are respectively the volumes of the required liquid VOCs and the testing chamber

Typically, a certain volume of liquid VOCs is injected into a micropipette testing room, where it is converted by heater into VOC gas vapor to make the chamber filled with targeted VOC gas. With the help of any one of the methods discussed earlier, the resistance level of a test under different VOC gases can be easily measured by the source meter. Next, the target VOC gas sensor response can be obtained based on the values of Ra and Rg. In addition, sensor response/acquisition time is a very important sensory parameter, defined sequentially to obtain 90% of the total resistance changes between the advertising process and the target gas reduction of VOCs (Wang et al., 2021). Very sensitive reaction, short response/recovery time, and low operating temperature contributes to the creation of a real-time gas detection system and VOCs, which are also desirable for the use of natural gas monitoring, as well as diagnostic for non-invasive diseases (Li et al., 2020a; Zhou et al., 2020).

5.9.4 Functionalized CNT-Based VOC Gas Sensor

There are two types of CNTs, SWCNT and MWCNT, and both have paid increasing attention to the VOC gas sector. However, pristine CNT-based gas sensors have a limited gas sensor response because of sp^2 bonds. Therefore, in order to design a highly efficient VOC gas sensor to meet the possible use of natural gas monitoring, non-invasive diagnoses, etc., CNTs are functionalized with suitable functional groups.

5.9.5 Functionalized Graphene-Based VOC Gas Sensors

Metal nanoparticles (NPs) have unique chemical and electronic properties in nanoscale and can retain their unique properties even in nanocomposites (Chen et al., 2021; Lee et al., 2021). This benefit encourages researchers to decorate metal NPs on the surface of graphene materials to obtain low-cost and effective VOC gas sensors. In particular, NPs of Au, Ag, and Pt have been extensively investigated to repair graphene structures, where the sensitivity response to targeted VOCs was significantly improved due to accelerated gas transfer between target gas and nerves. In addition, ternary compounds comprising graphene structures, noble NPs, and metal oxide have also been considered to improve the performance of targeted VOC gases.

5.9.6 BORON NITRIDE–BASED VOC GAS SENSORS

Boron nitride exhibits crystalline and electronic properties such as graphite and has high thermal conductivity and chemical and thermal stability (Zheng et al., 2019). Studies have suggested that a nanosheet of ultrathin hexagonal-boron nitride is optimized for atomic solidification. These ultrathin nanosheets were tested as a gas sensor and their gas sensing properties were reported. Improved gas sensors have shown rapid response and recovery with a prominent choice in C_2H_5OH. Hexagonal-boron nitride gas sensors also work at high operating temperatures and for NH_3 gas and C_2H_5OH (100 ppm). Moreover, by increasing the concentration of C_2H_5OH, sensitivity increases. In particular, the response/recovery time was very rapid, which means that the gas sensor has a high sensitivity (Lin et al., 2016).

5.9.7 FUNCTIONALIZED FULLERENE-BASED VOC GAS SENSORS

VOC sensor properties based on CNMs are given in Table 5.2 (Gakhar and Hazra, 2021). Carbon materials act as sensor by adsorption, which depend on number of parameters (Figure 5.19) (Zhang et al., 2017).

5.9.8 GRAPHDIYNE-BASED VOC GAS SENSORS

In 2010, the first large-scale graphdiyne film (3.6 cm²) was made by using a composite reaction in situ on a copper foil substrate. This was possible because of flexibility of sp bonds and their high structural strength (Li et al., 2010). Graphdiyne, unlike other carbon materials, has a 2D planer periodic structure which combines diacetylene (sp carbon atoms) with benzene ring moieties (sp² carbon atoms). It also has the advantages of direct bandgap, excellent company charging flow at ambient temperature, outstanding electrical conductivity (similar to silicon), and natural pores in the carbon structure such as non-natural carbon allotropes with good durability (Chen et al., 2019). In order to integrate graphdiyne material with various morphologies, a number of synthetic processes have been developed, including spontaneous growth, CVD, and explosive method (Xue et al., 2018). Graphic-based materials also have potential applications in different of fields, due to their unique chemical, physical, and electrical properties. Over the past seven years, many efforts have been made to extend the use of graphdiyne for catalysis, water processing, energy conservation, optoelectronic devices, and solar cells. It should be noted that the applications of graphdiyne gas sensing have recently grown rapidly and become a popular practice. In addition, graphdiyne-based electrochemical moisture sensors have been reported, showing higher response rate than carbon dioxide, due to sp hybridized carbons in graphdiyne material (Li et al., 2021). In addition, numerous DFT-based theoretical studies have been reported to reveal NO_x adsorption behavior on graphdiyne

TABLE 5.2
VOC Sensors Properties Based on Carbon–Metal Oxide Nanocomposite

S. No.	Material	Selective VOC	Operating Temp. (°C)	Response	Concentration (ppm)	Time Response/Time Recovery (s)
1	ZnO/rGO nanorods	C_2H_5OH	260	1.1–2.7	5–50	7/8
2	SnO$_2$/rGO film composite	C_3H_6O	Room Temp.	2.1–9.72	10–2000	107/95 146/141
3	WO$_3$-Graphene hemitubes	C_3H_6O	300	2.8–7	1–5	13.5/25 8.5/19
4	rGO/SnO$_2$ aerogel	C_3H_6O	Room Temp.	0.5–1.7	10–80 ppb	2.43/1.06
5	SnO$_2$/Graphene nanocomposite	C_6H_6	Room Temp.	1.87–94.3	5–100 ppb	2/8
6	rGO/α-Fe$_2$O$_3$ nanofiber	C_3H_6O	375	8.9	100	3/9
7	α/Fe$_2$O$_3$/Graphene nanostructure	C_2H_5OH	280	5–30	1–1000	10/12
8	ZnO quantum dot—Graphene	HCHO	Room Temp.	0.42–1.1	25–100	30/40
9	rGO/TiO$_2$ nanoparticles	HCHO	Room Temp.	0.8 ppmV	0.5 ppmV	112/126
10	ZnO nanosheets/GO	C_3H_6O	240	35.8	100	3/6
11	rGO/SnO$_2$ nanoparticles	C_2H_5OH	300	2.7–297.8	5–500	11/20
12	TiO$_2$ nanotubes/rGO hybrid	CH_3OH	Room Temp.	24–96.93	10–800	18/61
13	rGO/ZnO flower hybrid	HCHO	Room Temp.	2–5.5	2–10	39/50
14	Graphene loaded Ni-SnO$_2$	C_3H_6O	350	169.7	200	5.4
15	Graphene/ZnO	C_2H_5OH	340	97	200	5/20
16	rGO/Fe$_2$O$_3$ nanoparticles	C_2H_5OH	Room Temp.	1.86–6.5	1–100	10/50
17	SnO$_2$/Graphene	HCHO	260	35–44.3	100–500	--
18	GO loaded TiO$_2$ nanotubes	CH_3OH	Room Temp.	40	480	74/76
19	MoO$_3$-rGO Nanoflakes	C_2H_5OH	310	53–702	100–8000	6/54
20	Ag-ZnO-rGO	C_2H_2	150	2.5–35	1–1000	25/80
21	SnO$_2$/CNT	C_2H_5OH CH_3OH	Room Temp.	0.15–0.13	30	60/60
22	CNT/SnO$_2$	CH_3OH C_2H_5OH	250–300	65	100–1000	20/30
23	ZnO/MWNT nanorods	C_2H_5OH	Room Temp.	1.6–7.3	5–500	5–12/8–16

(Continued)

TABLE 5.2 (Continued)
VOC Sensors Properties Based on Carbon–Metal Oxide Nanocomposite

S. No.	Material	Selective VOC	Operating Temp. (°C)	Response	Concentration (ppm)	Time Response/Time Recovery (s)
24	SnO$_2$/MWNT	C$_3$H$_6$O	250	72	1	2/12
25	MWCNT/SnO$_2$	C$_2$H$_5$OH	300	1.83–5.8	10–1000	5/8
26	SnO$_2$/SWNT	C$_3$H$_6$O	Room Temp.	10–50	0.5–20	7/9
27	All—Fullerene	C$_2$H$_5$OH	–	208 Hz	3–5 Hz	–
28	C$_{60}$-ZnO	C$_2$H$_5$OH	Room Temp.	32	100	80/15
29	C$_{60}$-cryptand 22 coated quartz	CH$_3$OH C$_2$H$_5$OH	Room Temp.	73.2–750 Hz/mgl^{-1}	0.80 mg/l 0.70 mg/l	–
"30	C$_{60}$-encapsulated TiO$_2$ nanoparticles	HCHO	100	117	100	–
31	GO-WO$_3$ Hemitubes	Acetone	300	1.7 for 100 ppb	100	11.5/13.5
32	GO/SnO$_2$	HCHO	120	32 for 100 ppm	500	66/10
33	rGO Coated SiNWs	HCHO	300	6.4 for 10 ppm	35	30/10
34	PMMA/POSS/CNT	HCHO	Room Temp.	–	300	<5
35	Au@NGQDs/TiO$_2$	HCHO	150	9.1 for 100 ppb	40	18/20
36	Ti$_3$C$_2$T$_x$	Ethanol	Room Temp.	1.7 for 100 ppm	50	–
37	3D Mxene Framework	Acetone	Room Temp.	0.08 for 50 ppb	50	90/102
38	W$_{18}$O$_{49}$/Ti$_3$C$_2$T$_x$	Acetone	300	11.6 for 20 ppm	170	4.6/18.2
39	rGO/DF-PDI	TEA	Room Temp.	13.7 for 20 ppm	16	64/128
40	Graphene and ZnO	Acetone	Room Temp.	–	1.56	–
41	rGO/ZnSnO$_3$	HCHO	103	12.8 for 10 ppm	100	31/–
42	SnO$_2$/rGO	HCHO	160	138 for 100 ppm	Level	20/20
43	rGO/PMMA	HCHO	Room Temp.	–	100	150/180
44	Graphene/ZnO	HCHO	Room Temp.	–	180	36
45	RGO NS SnO$_2$ NF	Acetone	350	2 for 100 ppb	100	<198/<114"

Source: (Gakhar and Hazra, 2021)

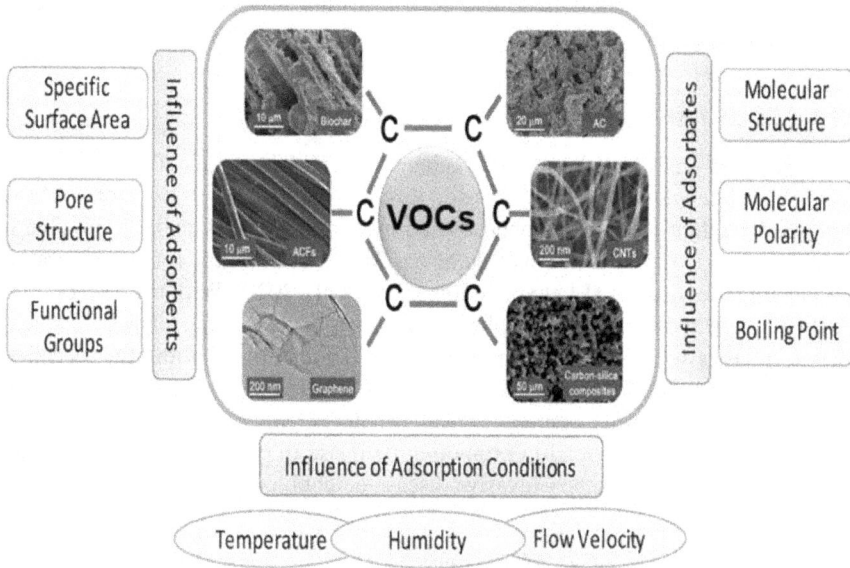

FIGURE 5.19   Carbon materials act as sensor by adsorption.

Source: (Zhang et al., 2017)

surfaces, indicating that graphdiyne has significant potential for use in gas sensory processing (Majidi and Nadafan, 2020; Wu et al., 2021).

5.9.9 OTHER CARBON NANOMATERIALS–BASED VOC GAS SENSORS

In fact, carbon-based nanomaterials are not limited to CNTs, graphene structures, and MXene materials, but can also be found in other forms such as carbon black (CB). Many CB particles are classified as clusters similar to grapes and most atoms are found in the surrounding environment, providing a specific area higher than 1000 m²/g (Zhao et al., 2020). As a result, CB materials are expected to achieve some success in targeted gas development and attract greater interest in the field of highly efficient VOC gas sensor design (Pargoletti and Cappelletti, 2020; Arduini et al., 2020; Daneshkhah et al., 2020). To achieve the distinction of various aldehyde analyzes by such a combination of PEI/CB, a multifactor machine learning algorithm based on key component analysis (PCA) was developed. In addition, a new form of carbon allotrope, graphene, has been successfully developed (Jambhulkar et al., 2020). Due to its unique electronic structure and excellent semiconductor performance, the new CNM is highly regarded and expected to operate in the field of semiconductors, electronics, and new energy. In addition, the composite structure of fullerene (C60) with CNT materials is enhanced by layer-by-layer spraying and detects a gas sensor that significantly increases VOCs (Majidi and Nadafan, 2020). A limited result showed that the sensor acquisition limit could reach sub-ppm (340 ppb) and set a high signal level (Jambhulkar et al., 2020). Indeed, many other carbon-based nanomaterials have also been investigated to create a more efficient sensor, including amorphous carbon, onion carbon, and carbon dot.

5.10 APPLICATIONS

Many excellent CNT-based sensors, graphene, Mxene (Guo et al., 2020b), and other CNMs have been developed and achieved through the successful discovery of VOC analysts. Excellent electrical properties and location of CNMs make it possible to achieve accurate and reliable analysis of gas VOCs even below room temperature and low target emissions (Li et al., 2020a; Chen et al., 2020). With the help of other highly efficient and effective technologies, VOC gas sensors using CNMs are able to operate in areas with high RH values. Additionally, combined with other excellent flexible materials, CNMs contribute to the VOC flexible gas sensor that maintains sensory performance after bending cycles. The development of these VOC gas sensors using CNMs is beneficial in promoting their use in potential applications such as natural gas monitoring, non-invasive disease detection, and environmental care.

5.10.1 Environmental Gas Monitoring

Many VOC gases are highly toxic and harmful to human health and the environment. HCHO obtained from building materials is a major domestic pollution. Even small quantity of HCHO can cause lot of problems. VOC gas sensor acts as a major component of the air quality monitoring system. H-SnO$_2$@rGO composite synthesized by combining SnO$_2$ and rGO exhibits an ultrahigh sensory response of 10–435 ppm HCHO gas (Hu et al., 2020). In addition, with the proliferation of the IoT (Internet of Things), real-time detection and online analysis of VOC are essential (Li et al., 2020a). In some cases, such as exposure to low voltage, the gas sensor needs to be set at a very low acquisition limit in order to detect low-level VOCs (Hou et al., 2020; Chen et al., 2020; Sun et al., 2020; Majidi and Nadafan, 2020). Benefiting from the important combination of

> rGO/DF-PDI heterojunction, the rGO/DF-PDI sensor showed 16 ppb LOD near TEA at ambient temperature. Additionally, the hydrophobic material of rGO provides the rGO/DF-PDI sensor with excellent protection from high humidity (up to 100% RH). A highly efficient rGO/DF-PDI sensor, low power consumption, and a well-tolerated high-density environment, were expected to provide a solution for obtaining TEA online in the real system. Recently, a combination of Ti$_3$C$_2$T$_x$/WSe$_2$
>
> (Li et al., 2020b)

with a simple treatment and skin removal procedure has been improved. Using the inkjet printing process, the Ti$_3$C$_2$T$_x$/WSe$_2$ hybrid was inserted into the PI substrate with Au interdigital electrodes and used as sensors. Finally, the gas sensor for wireless VOCs was detected by connecting a wireless monitoring system. The measured results showed that Ti$_3$C$_2$T$_x$/WSe$_2$ sensor has low LOD and noise level, high sensitivity, and fast response/recovery times, which paved the way for the development of the next generation of IoT VOC sensors (Chen et al., 2020).

5.10 NON-INVASIVE DISEASE DIAGNOSIS

Due to human activities, more than 200 types of VOC gases are found. These VOC gases are considered to be the biological markers of many diseases (Lee et al., 2020; Gilio et al., 2020). A gas sensor with a very short response/recovery time is consistent with the design of the real-time VOC sensor platform (Hyodo and Shimizu, 2020), which is expected to be used in diagnostic applications. The LOD sensor of VOCs must be low enough to detect VOCs generated (Choi et al., 2020). VOC sensors using CNMs can provide a very low LOD near VOC gas directed and summarized in Table 5.1. Many efforts have been made to construct VOC gas sensors using CNMs to achieve analytical detection. of follow-up VOCs at room temperature (Li et al., 2020a). Gas VOC sensors using CNMs have been considered for portable non-invasive diagnostic machines (Elakia et al., 2020; Hussain et al., 2020). It is not an easy task to find the best sensor with the right balance such as proper humidity, low LOD, and high sensitivity options at the same time. Therefore, further improvements in the sensory function of VOCs using CNMs are necessary to demonstrate effective biomarker detection and uncommon diagnosis of another disease.

5.11 CONCLUSIONS

NMs and functionalized NMs play important roles in different sectors. There is a variety of NMs which can be synthesized in different ways. Out of different NMs, CNMs are of immense importance. CNMs can be functionalized by different methods and are useful. In recent years, VOCs are released by different activities and are harmful for human health and the environment. It is found that CNMs can be a good sensor for VOCs. Sensing properties have been discussed.

REFERENCES

Abd-Elsalam Kamel A. (2020) Carbon nanomaterials: 30 years of research in agroecosystems, Carbon Nanomaterials for Agri-Food and Environmental Applications. https://doi.org/10.1016/B978-0-12-819786-8.00001-3

Andre Laurie, Desbois Nicolas, Gros Claude P., Brandes Stephane (2020) Porous materials applied to biomarker sensing in exhaled breath for monitoring and detecting non-invasive pathologies, Dalton Transactions, 49, 15161.

Arduini Fabiana, Cinti Stefano, Mazzaracchio Vincenzo, Scognamiglio Viviano, Amine Aziz, Moscone Danila (2020) Carbon black as an outstanding and affordable nanomaterial for electrochemical (bio) sensor design, Biosens. Bioelectron., 156, 112033.

Balasubramanian Kannan, Burghard Marko (2005) Chemically functionalized carbon nanotubes, Nano-Micro Small, 1(2), 180–192.

Barros Anerisede, Luisa Maria, Braunger, Furlande Rafael (2021) Oliveira, marystela ferreira, functionalized advanced carbon-based nanomaterials for sensing, Reference Module in Biomedical Sciences. https://doi.org/10.1016/B978-0-12-822548-6.00014-5

Bartelmess Juergen, Giordani Silvia (2014) Carbon nano-onions (multi-layer fullerenes): Chemistry and applications, Beilstein Journal of Nanotechnology, 5, 1980–1998. https://doi.org/10.3762/bjnano.5.207

Bartzis J., Wolkoff P., Stranger M., Efthimiou G., Tolis E., Maes F., Nørgaard A., Ventura G., Kalimeri K., Goelen E. (2015) On organic emissions testing from indoor consumer products' use, J. Hazard. Mater., 285, 37–45.

Chen J., Rao A. M., Lyuksyutov S., Itkis M. E., Hamon M. A., Hu H., Cohn R. W., Eklund P. C., Colbert D. T., Smalley R. E., Haddon R. C. (2001) Dissolution of full-length single-walled carbon nanotubes, J. Phys. Chem. B, 105, 2525–2528.

Chen Wei, Wang Zhenhe, Gu Shuang, Wang Jun, Wang Yongwei, Wei Zhenbo (2020) Hydrophobic amino-functionalized graphene oxide nanocomposite for aldehydes detection in fish fillets, Sensors and Actuators B: Chemical, 306, 127579.

Chen Wei Cheng, Jing Shiuan Niu, Liu I-Ping, Chen Hong Yu, Cheng Shiou Ying, Lin Kun Wei, Liu Wen Chau (2021) Hydrogen sensing properties of a novel GaN/AlGaN Schottky diode deco-rated with palladium nanoparticles and a platinum thin film, Sensors and Actuators B: Chemical, 330, 129339.

Chen Winston Yenyu, Jiang Xiaofan, Lai Sz Nian, Peroulis Dimitrios, Stanciu Lia (2020) Nanohybrids of a MXene and transition metal di-chalcogenide for selective detection of volatile organic compounds, Nature Communications, 11, 1, 10.

Chen Zhi, Lin Tao, Cheng Fang, Su Chenliang, Loh Kian Ping (2019) Hydrogen bond guided synthesis of close-packed one-dimensional graphdiyne on the Ag (111) surface, Chemical Science, 47, 10849.

Choi Junghoon, Kim Yong-Jae, Cho Soo Yeon, Park Kangho, Kang Hohyung, Kim Seon Joon, Jung Hee Tae (2020) In situ formation of multiple Schottky barriers in a Ti_3C_2 MXene film and its application in highly sensitive gas sensors, Advanced Functional Materials, 30, 2003998.

Daneshkhah Ali, Vij Shitiz, Siegel Amanda P., Agarwal Mangilal (2020) Polyetherimide/carbon black composite sensors demonstrate selective detection of medium-chain aldehydes including nonanal, Chemical Engineering Journal, 383, 123104.

Dariyal Pallvi, Sharma Sushant, Chauhan Gaurav Singh, Singh Bhanu Pratap, Dhakate Sanjay R., (2021) Recent trends in gas sensing via carbon nanomaterials: Outlook and challenges, Nanoscale Adv., 3, 6514–6544.

Darwish, M., Mohammadi, A. (2018) Functionalized nanomaterial for environmental techniques, In Nanotechnology in Environmental Science. Chaudhery M. H.; Mishra A. K. Eds., Wiley-VCH Verlag GmbH & Co. KGaA., Weinheim, Germany.

Das Tanmay, Mohar Mrittika (2020) Development of a smartphone-based real time cost-effective VOC sensor, Heliyon, 6, e05167.

Elakia Manoharan, Gobinath Marappan, Sivalingam Yuvaraj, Palani Elumalai, Ghosh Soumyajit, Nutalapati Venkatramaiah, Surya Velappa Jayaraman (2020) Investigation on visible light assisted gas sensing ability of multi-walled carbon nanotubes coated with pyrene based organic molecules, Physica E: Low-Dimensional Systems and Nanostructures, 124, 114232.

Fernando K. A. S., Lin Y., Sun Y. P. (2004) High aqueous solubility of functionalized single-walled carbon nanotubes, Langmuir, 20, 4777.

Feynman R. P. (1960) There's plenty of room at the bottom, Eng. Sci., 23, 22–36.

Gakhar Teena, Basu Sukumar, Hazra Arnab (2022) Carbon-metal oxide nanocomposites for selective detection of toxic and hazardous volatile organic compounds (VOC): A review, Green Analytical Chemistry, 1, 100005.

Gakhar Teena, Hazra Arnab (2021) C_{60}-encapsulated TiO_2 nanoparticles for selective and ultrahigh sensitive detection of formaldehyde, Nanotechnology, 32, 505505.

Georgakilas Vasilios, Tiwari Jitendra N., Kemp K. Christian, Perman Jason A., Bourlinos Athanasios B., Kim Kwang S., Zboril Radek (2016) Noncovalent functionalization of graphene and graphene oxide for energy materials, biosensing, catalytic, and biomedical applications, Che. Rev., 116(9), 5464–5519.

Gilio Alessia Di, Catino Annamaria, Lombardi Angela, Palmisani Jolanda, Facchini Laura, Mongelli Teresa, Varesano Niccolo, Bellotti Roberto, Galetta Domenico, de Gennaro Gianluigi, Tangaro Sabina (2020) Breath analysis for early detection of malignant pleural mesothelioma: volatile organic compounds (VOCs) determination and possible biochemical pathways, Cancers, 12, 1262.

Guo Huihui, Guo Aohui, Gao Yang, Liu Tingting (2020a) Influence of external swelling stress on the frequency characteristics of a volatile organic compound (VOC) sensor based on a polymer-coated film bulk acoustic resonator (FBAR), Instrumentation Science & Technology, 48, 431.

Guo Wenzhe, Surya Sandeep G., Babar Vasudeo, Ming Fangwang, Sharma Sitansh, Alshreef Husam N., Schwingenschlogl Udo, Salama Khaled N. (2020b) Selective toluene detection with Mo_2CT_xMXene at room temperature, ACS Applied Materials & Interfaces, 12, 57218.

Hailin W., Lie N., Jing L., Yufei W., Gang W., Junhui W., Zhengping H. (2013) Characterization and assessment of volatile organic compounds (VOCs) emissions from typical industries, Chin. Sci. Bull., 58, 724–730.

Hamon M. A., Chen J., Hu H., Chen Y. S., Itkis M. E., Rao A. M., Eklund P. C., Haddon R. C. (1999) Dissolution of single-walled carbon nanotubes, Adv. Mater., 11, 834.

Hirsch A. (2010) The era of carbon allotropes, Nat. Mater, 9, 868–871.

Hou Xinghui, Wang Zhaowu, Fan Guijun, Ji Haipeng, Yi Shasha, Li Tao, Wang Yu, Zhang Zongtao, Yuan Lei, Zhang Rui, Sun Jing, Chen Deliang (2020) Hierarchical three-dimensional MoS_2/GO hybrid nanostructures for triethylamine-sensing applications with high sensitivity and selectivity, Sensors and Actuators B: Chemical, 128, 236.

Hu Jicu, Chen Mingpeng, Rong Qian, Zhang Yumin, Wang Huapeng, Zhang Dongming, Zhao Xinbo, Zhou Shiqiang, Zi Baoye, Zhao Jianhong, Zhang Jin, Zhu Zhongqi, Liu Qingju (2020) Formaldehyde sensing performance of reduced graphene oxide-wrapped hollow SnO_2 nanospheres composites, Sensors and Actuators B: Chemical, 307, 127584.

Hussain Tanveer, Sajjad Muhammad, Singh Deobrat, Bae Hyeonhu, Lee Hoonkyung, Larsson J. Andreas, Ahuja Rajeev, Karton Amir (2020) Sensing of volatile organic compounds on two-dimensional nitrogenated holey graphene, graphdiyne, and their heterostructure, Carbon, 163, 213.

Hyodo Takeo, Shimizu Yashiro (2020) Adsorption/combustion-type micro gas sensors: Typical VOC-sensing properties and material-design approach for highly sensitive and selective VOC detection, Analytical Science, 36(4), 401.

Iijima S. (1991) Helical microtubules of graphitic carbon, Nature, 354(6348), 56–58.

Jariwala Dhruvil Hiteshkumar, Patel Dhrumi, Wairkar Sarika (2020) Surface functionalization of nanodiamonds for biomedical applications, Materials Science and Engineering: C, 113, 110996.

Jambhulkar Sayli, Xu Weiheng, Franklin Rahul, Ravichandran Dharneedar, Zhu Yuxiang, Song Kenan (2020) Integrating 3D printing and self-assembly for layered polymer/nanoparticle microstructures as high-performance sensors, Journal of Material Chemistry C, 8, 9495.

Kharisov B. I., Kharissova, O. V. (2019). Carbon Allotropes: Metal-Complex Chemistry, Properties and Applications, Springer, Cham, Switzerland.

Koppmann R. (2007) Volatile Organic Compounds in the Atmosphere, Blackwell Publishing Ltd., Oxford, UK.

Kroto H. W., Heath J. R., O'Brien S. C., Curl R. F., Smalley R. E. C. (1985) buckminster-fullerene, Nature 318, 162–163.

Kus Funda, Altinkok Cagatay, Zayim Esra, Erdemir Serkan, Tasaltin Cihat, Gurol Ilke (2021) Surface acoustic wave (SAW) sensor for volatile organic compounds (VOCs) detection with calix-4-arene functionalized Gold nanorods (AuNRs) and silver nanocubes (AgNCs), Sensors and Actuators B: Chemical, 330, 129402.

Lee Janghyeon, Lee Junsuk, Lim Si Hyung (2020) Micro gas preconcentrator using metal organic framework embedded metal foam for detection of low-concentration volatile organic compounds, J. Hazard Mater., 392, 122145.

Lee Jinho, Jung Youngmo, Sung Seung Hyun, Lee Gilho, Kim Jungmo, Seong Jin, Shim Young Seok, Jun Sheong Chan, Jeon Seakwoo (2021) High-performance gas sensor array for indoor air quality monitoring: The role of Au nanoparticles on WO_3, SnO_2, and NiO based gas sensors, Journal of Materials Chemistry, 9, 1159.

Leidinger M., Sauerwald T., Conrad T., Reimringer W., Ventura G., Schütze A. (2014) Selective detection of hazardous indoor VOCs using metal oxide gas sensors, Procedia Eng., 87, 1449–1452.

Li Guoxing, Li Yuliang, Liu Huibiao, Guo Yanbing, Li Yongjun, Zhu Daoben (2010) Architecture of graphdiyne nanoscale films, Chemical Communication, 46, 3256.

Li Hai-Yang, Zhao Shu-Na, Zang Shuang Quan, Li Jing (2020a) Functional metal-organic frameworks as effective sensors of gases and volatile compounds, Chemical Society Reviews, 17, 6364.

Li Peipei, Cao Changyan, Song Weiguo (2021) Graphdiyne: A highly sensitive material for ppb-level NO2 gas sensing at room temperature, Chemical Research in Chinese University, 37, 1317.

Li Shanshan, Zhao Chuanrui, Zhou Shauang, Zhang Yan, Zhu Peihua, Yu Jinghua (2020b) Non-covalent interaction-driven self-assembly of perylene diimide on rGO for room-temperature sensing of triethylamine with enhanced immunity to humidity, Chemical Engineering Journal, 385, 123397.

Li W. Z., Xie, S. S., Qian, L. X., Chang, B. H., Zou, B. S., Zhou, W. Y., Zhao, R. A., Wang, G. (1996) Large scale synthesis of aligned carbon nanotubes, Science, 274, 1701–1703.

Likhite Rugved, Banerjee Aishwaryadev, Majumder Apratim, Karkhanis Mohit, Kim Hanseup, Mastrangelo Carlos H. (2020) VOC sensing using batch-fabricated temperature compensated self-leveling microstructures, Sensors and Actuators B: Chemical, 311, 127817.

Lin Liyang, Liu Tianmo, Zhang Yu, Sun Rong, Zeng Wen, Wang Zhongchang (2016) Synthesis of boron nitride nanosheets with a few atomic layers and their gas-sensing performance, Ceramics International, 42, 971.

Luo Shao Xiong Lennon, Lin Che Jen, Ku Kang Hee, Yoshinaga Kosuke, Swager Timothy M. (2020) Pentiptycene polymer/single-walled carbon nanotube complexes: Applications in benzene, toluene, and o-xylene detection, ACS Nano, 14, 7297.

Majidi Roya, Nadafan Marzieh (2020) Detection of exhaled gas by γ-graphyne and twin-graphene for early diagnosis of lung cancer: a density functional theory study, Physics Letters A, 384, 126036.

McGinn Christine K., Lamport Zachary A., Kymissis Ioannis (2020) Review of gravimetric sensing of volatile organic compounds, ACS Sensors, 5, 1514.

Mostofizadeh A., Li Y., Song B., Huang Y. (2011) Synthesis, properties, and applications of low-dimensional carbon-related nanomaterials, J. Nanomater., 685081.

Mykhailiv Olena, Zubyk Halyna, Plonska-Brzezinska Marta E. (2017) Carbon nano-onions: Unique carbon nanostructures with fascinating properties and their potential applications, Inorganica Chimica Acta, 468, 49–66.

Panda Atanu, Arumugasamy Shiva Kumar, Lee Jihyen, Son Younghu, Yun Kyusik, Venkateswarlu Sada, Yoon Minyoung (2021) Chemical-free sustainable carbon nano-onion as a dual-mode sensor platform for noxious volatile organic compounds, Applied Surface Science, 537, 147872.

Pargoletti Eleonora, Cappelletti Giuseppe (2020) Breakthroughs in the design of novel carbon-based metal oxides nanocomposites for VOCs gas sensing, Nanomaterials, 10, 1485.

Purohit Rajesh, Purohit Kuldeep, Rana Saraswati, Rana R.S., Patel Vivek (2014) Carbon nanotubes and their growth methods, Procedia Materials Science, 6, 716–728.

Ruzsanyi Veronika, Wiesenhofer H., Ager C., Herbig J., Aumayr G., Fischer M., Renzler M., Ussmueller T., Lindner K., Mayhew C. (2021) A portable sensor system for the detection of human volatile compounds against transnational crime, Sensors and Actuators B: Chemical, 328, 129036.

Sabourin P.J., Bechtold W.E., Henderson R.F. (1988) A high pressure liquid chromatographic method for the separation and quantitation of water-soluble radiolabeled benzene metabolites, Anal. Biochem., 170, 316–327.

Saleh Tawfik Abdo (2020) Nanomaterials: Classification, properties, and environmental toxicities, 20, 101067.

Saleh Tawfik Abdo, Gupta Vinod Kumar (2016) Chapter 4: Synthesis, classification, and properties of nanomaterials, Nanomaterial and Polymer Membranes, 83–133. http://dx.doi.org/10.1016/B978-0-12-804703-3.00004-8

Speranza Giorgio (2021) Carbon nanomaterials: Synthesis, functionalization and sensing applications, Nanomaterials, 11, 967.

Stewart Katherine, Limbu Saurav, Nightingale James, Pagano Katia, Park Byoungwook, Hong Soonil, Lee Kwanghee, Kwon Sooncheol, Kim Ji Seon (2020) Molecular understanding of a p-conjugated polymer/solid-state ionic liquid complex as a highly sensitive and selective gas sensor, Journal of Materials Chemistry C, 8, 15268.

Subbiah R., Veerapandian M., Yun K.S. (2010) Nanoparticles: Functionalization and multifunctional applications in biomedical sciences, Current Medicinal Chemistry, 17(36), 4559–4577.

Sun Shibin, Wang Mingwei, Chang Xueting, Jiang Yingchang, Zhang Dongzhi, Wang Dongsheng, Zhang Yuliang, Lei Yanhua (2020), $W_{18}O_{49}/Ti_3C_2T_x$ Mxene nanocomposites for highly sensitive acetone gas sensor with low detection limit, Sensors and Actuators B: Chemical, 304, 127274.

Teixeira L. S. G., Leão E. S., Dantas A. L. F., Pinheiro H. S. L. C., Costa A. C. S., de Andrade J. B. (2004) Determination of formaldehyde in Brazilian alcohol fuels by flowinjection solid phase spectrophotometry, Talanta, 64, 711–715.

Tiwari, J. N., Tiwari, R. N., Kim, K. S. (2012) Zero-dimensional, one-dimensional, two-dimensional and three-dimensional nanostructured materials for advanced electrochemical energy devices, Prog. Mater. Sci., 57, 724–803.

Tiwari Santosh K., Sahoo Sumanta, Wang Nannan, Huczko Andrzej (2020) Graphene research and their outputs: Status and prospect, Journal of Science: Advanced Materials and Devices, 5, 10–29.

Tiwary A. (2009) J. Colls, Air Pollution: Measurement, Modelling and Mitigation, CRC Press, Boca Raton, US.

Ugarte, D. (1992) Curling and closure of graphitic networks under electron-beam irradiation, Nature, 359, 707–709. doi:10.1038/359707a0

Vesely P., Lusk L., Basarova G., Seabrooks J., Ryder D. (2003) Analysis of aldehydes in beer using solid-phase microextraction with on-fiber derivatization and gas chromatography/mass spectrometry, J. Agric. Food Chem., 51, 6941–6944.

Wang Chao, Sun Jingwen, Sun Yuntong, Tan Zongyao, Xu Xuran, Fu Yongsheng, Feng Zhangqi, Zhu Junwu (2021) Fabrication of cubic Co_3O_4-hexagonal ZnO disk/rGO as a two-phase benzaldehyde sensor via a sequential nucleation strategy, Sensors and Actuators B: Chemical, 330, 129384.

Wu Yang, Chen Xingzhu, Weng Kaiyi, Arramel Jiang Jizhou, Ong Wee-Jun, Zhang Peng, Zhao Xiujian, Li Neng (2021) Highly sensitive and selective gas sensor using hetero-atom doping graphdiyne: A DFT study, Advanced Electronic Materials, 7.

Xue Yurui, Huang Bolong, Yi Yuanping, Guo Yuan, Zuo Zicheng, Li Yongjun, Jia Zhiyu, Liu Huibiao, Li Yuliang (2018) Anchoring zero valence single atoms of nickel and iron on graphdiyne for hydrogen evolution, Nature Communication, 9, 1460.

Yin Feifei, Yue Wenjing, LiYang, Gao Song, Zhang Chunwei, Kan Hao, Niu Hongsen, Wang Wenxiao, Guo Yunjian (2021) Carbon-based nanomaterials for the detection of volatile organic compounds: A review, Carbon, 180, 274–297.

Yudasaka M., Komatsu T., Ichihashi T., Iijima S. (1997) Single wall carbon nanotube formation by laser ablation using double-targets of carbon and metal, Chem. Phys. Lett, 278, 102–106.

Zeng T. L. W., Wang Z., Tsukimoto S., Saito M., Ikuhara Y. (2009) Selective detection of formaldehyde gas using a Cd-doped TiO2-SnO2 sensor, Sensors, 9, 9029.

Zhan Jing, Lei Zhendong, Zhang Yong (2022) Non-covalent interactions of graphene surface: Mechanisms and applications, Chem, 8, 947–997.

Zhang Xueyang, Gao Bin, Creamer Anne Elise, Cao Chengcheng, Li Yuncong (2017) Adsorption of VOCs onto engineered carbon materials: A review, Journal of Hazardous Materials, 338, 102–123.

Zhao Wei Na, Yun Na, Dai Zhen Hua, Ye Fei Li (2020) A high-performance trace level acetone sensor using an indispensable $V_4C_3T_x$-MXene, RSC Advanced, 10, 1261.

Zheng Xuan, Wang Guangjin, Huang Fei, Liu Hai, Gong Chunli, Wen Sheng, Hu Yuanqiang, Zheng Genwen, Chen Dongchu (2019) Liquid phase exfoliated hexagonal boron nitride/graphene heterostructure based electrode toward asymmetric supercapacitor application, Frontiers in Chemistry, 7, 544.

Zhou Xinyuan, Xue Zhenjie, Chen Xiangyu, Huang Chuanhui, Bai Wanqiao, Lu Zhili, Wang Tie (2020) Nanomaterial-based gas sensors used for breath diagnosis, Journal of Materials Chemistry, B 8, 3231.

Zhu Today Shuang, Meng Huan, Gua Zhanjun, Zhao Yuliang (2021) Research trend of nano-science and nanotechnology: A bibliometric analysis of Nano, Nano Today, 39, 101233.

6 Functional Nanomaterials for Photocatalysis Applications

Kefayat Ullah, Bakht Mand Khan,
and Won-Chun Oh

CONTENTS

6.1 INTRODUCTION

Nowadays, the world is facing two major issues as a result of the significant increase in global population, which is associated with massive industrial growth and usage of natural resources. The first is concerned with environmental protection and remediation, and the second is mainly related to energy storage and control, as well as alternative conversion [1, 2]. Several techniques to address these issues have been proposed. Among them, applying photocatalysis technology to transform solar energy into chemical energy is a viable green strategy to address or reduce future energy needs and

DOI: 10.1201/9781003263852-6

environmental concerns [3]. Photocatalysis is a method that makes use of light and semiconductors. It is very stable, non-toxic, and corrosion-resistant, and it does not produce secondary pollution. Photocatalysis has developed as one of the highly promising, effective, and low-cost technologies for not only generating energy but also removing pollutants from industrial wastes [4]. According to generally recognized theory, the basic concept of photocatalysis may be described. When photocatalytic material is subjected to light with an energy equal to or greater than that of the photocatalyst's bandgap, electron-hole pairs are formed. Formed electron-hole pairs disintegrate into electrons (e) in the conduction band and holes (h^+) in the valence band. The electrons and holes cause the reduction and oxidation of molecules adsorbed on the surface of the photocatalytic material. However, electron-hole recombination occurs often, which may result in the absence of oxidation and reduction events on the surface of photocatalytic material. Increased or decreased reaction rate is frequently related to increased or suppressed electron-hole recombination, accordingly [5].

Photocatalysis has demonstrated its enormous application potential in renewable energy, environmental remediation, and a variety of other sectors. As a result, researchers are presently focusing on developing new photocatalytic materials. The three basic principles of photocatalysis are CO_2 reduction reaction (CO_2RR), photocatalytic hydrogen evolution reaction (HER), and photocatalytic degradation reaction [6]. Photocatalysis is a green and environmentally beneficial technique since it can transform solar energy into different sorts of chemical energy. As a result, finding highly effective photocatalysts is crucial in the field of photocatalysis. Until now, several various types of functional nanomaterials—such as metal oxides, metal sulfides, metal phosphides, and carbon-based composites—have now been produced and used as photocatalysts in energy conversion technology [7]. Significant development has been achieved in the production of photocatalytic nanomaterials in recent years. These nanomaterials have unique properties including large surface areas, an abundance of active sites, controllable architectures, and tunable bandgaps, all of which are extremely useful for photocatalysis. To be more specific, different nanomaterials have been utilized as photo redox-active constituents in dual catalysts due to their superior light absorption and charge-separation capabilities [8]. Solar energy may be converted into essential chemical fuels, which include clean hydrogen by visible-light semiconductor photocatalytic technology, by water splitting, and by CO_2 photoreduction [9]. It is regarded as one of the most critical technologies for addressing energy shortages and environmental concerns [10].

In this chapter, we mainly focus on different multifunctional nanomaterials including graphene, graphitic carbon nitride, zinc oxide (ZnO) and titanium dioxide (TiO_2) for photocatalysis applications such as hydrogen production, degradation of organic dyes and carbon dioxide (CO_2) reduction.

6.2   NANOMATERIALS FOR PHOTOCATALYSIS APPLICATION

6.2.1   Graphene-Based Photocatalysis

Graphene is a hexagonal lattice of single layer sp^2 carbon atoms with significant catalytic performance in photocatalysis due to its superior electron capturing and transport conductivity, huge specific surface area, and its perfect interaction with other catalyst particles [11]. It has outstanding properties including excellent thermal conductivity of about 5000 W m^{-1} K^{-1}, quite high (97.7%) transparency to visible light, large carrier mobility of approximately 200,000 cm^2 V^{-1} s $^{-1}$, ambipolar electric field effect, large surface area, and quantum Hall effect at room temperature [12, 13]. Graphene has considered to be a zero bandgap semiconductor, which means that its band structure is linearly dispersed and the charge carriers behave as massless Dirac fermions in a particular k-point in the first Brillouin zone [14]. Because graphene has a zero bandgap, it absorbs light in a wide variety of spectra ranging from infrared to ultraviolet, allowing it to be used in electro-optical systems/devices [15]. Because of the conduction and valence band overlap, graphene behaves zero-gap semiconductor with large carrier mobility at the speed of about 106 m s^{-1} (relativistic speed) [16]. Graphene is considered to be gapless material because the conduction and valence band are symmetrical and meet at the same Dirac point [17], and it has unique two-dimensional (2D) material with one-atom-thick layers of carbon atoms, absorbing 2.3% of the total incident light [18]. Because of its superior electrical characteristics, graphene may efficiently increase charge separation as well as inhibit recombination of excited carriers produced by photocatalysts [19]. Graphene and modified graphene are introduced in several fields of catalysis because of their excellent characteristics [20]. There are several benefits to graphene-based nanocomposites. When graphene is applied to or deposited on a photocatalyst, it improves photocatalytic activity by enhancing charge separation and inhibiting charge-carrier recombination [21].

6.2.1.1 Graphene-Based Photocatalytic Hydrogen Production

Hydrogen produced from water is a clean, safe, and suitable option for long-term energy generation [22]. Hydrogen is now considered to be the fuel of the near future. Hydrogen is a sustainable and environmentally friendly form of energy as compared to carbon fuel. On an industrial scale, there are several techniques for manufacturing hydrogen. However, all known technologies have a significant energy consumption, making large-scale hydrogen synthesis difficult. The creation of hydrogen through the photocatalytic splitting of water is technologically easy, and the resulting gases are friendly to the environment [23].

Graphene has been widely researched for photocatalytic water splitting because of its exceptional features including high electron mobility and large specific surface area, which make graphene a well-suited nanomaterial for the creation and separation of charge carrier. However, apart from being a transfer channel as well as co-catalyst, graphene may also be utilized as a good photocatalyst. The conduction band minimum (CBM) of reduced graphene oxide has been proved theoretically and experimentally to be appropriate for hydrogen production, although graphene offers some advantages for water splitting, which has several practical limitations, including low catalytic activity and lack of bandgap. However, graphene is combined with other materials to improve its photocatalytic performance for water splitting [24]. In photocatalysts, graphene, which is commonly referred to as GO and rGO, seems to have the potential to tune the bandgap energy of semiconductor materials and suppress recombination of electron-hole pairs, resulting in a significantly enhanced photocatalytic activity when exposed to a given light [25].

In photocatalytic water splitting, oxidation/reduction occurrences occur in a photocatalyst simultaneously as it is exposed. Semiconductor-based materials are commonly used as photocatalysts. The photocatalytic water splitting process consists of four main steps (Figure 6.1). (1) The photogeneration of e/h+ pairs in the semiconductor because of light absorption is the first step. (2) The photogenerated electrons and holes migrate through the photocatalyst to reach the exterior surface in the second stage. (3) Following that, electrons and holes that come into touch with water undergo oxidation/reduction processes. (4) Finally, throughout electron/hole migration or at the photocatalyst surface, unfavorable charge recombination might occur, leading to charge annihilation [26].

Shanshan Qiao and coworkers prepared Co/CoO combined on N-doped graphene by calcinating ZIF-67 and using a simple preparation method, which showed great performance (water oxidation about 10.22% at 450 nm, oxygen rate of production about 543,198 mmol^{-1} g^{-1} h^{-1}, and hydrogen rate of production about 330 mmol^{-1} g^{-1} h^{-1}). The core-shell architecture of Co/CoO has a synergic activity with graphene, which is advantageous for light absorption and charge separation. However, repeated water oxidation, as well as water reduction studies, showed that such catalyst was extremely stable throughout the reaction process [27].

Gouri Sankar Das et al. developed a simple and clean method to make nitrogen-combined functionalized graphene nanosheets (N-fGNS), which they then applied for water splitting through visible light. The prepared N-fGNS shows H$_2$ production about 1380 and O$_2$ production about 689 M g^{-1} h^{-1}, effectively without co-catalysts when directly exposed to light. The nitrogen doping and plentiful surface defects are accountable for N-fGNS photocatalytic water splitting effectiveness. Because of the many surface imperfections, the bandgap was narrowed, separation of charge carrier was improved, and the lifetime of charge carrier was also increased. The photocatalytic characteristics, cycle stability, and storage stability of N-fGNS all were outstanding. As a result, the N-fGNS might be a potential single photocatalyst for water splitting [28].

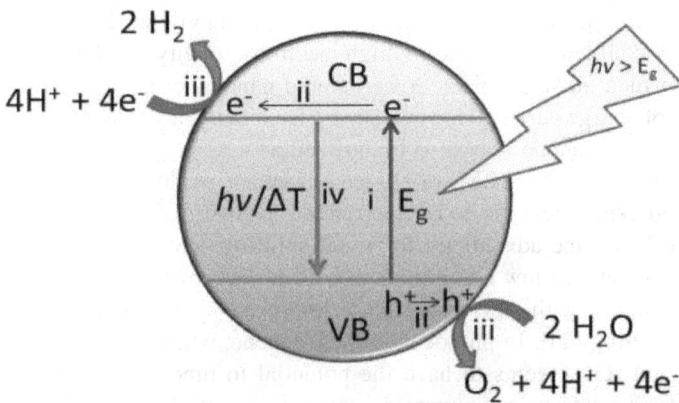

FIGURE 6.1 Fundamental mechanism of photocatalytic water splitting reactions [26].

6.2.1.2 Graphene-Based Photocatalytic Degradation of Dyes

The urgent demand for clean water has grown because of the increasing water pollution. Consequently, the new purification method is both environmentally friendly as well as the most researched procedure at the moment. Our future is largely dependent on understanding these needs and developing effective, long-term, and environmentally friendly water treatment technology. The negative impact of wastewater treatment on the environment—including human health and hygiene, as well as that of other species—demands the search for environmentally friendly, energy-efficient, and cost-effective water purifying solutions. Textile wastewater streams are loaded with aromatic chemicals and organic dyes, causing a hazardous danger to the environment, and must be cleaned before reuse or release. Several materials have been investigated for use as photocatalysts for pollutant degradation that solve the problems in water treatment technologies [29].

Graphene is one of the 2D nanomaterials that is widely used for adsorption and photodegradation because of its tunable structure an large surface area [30]. Graphene and its derivatives are interesting choices for photocatalytic degradation contaminants in wastewater because of their exceptional characteristics. Since graphene has a zero bandgap, it is generally used in conjunction with other semiconductor catalysts. Due to its remarkable electronic, mechanical, and optical properties, it can improve photocatalytic properties by functioning as an acceptor and transfer, lowering the recombination process of e/h+ pairs generated, as well as increasing the light absorption when combined with other materials [31].

Hamed Safajou and his colleagues synthesized RGO/Cu nanocomposite via reducing graphene oxide and Cu_2 ions with spearmint extracted as reductant and stabilizing agents. Figure 6.2 shows the photocatalytic activity of the prepared composite. The photocatalytic behavior of RGO/Cu nanocomposites demonstrated that it is an effective catalyst for the degradation of organic contaminants. This chemical can remove 91.0% of Rhodamine B (RhB), as well as 72.0% of methylene blue (MB) [32].

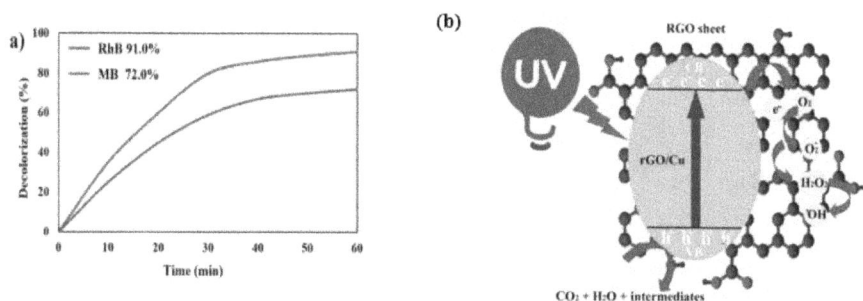

FIGURE 6.2 (a) The photocatalytic activity of the prepared RGO/Cu nanocomposite and (b) The photocatalytic mechanism of the obtained RGO/Cu nanocomposite in UV-visible light [32].

6.2.1.3 Graphene-Based Photocatalytic CO_2 Reduction

Photocatalytic carbon dioxide reduction is an excellent method for reducing greenhouse gas emissions while also meeting the need for renewable fuels. As a result, numerous researchers have in recent decades dedicated themselves to studying various semiconductor composites toward CO_2 photocatalytic conversion [33]. Graphene and its derivatives have been the focus of intense study in recent decades because of its large surface area and unique tunable features. Graphene, in particular, has been intensively researched, with applications in both electrocatalyst and photocatalytic CO_2 reduction being considered [34]

For photocatalytic applications, Velasco-Hernandez et al. applied a sol-gel technique to create TiO_2 doped graphene oxide nanocomposite (GO/TiO_2) thin films on top of glass substrates through using dip-coating process. XRD, Raman, UV-VIS, and electron microscopy spectroscopy were used to investigate the optical, structural, and morphological features of the graphene oxide powder, as well as GO thin films, which were produced using a modified Hummers' technique. After heating at 450°C, the thin films' energy bandgap was calculated to be in the region of 3.38–3.45 eV; however, increasing the amount of GO decreased the transmittance as well as Raman intensity considerably. The photocatalytic performance of obtained GO/TiO_2 nanostructures was increased due to the addition of GO, as measured by methylene blue blenching; moreover, the films indicated prospective uses in the CO_2 reduction process, having a production of methanol about 68.443 $\mu mol\ cm^{-2}$ was obtained after five hours as in UV light [35].

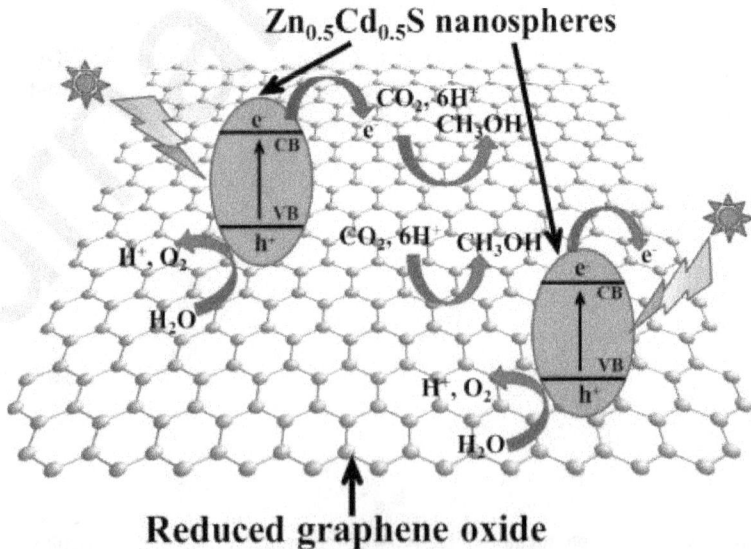

FIGURE 6.3 Charge transfer mechanism of 2G/ZCS composite material for photocatalytic CO_2 reduction [36].

A simple dual-step hydrothermal technique at a temperature of 180°C was used to create unique hierarchical structure graphene–$Zn_{0.5}Cd_{0.5}S$ (xG-ZCS) nanocomposites. The synthesized composites showed improved photocatalytic CO_2 reduction into methanol conversion (CH_3OH). The largest quantity of CH_3OH was formed by the 2G-ZCS (2 wt% graphene-$Zn_{0.5}Cd_{0.5}S$) composite (about 1.96 µmol g^{-1} h^{-1}), which is over 98 times more than pure ZCS nanospheres ($Zn_{0.5}Cd_{0.5}S$). The improved photocatalytic CO_2 reduction might be due to graphene's role as an excellent electron absorber and carrier, lowering charge transportation reconsolidation and improving catalytic properties [36].

6.2.2 Graphitic Carbon Nitride (g-C₃N₄)–Based Photocatalysis

The metal-free polymeric catalyst g-C_3N_4 is made out of triazine units that repeated. It has a bandgap of approximately 2.56–2.70 eV, it is promising to be a very effective photocatalytic catalyst [37, 38]. It is one of the most widely recognized materials for photocatalysis applications because of its extremely effective performance and environmentally friendly nature. Owing to some of its limitations, research is being conducted to enhance its practicality and cost-effectiveness by increasing its photocatalytic activity. For creating heterostructures, many strategies have been studied, including morphological modification by altering temperature, modifying g-C_3N_4 with metals, and doping [39]. It has large activity and effective visible-light absorbance, graphitic carbon nitride (g-C_3N_4) has received a lot of attention. Because of the strong covalent link between nitrogen and carbon atoms in the conjugate layer structure, it includes just Earth-abundant elements such as carbon and nitrogen, and it also has an excellent chemical as well as thermal stability [40]. It also has a modest bandgap energy of about 2.7 eV (460) nm, which allows it to absorb visible light and has favorable CB and VB edges locations for photocatalysis. As a result, g-C_3N_4 has developed into one of the most excellent materials for the photocatalysis field as an eco-friendly photocatalyst. Furthermore, the widespread use of g-C_3N_4 as a photocatalyst is due to its ability to be obtained from inexpensive and easily available precursors using a relatively easy synthesis process. Moreover, g-C_3N_4 has a novel delocalized conjugated composition that includes graphitic stacking of C_3N_4 layers joined by tertiary amines, resulting in excellent electrical conductivity [3].

6.2.2.1 g-C₃N₄–Based Photocatalytic Hydrogen Production

Because of its outstanding performance, g-C_3N_4 has been widely utilized in the area of photocatalytic production of hydrogen [41], and it has been considered one of the best photocatalysts for hydrogen production due to its natural availability, easy synthesis, and excellent thermal and photochemical stability [42]. However, the photocatalytic performance of the directly polymerized g-C_3N_4 material is far from desirable because of its small surface area, fewer active sites, and fast recombination, as well as limited visible-light absorption range [43]. To increase the photocatalytic performance of g-C_3N_4 materials, different strategies are used such as modifying its morphological structure, doping its metallic/non-metallic elements, heterostructure construction by Z-scheme, protonation, and inducing element defects or voids, and some other technologies may be applied [44–46].

A simple three-step procedure was used to synthesize graphitic carbon nitride (g-C_3N_4) nanosheets attached with amorphous NiS metallic Ni interface layers co-catalyst: packing of Ni(OH)$_2$ nanosheets, greater H$_2$ reduction, and further deposition of amorphous NiS nanostructures (Figure 6.4). The results showed that both strong metallic Ni interface layers and amorphous NiS may be used as electron co-catalysts to significantly increase visible-light hydrogen production above that of g-C_3N_4 semiconductors. The optimized g-C_3N_4–based photocatalyst including 1.0% NiS and 0.5 wt% Ni exhibited the maximum hydrogen production of about 515 μmol g^{-1} h^{-1}, and it was about 4.6 and 2.8 times more than those observed on binary g-C_3N_4_0.5% and g-C_3N_4_1.0% Ni, respectively. Metallic Ni interface layers appear to serve multifunctional functions in boosting visible-light H$_2$ production by first collecting photogenerated electrons from semiconductor g-C_3N_4 material and then accelerating the surface H$_2$-production reaction kinetics on amorphous NiS co-catalysts [47].

S-type heterojunction photocatalyst based on g-C_3N_4/α-Fe$_2$O$_3$ containing Co$_3$S$_4$ nanoparticles was used for the hydrogen production process. The obtained composite shows an excellent photocatalytic performance and good stability for hydrogen production: g-C_3N_4/α-Fe$_2$O$_3$/Co$_3$S$_4$ exhibits a large amount of hydrogen production about 191.41 mol—that is, approximately 30 times as compared to pure Co$_3$S$_4$ of 6.38 mol. The EY molecules enhance the light absorption performance of the prepared samples and also acts as electron donors which highly dispersed on Co$_3$S$_4$ nanoparticles could provide a large quantity of reduction sites; also, the designed S-scheme heterojunction utilizes unnecessary electrons and holes as in the hydrogen evolution system, and uses the composite material's solid redox potential to enhance the photogenerated carrier separation [48].

6.2.2.2   g-C_3N_4–Based Photocatalytic Degradation of Dyes

Photocatalytic reduction of carbon dioxide (CO$_2$) to obtain energy fuels seems to be a promising technique for overcoming the two most significant challenges: fossil fuel

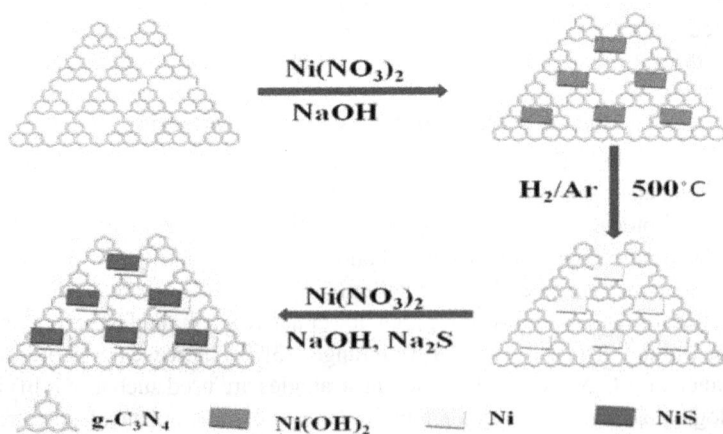

FIGURE 6.4   Synthesis of ternary g-C_3N_4-Ni-NiS photocatalysts [47].

dependency and pollution. Given its fascinating features, such as a moderate band-gap having a high CB potential layered structure, outstanding stability, and low cost, graphitic carbon nitride (g-C_3N_4) has been intensively researched for its potential material for CO_2 reduction among various available photocatalysts [49]. Graphitic carbon nitride (g-C_3N_4) is considered to be an effective catalyst for CO_2 photoreduction due to its moderate bandwidth, excellent stability and metal-free property [50]. g-C_3N_4 has now been widely described as a suitable photocatalyst in the degradation of organic pollutants as a metal-free polymeric semiconductor [51].

Fahad A. Alharthi et al. synthesized graphitic carbon nitride loaded zinc oxide g-C_3N_4/ZnO (Zn-Us) nanocomposites by using the various amounts of urea. Following UV light irradiation, the obtained nanocomposites were applied as photocatalysts for the degradation of the photosensitive dyes rose bengal (RB) and methylene blue (MB), and a suitable photomechanism was postulated. Using these photocatalytic properties, urea-derived g-C_3N_4/ZnO nanocomposite photocatalysts were shown to exhibit good photodegradation performance against RB and MB at four hours, correspondingly. The degradation rate of synthesized Zn-Us was revealed to be 90% for both dyes during the provided testing circumstances. Anionic RB dye degrades more aggressively on the surface of produced photocatalysts than cationic MB dye. The findings can be efficiently used to future potential implementation in wastewater treatment [52].

J. Zhang et al. created a ZnO/g-C_3N_4/GO nanocomposite by first modifying porous carbon nitride (g-C_3N_4) with zinc oxide (ZnO) nanoparticles and thereafter loading graphene oxide (GO) nanosheets (Figure 6.5). A series of experiments were performed on this synthesized ternary and binary composites with varying weight ratios. Whenever the g-C_3N_4 weight ratio reached 40%, the binary compound ZnO/g-C_3N_4 moved its absorption edge toward lower energy and exhibited the highest photocatalytic activity. Furthermore, the inclusion of GO significantly reduced photocatalytic charge recombination and increased absorption rate in the visible-light spectrum.

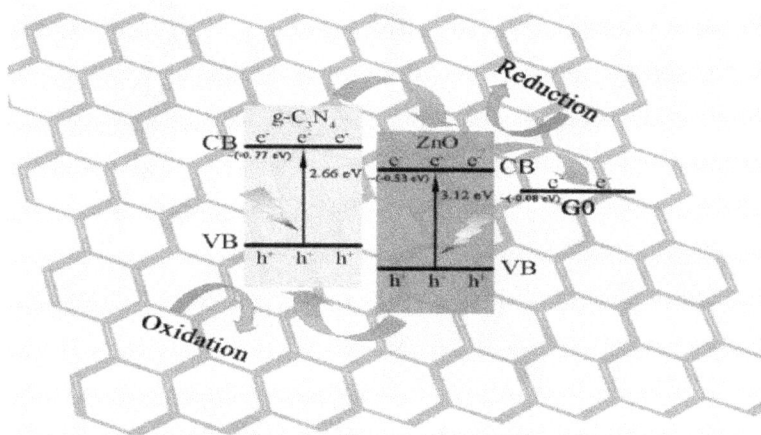

FIGURE 6.5 Charge transfer pathway and the working principle of a heterojunction [53].

The ZnO/g-C$_3$N$_4$ (40%)/GO (15%) showed 98% degrading efficiency in 100 minutes, which was approximately 2.23 times faster than ZnO/g-C$_3$N$_4$ (40%). The conduction and valence bands potentials of the obtained material was estimated using VB-XPS, and when paired with the free radical capturing experiment, demonstrated the most efficient free radical throughout this experiment and demonstrated the reaction rate of the photocatalytic process. The photocurrent, as well as EIS findings, examined the material's strong responsivity and low photogenerated electron and hole recombination rate. In conclusion, the interface impact, as well as synergistic effect of such ternary heterojunction, considerably increased the photocatalytic degradation of hazardous dyes [53].

6.2.2.3 g-C$_3$N$_4$–Based Photocatalytic CO$_2$ Reduction

Graphitic carbon nitride (g-C$_3$N$_4$) is considered to be an effective catalyst for photocatalytic CO$_2$ reduction due to its strong stability, moderate bandwidth and metal-free property [50]. However apart from its low cost, the advantages of g-C$_3$N$_4$ based CO$_2$ photoreduction are as follows: (1) it is nitrogen-rich, as well as an emerging material for CO$_2$ activation; (2) the band architecture of g-C$_3$N$_4$ is appropriate for photocatalytic reduction of CO$_2$ to different value-added compounds; (3) its 2D layered structure might enable electron transfer toward the chemical surface adsorption sites; and (4) it is complete of capability for its greatly tunable structure: via changing polymerization conditions and precursors, g-C$_3$N$_4$ with various band edges may be obtained [54].

T. Beenish et al. successfully prepared WO$_3$/g-C$_3$N$_4$ composite for photocatalytic reduction of CO$_2$ applications and tested at visible-light irradiation. Effective charge

FIGURE 6.6 FE-SEM analyses of (a) g-C$_3$N$_4$, (b) WO$_3$/g-C$_3$N$_4$ material, and (c) TEM analyses of WO$_3$/g-C$_3$N$_4$ material, and (d) UV-visible analyses of WO$_3$, g-C$_3$N$_4$ and WO$_3$/g-C$_3$N$_4$ photocatalysts [55].

separation was reported in a Z-scheme–prepared WO_3/g-C_3N_4 composite, allowing for greater CO generation via photocatalytic dry as well as bireforming processes. The maximum CO production for WO_3/g-C_3N_4 composite is approximately 310 µmol gcat^{-1}h^{-1}. The performance of several reforming processes—including dry reforming of CH_4, bireforming of CH_4, and reverse water-gas reactions—was further examined. Under the identical reaction circumstances, most CO was generated using BRM as well as RWGS reactions. It is possible to infer that the WO_3/g-C_3N_4 nanocomposite is a promising photocatalytic material for photocatalytic CO_2 reduction in visible-light illumination and can be used in additional solar energy-supported applications because of improved charge separation and greater visible-light absorption [55].

6.2.3 ZnO-Based Photocatalysis

Zinc oxide (ZnO) is a well-studied multifunctional semiconductor material due to its remarkable features such as biocompatibility, strong chemical stability, unique electronic structure, biocompatibility, and low manufacturing cost, and it has been widely used in photocatalysis and many energy conversion applications [56–58]. These characteristics enable ZnO to play a significant role in the field of photocatalysis and solar cell technologies. Because of its excellent carrier mobility, as well as outstanding optical properties, ZnO is a possible photocatalytic material having an appropriate bandgap that has received a lot of attention [59]. At 300 K, ZnO has a broad direct bandgap about 3.30–3.37 eV, and has a binding energy of 60 meV. These two properties make ZnO an interesting choice in a broad variety of optical and electronic applications [60, 61].

6.2.3.1 ZnO-Based Photocatalytic Hydrogen Production

Photocatalytic hydrogen production is a unique, environmentally beneficial, and advantageous technology for producing sustainable and clean energy by light energy [62]. Particularly regarding the water splitting procedure for hydrogen production, zinc oxide has shown outstanding photocatalytic characteristics comparable to those identified for TiO_2, that may be contributed to ZnO high electron mobility; that is approximately 100-fold greater than those of TiO_2, resulting in decreased the rate of charge-carrier recombination. Furthermore, the lower charge transfer resistance—as well as the larger surface area of ZnO different morphological features—contribute to its tremendous potential in photoelectrochemical applications [63–66].

Yang et al. used a one-step hydrothermal process to create ZnO flowers. A solution recrystallization technique was used to attach the ZnS nanoparticles uniformly on the surface of ZnO, resulting in a ZnO/ZnS core-shell composite. An additional rise in thiourea causes the Kirkendall effect, which results in the creation of a ZnS hollow architecture. The flower-like architecture has remained constant throughout the procedure. The effects of various ZnS loadings on the architecture, crystallization characteristics, photoelectric response, specific surface area, and catalytic performance of ZnO were studied. Under synthetic light irradiation, the optimal ZnO/ZnS composite demonstrated good photocatalytic activity about 757.07 µmol g^{-1} of hydrogen. The findings suggest that placing enough ZnS nanoparticles above zinc oxide nanorods significantly increases the charge transfer mechanism. The charge transfer and large specific surface area of the catalyst boost photocatalytic hydrogen production activity [67].

X. Liu et al. developed nitrogen-rich $ZnO/g-C_3N_4$ nanocomposite with varying Zn stacking mass contents to test visible-light photocatalytic efficiency for hydrogen production as well as NO photo-oxidation. The synthesis of the N-doped $ZnO/g-C_3N_4$ nanocomposite was verified by characterization data. The experimental findings show that N dopant can improve the photocatalytic efficiency of $ZnO/g-C_3N_4$ nanocomposite for hydrogen production. With visible-light irradiation, N-ZnO-CN composite has the greatest photocatalytic activity for hydrogen production rate about 0.78 mmol h^{-1} g^{-1}, which is approximately 77% greater than $g-C_3N_4$. The photo-oxidation NO extraction efficiency of N-ZnO-CN is about double that of $g-C_3N_4$. Because photogenerated electrons may readily be moved from $g-C_3N_4$ to the CB of ZnO via potential difference, this mechanism is a sustainable mechanism for the photocatalytic H_2 production process. The results demonstrate that the Z-scheme technique of N-doped $ZnO/g-C_3N_4$ nanocomposite is beneficial for NO photo-oxidation since the Type II strategy is unfavorable for O_2 production [68].

H. Yoo et al. produced ZnO/CuS nanocomposites in two steps using simple precipitation procedures, and then heat-treated them to generate ZnO/Cu_2O-CuO heterojunction photocatalysts. After heat sintering, the photocatalytic efficiency of the produced materials for bringing cuprous and cupric oxides was considerably increased for hydrogen generation, with the greatest value of about 1092.5 µmol g^{-1} h^{-1} at the optimal copper precursor amounts of 33% and optimum cuprous oxides amount of 14.3%. This extremely enhanced photocatalytic behavior of the ZnO/Cu_2O-CuO structure could be contributed by three major factors: (1) greatly enhanced light absorption having to introduce copper elements; (2) reduced e/h^+ recombination in every cuprous and cupric oxide region with the help of Z-scheme mechanism in both cuprous and cupric oxides; and (3) effective charge flows directly in CB from Cu_2O to zinc oxide energy levels [69].

6.2.3.2 ZnO-Based Photocatalytic Degradation of Dyes

ZnO has a good optical capacity, low cost, availability, and chemical stability, and it is widely used to degrade organic compounds including ciprofloxacin, acid res, rhodamine B, and methylene blue [70]. Under UV light (UV) visible irradiation, titanium dioxide demonstrates an exceptional ability to degrade organic pollutants via oxidation/reduction driven by excited electron-hole pairs between both the valence and conduction bands [71]. Figure 6.7 shows a schematic illustration of the photocatalytic mechanism of organic pollutants degradation using silver-ZnO-MWCNT as a photocatalyst. In this mechanism, the production of a hole-electron following incoming light photons might cause a charge transfer in the photocatalyst, facilitating the synthesis of O_2 radicals. Organic pollutants could be degraded by the generated radicals [72].

Demirci et al. produced ZnO particles using the sol-gel technique at 500°C for two hours at various heating rates. The degradation of methylene blue was used to assess the influence of heating rate on the photocatalytic performance of ZnO particles. The non-isothermal rates and thermodynamic characteristics were also calculated. The

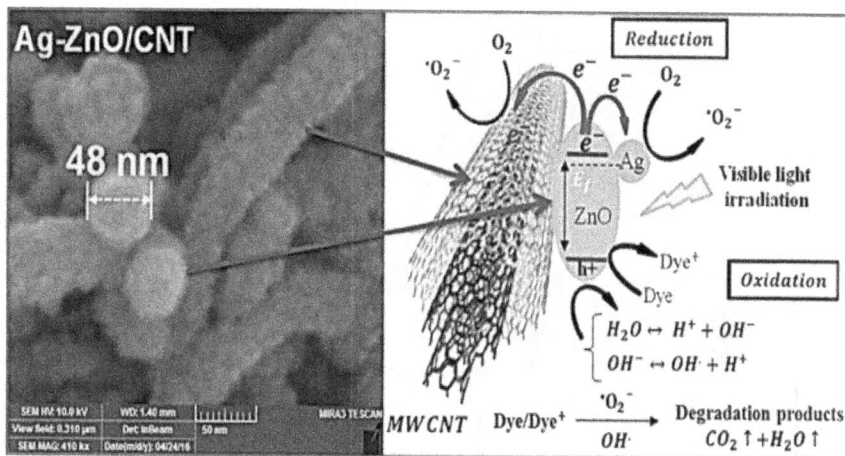

FIGURE 6.7   SEM picture of the Ag-ZnO-MWCNT photocatalyst material and dye photo-degradation mechanism [72].

varying heating rates had little effect on the surface structure of ZnO particles. It was discovered that the heating rate had to have a significant impact on the lowering of bandgap, as well as photocatalytic performance. The bandgap of the zinc oxide particles ranged from 3.10–3.17. The maximum photocatalytic performance was found in a ZnO sample produced at 1°C/min. Its comparative photocatalytic degradation efficiency and kinetic constant reported 92.7% and as 1.069×10^{-2} min^{-1}, respectively. The findings might be attributed to low bulk defects, large surface oxygen vacancies, and a narrow energy bandgap. In addition, ZnO photocatalysts demonstrated high stability after four series of testing. This study proposes a new technique for improving the photocatalytic performances of ZnO photocatalysts for organic pollutant removal [73].

Atchudan et al. studied the solvothermal production of ZnO–GO nanocomposites by doping GO with ZnO nanoparticles. Before solvothermal impregnation, the GO and ZnO nanoparticles were produced by Hummers' method, as well as thermal oxidation procedures (Figure 6.8), respectively. Because of the enhanced light adsorption and lower electron-hole recombination caused by the creation of a ZnO-GO nanocomposite, these composites have been efficient in the photodegradation of methylene blue dye at UV light irradiation—to be specific, 98.5% degradation after 15 minutes of irradiation [74].

The ZnO/GO nanocomposites were effectively produced by using a precipitation technique using GO layers as the catalytic carrier. The hydroxyl groups of GO serve as anchored sites for the formation of cross-linked zinc oxide nanoflakes on the surface of GO. It is possible to identify that the ZnO has been effectively coated on the surface of GO based on extensive characterization using XRD, XPS, and SEM. The produced nanocomposite has also been used to study the photodegradation of MB

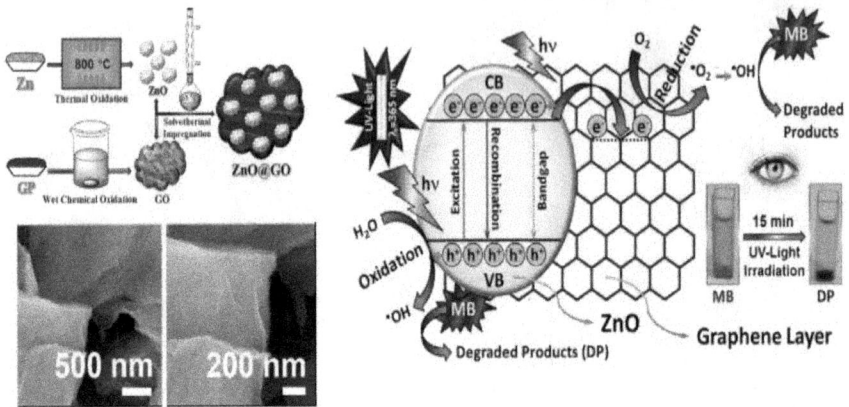

FIGURE 6.8 Schematic illustration of the preparation and SEM images of ZnO–GO nanocomposites and suggested photocatalytic process for the photodegradation of dye (methylene blue dye) by using the obtained ZnO–GO nanocomposites under ultraviolet light. [74]

dye from water. We discovered that processed ZnO/GO nanocomposites have more efficient photocatalytic properties than pure ZnO. The degradation rate at optical circumstances approaches 97.6% in just 90 minutes, whereas the first-rate constant, was 0.04401 min⁻¹. This efficient photocatalyst is an excellent choice for photocatalytic degradation of dyes [75].

Wu et al. used a facile method to synthesize $ZnO/Fe_3O_4/g-C_3N_4$ nanocomposites as a photocatalysis. With all able to prepare photocatalysts, the $ZnO/Fe_3O_4/g-C_3N_{4-}$ 50% nanocomposite photocatalyst demonstrated the most effective photocatalytic performance, owing to enhanced light absorption characteristics leading as from a heterojunction framework between $g-C_3N_4$ and ZnO, as well as synergistic effect of its internal electric field and the corresponding band structure framework. Furthermore, the rate of the photogenerated carriers remained large. The degradation rates of orange (G), monoazo dyes, alizarin yellow R (AYR), and methyl blue (MO) over $ZnO/Fe_3O_4/g-C_3N_{4-}$ 50% was 97.87%, 98.05%, and 83.35%, respectively. This was attributed to the quantity of the dye molecules adsorbed on the photocatalyst, as well as the geometry of the azo dye molecules that had to influence the degradation. The rates of MO breakdown on the composite photocatalyst were consistent with first-order kinetics. Moreover, the inclusion of Fe_3O_4 increased the photocatalyst's stability as well as its recyclability substantially [76].

Park and his coworkers used an electrospinning technology followed by a hydrothermal procedure to create new photocatalytic $ZnO-TiO_2$–combined carbon nanofibers (ZnO/TiO₂/CNFs) nanocomposite. Initially, TiO_2 nanoparticles doped with carbon nanofibers (TiO₂/CNFs) were created via electrospinning and carbonization procedure. Following that, the zinc oxide particles were hydrothermally developed into TiO₂/CNFs. State-of-the-art methods were used to investigate the morphology, structure, and chemical composition. The photocatalytic activity of the nanocomposite was investigated utilizing three sequential cycles of degrading organic dye such as methylene blue (MB) with UV irradiation. It was discovered that ZnO/TiO$_2$/CNFs nanocomposite had greater MB removal characteristics than other compositions, which might be attributed to synergistic effects caused by carbon CNFs as well as the metal oxides used (ZnO and TiO$_2$). Carbon fiber adsorption, along with matching band potentials of TiO$_2$ and zinc oxide, work together to improve the total photocatalytic activities of the nanocomposite. The study's findings suggested that it might be a cost-effective and ecologically friendly photocatalyst [77].

6.2.3.3 ZnO-Based Photocatalytic CO$_2$ Reduction

Much effort has been focused on the search for appropriate materials for photocatalytic CO$_2$ reduction. Among all the possibilities, semiconductors are the greatest potential and popular photocatalysts. Especially ZnO has received a lot of attention because to its acceptable bandgap (Eg), affordability, and high exciton binding energy [78–80]. Different lighting sensitizers have been coupled with ZnO to build heterojunction structures in order to expand ZnO light absorption to the visible-light range [81].

Qingfeng Ge et al. effectively synthesized a hollowed structured p-n heterojunction catalyst, polydopamine (PDA)-ZnO/Co$_3$O$_4$, by pyrolyzing bimetallic ZnCo-ZIFs and then modifying them with PDA. A good separation ratio of photogenerated e/h$^+$ pairs was obtained by adjusting the Zn/Co ratio. Moreover, the adsorption rate of CO$_2$ may be enhanced by adjusting the quantity of PDA on the interface of ZnO/Co$_3$O$_4$. As a result, without any of the photosensitizer and extra sacrificial ingredient, PDA/ZnO/Co$_3$O$_4$ demonstrated a CO generation rate of 537.5 mol/g/h, as well as a CO selectivity of 97.7%. A viable method for photocatalytic carbon dioxide reduction mostly on PDA/ZnO/Co$_3$O$_4$ was proposed using band structure analysis on X-ray photoelectron spectroscopy, photoelectrochemical characterization, and density functional theory computation. The current work presents an efficient method for fabricating highly effective photocatalysts for carbon dioxide reduction relying on semiconducting metal oxides with outstanding characteristics for absorbing light and CO$_2$, as well as separating charge carriers [82].

Fan Zhang et al. used a solvothermal approach combined with a chemical reduction procedure to create an Ag-Cu$_2$O/ZnO photocatalyst. The optimized Ag-Cu$_2$O/ZnO photocatalyst showed substantially greater activity for reduction of CO$_2$ to CO than pure ZnO, binary Cu$_2$O/ZnO, or Ag/ZnO composites due to manipulation of the multifunctional constituents (Ag and Cu$_2$O). This has been discovered that coated Cu$_2$O may improve CO$_2$ adsorption process on the surface of catalyst and that the created Z-scheme charge transport channel between ZnO and Cu$_2$O permits more effective charge separation, thereby boosting catalyst photocatalytic properties. As the Ag nanoparticles work as a co-catalyst more to absorb the excess of electrons collected on the surface of Cu$_2$O, that can alleviate Cu$_2$O self-photoreduction, considerably increasing the photocatalytic activity as well as the stability of Ag-Cu$_2$O/ZnO NRs. We believe that our research will open the path for the logical design of multicomponent semiconductor-based photocatalytic CO$_2$ reduction processes [79].

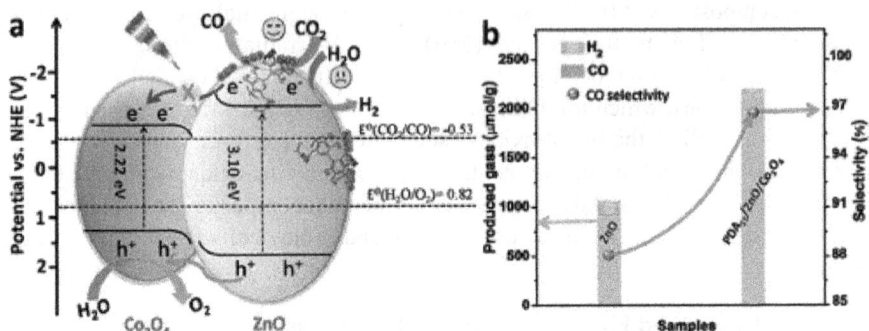

FIGURE 6.9 Photocatalytic CO_2 reduction process and charge transfer path for PDA/ZnO/ Co_3O_4 (b) CO and hydrogen generation rates of ZnO and PDA/ZnO/Co_3O_4 with various concentrations of PDA into the DMF/H_2O solution under UV light irradiation. [82]

6.2.4 TiO_2 Photocatalysis

TiO_2 has been actively researched in photocatalysis sectors during the previous century as a typical and widely utilized semiconductor material. The fundamental principles of TiO_2 photocatalysis, as well as photocatalytic efficiency increase and reaction system enhancement, have been extensively researched. Apart from hydrogen production and pollutant degradation, TiO_2 photocatalysis has been used in a variety of practical applications, including water/wastewater disinfection and super hydrophilic self-cleaning, bone-implant fixation, as well as the generation of fine organic compounds [8]. TiO_2 is now the most researched and frequently utilized photocatalytic degradation material. However, TiO_2 has a quick e/h$^+$ pair recombination as well as a large bandgap about Eg = 3.1–3.2 eV, that can only be stimulated by UV light. To solve these problems, semiconductor-coupling provides a promising structure-properties engineering option for increasing TiO_2 photocatalytic activity by reducing e/h+ recombination rate while increasing visible-light absorption [83]. TiO_2 is regarded as a multifunctional material due to its physicochemical features, which include the formation of charge carriers (electrons and holes) during UV light absorption, correlating to the bandgap as well as greater refractive index. As a result, it is a excellent material for different applications including CO_2 reduction, solar water splitting, and photocatalytic degradation [84]. Without a doubt, the development of nanocomposites is a potential strategy for increasing the photocatalytic activity of photocatalysts. TiO_2-based nanocomposites are developed to improve photon usage efficiency in the existence of UV and visible light [85].

6.2.4.1 TiO_2-Based Photocatalytic Degradation of Dyes

Organic compounds produced by many sectors contribute to a variety of harmful contaminants in wastewater. Photocatalysis is among the most efficient technologies for removing organic contaminants from wastewater. TiO_2 photocatalysts are innovative materials with exceptional absorption behavior against organic compounds in

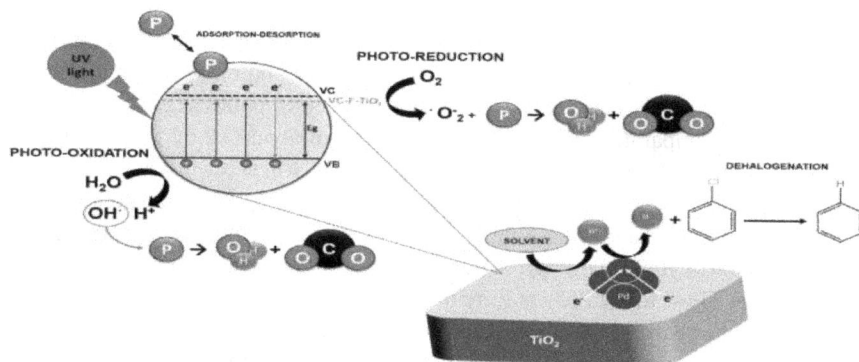

FIGURE 6.10 Proposed technique for pollutant degradation as well as dehalogenation using TiO_2 materials. [89]

wastewater because of their amazing qualities such as nontoxicity, strong photocatalytic degradation capability, and superior thermal and chemical stabilities [86–88].

TiO_2 photocatalytic degradation of organic dyes (methylene blue) under ultraviolet irradiation about 365 nm, 36 W, and having irradiated intensity about 1.02×10^3 mW cm^{-2} was used to photocatalyzed the breakdown of MB, resulting in the generation of radicals OH (through photo-oxidation) as well as O_2 (by photoreduction) (Figure 6.9). These generated radicals react with the MB that is adsorbed upon the surface of photocatalyst to break down the organic component to CO_2 and water [89].

Md. Tamez Uddin et al. used the polyol approach to create heterostructure ZnO/SnO_2 composite materials. As-synthesized samples were tested for optical, structural, electrical, morphological, optical, and photocatalytic properties. For such degradation of MB dye with UV light irradiation, the obtained heterojunction photocatalyst demonstrated greater photocatalytic efficiency than pure SnO_2, ZnO, or reference commercial TiO_2. The enhanced separation of photogenerated electron-hole caused by band offsets created at the surface of SnO_2/ZnO matrix composites was related to the greater catalytic efficiency of SnO_2/ZnO nanocomposites. Moreover, the heterostructure SnO_2/ZnO photocatalysts might be readily recycled without losing catalytic activity, demonstrating the catalysts' stability and repeatability. This notion of semiconducting heterostructure nanocatalysts with enhanced photocatalytic activity could find practical use in wastewater treatment systems, as well as the removal of unwanted organics from the environment in the future [90].

6.2.4.2 TiO_2-Based Photocatalytic Hydrogen Production

The most researched photocatalytic technology is photocatalytic water splitting to produce hydrogen, an excellent and extremely promising future fuel [2]. Photocatalytic solar hydrogen generation, which generates hydrogen from water photolysis, is one of the most appealing eco-friendly techniques of energy harvesting [81]. Because of its outstanding properties, such as abundance, cheap cost, great photochemical stability, and nontoxicity, titanium dioxide (TiO_2) has been widely researched for numerous photocatalytic applications in the energy and environmental sectors. Instead of its great performance, pure TiO_2 efficiency in photocatalysis is reduced because of its comparatively large bandgap energy of about 3.2 eV and short lifespan of photoexcited charge carriers [91–95]. Sujuan Wu et al. produced mesoporous anatase-TiO_2 and discovered numerous lattice deformation and point defects in pure mesoporous TiO2. These flaws might be partially eliminated by increasing crystallinity throughout the hydrothermal process by introducing Cl ions. In addition, effective Ti3+ atomic defects were introduced in highly crystalline mesoporous TiO_2 to enhance charge-separation efficiency, solar absorption and also dramatically boosted hydrogen generation under simulating solar irradiation [96].

B. Rusinque et al. used near-UV light to create a palladium combined TiO_2 photocatalyst before employing it. The photocatalysts having varied metal loadings about 0.25–5.00 wt% Pd were prepared using a sol-gel approach. Water splitting was used to generate hydrogen by light radiation, with ethanol acting an organic scavenger. It was discovered that the mesoporous 0.25 wt% Pd-TiO2 with a 2.5 leV bandgap exhibited the greatest hydrogen generation performance using visible light, with a quantum yield of 1.58% [97].

A molten salt template-assisted pyrolytic technique was employed to generate NiO-TiO_2-x nanoparticles attached above the carbon nanosheets, which were then used as photocatalytic hydrogen generation catalysts. The pyrolysis of oleylamine resulted in the formation of carbon nanosheets as well as the formation of defective TiO_2. Carbon nanosheets formed because of this process assisted in the distribution as well as stability of titanium dioxide nanoparticles. Under UV light, the addition of NiO to TiO_2-x aided in the separation of e/h+ pairs. These benefits enable NiO-TiO_2-x/C composites to function as extremely good photocatalytic material for hydrogen production. This study describes a simple method

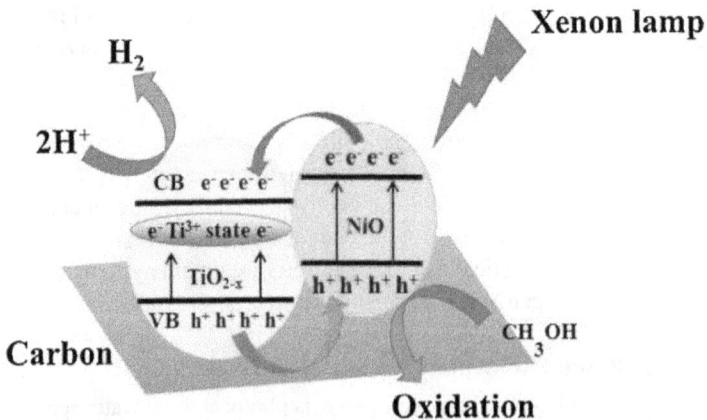

FIGURE 6.11 Schematic illustration of the proposed photocatalytic process of NiO-TiO_2-x/C nanocomposites at xenon lamp illumination. [98]

for producing NiO-TiO$_2$ anchored lying on carbon nanosheets with strong photo-catalytic activity [98].

Hongbing Ji et al. used an improved sol-gel process to create TiO$_2$/BiYO$_3$ photocatalysts. The photocatalytic hydrogen production performance of the obtained TiO$_2$/BiYO$_3$ composite was ten times greater as compared to pure TiO$_2$ and 57 times greater as compared to BiYO$_3$. The heterostructure formed by BiYO$_3$ and TiO$_2$ facilitated the transmission of photogenerated electrons, as well as the separation of photogenerated carriers, resulting in improved photocatalytic activity. The results show that TiO$_2$/BiYO$_3$ composite materials are potential photocatalysts for the production of hydrogen from water [99].

6.2.4.3 TiO$_2$-Based Photocatalytic CO$_2$ Reduction

The steady increasing of greenhouse gas (CO$_2$) concentrations in the atmosphere is now a global environmental problem. To address this environmental challenge, a major research effort has been devoted toward CO$_2$ capture and storage (CCS) to reduce emissions [100]. Titanium dioxide (TiO$_2$) is ecologically beneficial, chemically stable, and abundant on Earth, and it has been identified for CO$_2$ reduction [101]. However, the usefulness of separate component photocatalysts, such as TiO$_2$, is restricted by challenges in producing both high redox ability as well as a wide light response ranges. Among the several ways for improving photocatalytic performance, creating a heterojunction is usually recommended due to its benefits in merging unique advantages of every component as well as separating photogenerated electron-hole pairs [102–104].

Zhang et al. used hydrothermal methods to create CdS/TiO$_2$ HS with increased photocatalytic CO$_2$ reduction efficiency. The following parameters can be attributed to the improved photocatalytic CO$_2$ reduction efficiency: (1) hollow architecture helps to promote light absorption and provides extra active sites for photocatalytic reduction of CO$_2$; (2) intimate heterojunction improves charge-carrier separation even in CdS/TiO2 HS; and (3) step-scheme photocatalytic process provides CdS/TiO$_2$ HS to increased oxidation/reduction capability, which serves as a primary determinant for superior photocatalytic CO$_2$ reduction behavior [105].

Dehkordi et al. constructed a hierarchical g-C$_3$N$_4$/TiO$_2$ composite sphere exhibiting outstanding photocatalytic performance designed for CO$_2$ reduction using visible light by attaching TiO$_2$ shell on the surface of g-C$_3$N$_4$ shell (HCNS). This designed photocatalyst was created in two steps. Initially, nanocoating was performed using templating approach, following the reaction kinetics coating technique. Mesoporous silica spheres and cyanamide were utilized as a sacrificial templates and g-C$_3$N$_4$ precursors, correspondingly, for the manufacture of hollow g-C$_3$N$_4$ spheres that were then modified using TiO$_2$ nanosheets. The extraordinary photocatalytic performance of the obtained nanocomposite because of its unique structure, which provides attributes like light absorption, and many light reflection, as well as an improved active site. Because of the synergistic impact of g-C3N4 titanium dioxide, the light absorption zone was thus increased, resulting in improved photocatalytic performance. CO$_2$ photoreduction performance of the obtained HCNS/TiO$_2$ nanomaterial produced about 11.3 µmol gcat^{-1}h^{-1}; that is, approximately five times greater as compared to the values for pure g-C$_3$N$_4$ [106].

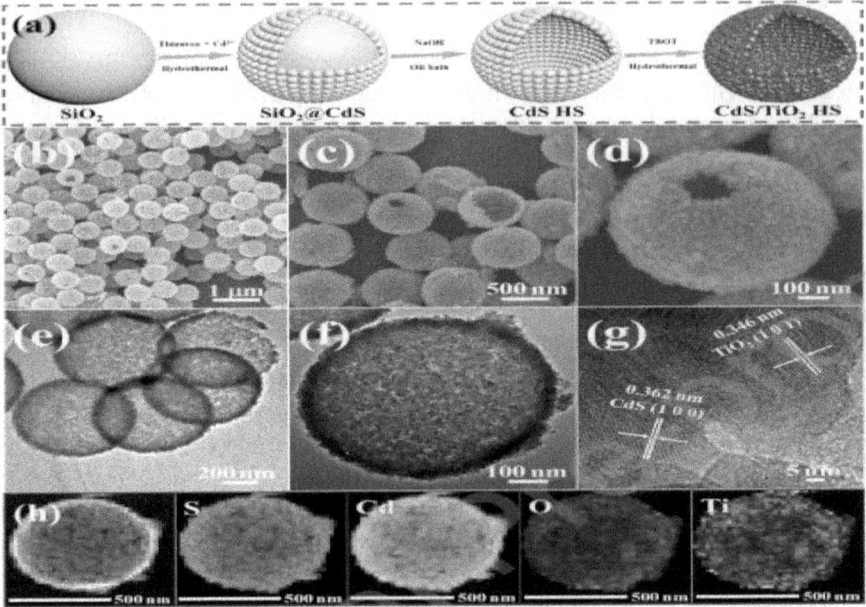

FIGURE 6.12 (a) Synthesis process of CdS/TiO$_2$ HS. (b) FESEM pictures of CdS HS (b) and (c,d) CdS/TiO$_2$ HS (e,f) TEM picture, (g) HRTEM picture, (h) EDS elemental analysis of CdS/TiO$_2$ HS [105].

G. Li et al. synthesized defective TiO$_2$ nanoparticles having a hierarchically porous structure using a hydrothermal procedure supported by an organic surfactant, while Ag quantum dot used as a co-catalyst was evenly stabilized on the TiO$_2$ surface using in situ photo deposition techniques. When compared to commercial TiO$_2$, the produced samples TiO$_2$ and Ag/TiO$_2$ demonstrate outstanding as well as consistent photocatalytic CO$_2$ reduction capability. The super photocatalytic efficiency can be contributed to the configuration of hierarchically porous structures, the creation of defects, and the modification of Ag quantum dots, which can greatly enhance adsorption capacities and absorption efficiency of visible-light, and significantly speed up the separation of photogenerated charges. Furthermore, the synergistic impact of Ag and TiO$_2$ can boost CO$_2$ adsorption, as well as the separation of e/h$^+$ pairs across the Schottky barrier, resulting in enhanced photocatalytic CO$_2$ reduction efficiency [107].

6.3 CONCLUSION

The present developing field of materials research is to hunt new materials for photocatalysis applications. The primary characteristics of charge separation, visible-light absorption, and interfacial photochemical reactions perform a role in material improvement. The photocatalytic activity begins with light and catalyst nanomaterials. Semiconductors are the most useful photocatalyst in nanomaterials because of their tunable bandgap in the visible region. Our interest is to collect and compare several environmental friendly and affordable photocatalyst materials. In the energy sector, the production of hydrogen by photocatalytic techniques is an effective approach to produce and utilize hydrogen in different fields of energy harvesting technologies. In the case of pollution and environmental remediation, the photocatalytic process is the most eco-friendly, affordable, and efficient method of combating heavy metals, microbial, organic, and inorganic contaminants. Several parameters influence catalytic activity, including recombination rate and e/h$^+$ pair separation, catalytic structure, pH, and incoming light intensity. In conclusion, graphene-based nanomaterials, carbon nitride, TiO_2, and ZnO are considered to be among the best materials for photocatalysis applications.

REFERENCES

[1] Djellabi, R., et al., *A review of advances in multifunctional XTiO3 perovskite-type oxides as piezo-photocatalysts for environmental remediation and energy production.* 2022. **421**: p. 126792.

[2] Sun, K., et al., *Incorporating transition-metal phosphides into metal-organic frameworks for enhanced photocatalysis.* 2020. **59**(50): p. 22749–22755.

[3] Ismael, M.J.J.O.A. and Compounds, *A review on graphitic carbon nitride (g-C3N4) based nanocomposites: Synthesis, categories, and their application in photocatalysis.* 2020: p. 156446.

[4] Pedanekar, R., S. Shaikh, and K.J.C.A.P. Rajpure, *Thin film photocatalysis for environmental remediation: A status review.* 2020. **20**(8): p. 931–952.

[5] Rueda-Marquez, J.J., et al., *A critical review on application of photocatalysis for toxicity reduction of real wastewaters.* 2020. **258**: p. 120694.

[6] Li, X., et al., *Applications of MXene (Ti 3 C 2 T x) in photocatalysis: A review.* 2021. **2**(5): p. 1570–1594.

[7] Qian, Y., F. Zhang, and H.J.A.F.M. Pang, *A review of MOFs and their composites-based photocatalysts: Synthesis and applications.* 2021. **31**(37): p. 2104231.

[8] Qiu, X., et al., *Applications of nanomaterials in asymmetric photocatalysis: Recent progress, challenges, and opportunities.* 2021. **33**(6): p. 2001731.

[9] Shen, R., et al., *Nanostructured CdS for efficient photocatalytic H 2 evolution: A review.* 2020: p. 1–36.

[10] Lai, C., et al., *Future roadmap on nonmetal-based 2D ultrathin nanomaterials for photocatalysis.* 2021. **406**: p. 126780.

[11] Ma, M., et al., *Rational design, synthesis, and application of silica/graphene-based nanocomposite: A review.* 2021. **198**: p. 109367.

[12] Ma, R., et al., *Multidimensional graphene structures and beyond: Unique properties, syntheses and applications.* 2020. **113**: p. 100665.

[13] Mamba, G., et al., *State of the art on the photocatalytic applications of graphene based nanostructures: From elimination of hazardous pollutants to disinfection and fuel generation.* 2020. **8**(2): p. 103505.

[14] Ulian, G., D. Moro, and G.J.C.S. Valdre, *Electronic and optical properties of graphene/molybdenite bilayer composite.* 2021. **255**: p. 112978.

[15] Chegel, R. and S.J.S.R. Behzad, *Tunable electronic, optical, and thermal properties of two-dimensional germanene via an external electric field.* 2020. **10**(1): p. 1–12.

[16] Su, H. and Y.H. Hu, *Recent advances in graphene-based materials for fuel cell applications.* 2021. **9**(7): p. 958–983.

[17] Chen, X., et al., *Recent progress in graphene terahertz modulators.* 2020. **29**(7): p. 077803.

[18] Wang, J., et al., *Optoelectronic and photoelectric properties and applications of graphene-based nanostructures.* 2020. **13**: p. 100196.

[19] Lim, Y., et al., *Low dimensional carbon-based catalysts for efficient photocatalytic and photo/electrochemical water splitting reactions.* 2020. **13**(1): p. 114.

[20] Fadlalla, M.I., et al., *Emerging energy and environmental application of graphene and their composites: A review.* 2020. **55**(17): p. 7156–7183.

[21] Tahir, M.B., et al., *Recent advances on photocatalytic nanomaterials for hydrogen energy evolution in sustainable environment.* 2020: p. 1–19.

[22] Singh, N., et al., *Graphene-supported TiO 2: Study of promotion of charge carrier in photocatalytic water splitting and methylene blue dye degradation.* 2020. **3**(1): p. 127–140.

[23] Bakbolat, B., et al., *Recent developments of TiO2-based photocatalysis in the hydrogen evolution and photodegradation: A review.* 2020. **10**(9): p. 1790.

[24] Ayyub, M.M., R. Singh, and C.N.R.J.S.R. Rao, *Hydrogen generation by solar water splitting using 2D nanomaterials.* 2020. **4**(8): p. 2000050.

[25] Guan, G., et al., *Hybridized 2D nanomaterials toward highly efficient photocatalysis for degrading pollutants: Current status and future perspectives.* 2020. **16**(19): p. 1907087.

[26] Albero, J., D. Mateo, and H.J.M. García, *Graphene-based materials as efficient photocatalysts for water splitting.* 2019. **24**(5): p. 906.

[27] Qiao, S., et al., *Core-shell cobalt particles Co@ CoO loaded on nitrogen-doped graphene for photocatalytic water-splitting.* 2020. **45**(3): p. 1629–1639.

[28] Das, G.S., et al., *Sustainable nitrogen-doped functionalized graphene nanosheets for visible-light-induced photocatalytic water splitting.* 2020. **56**(51): p. 6953–6956.

[29] Chaudhary, R.G., et al., *Graphene-based materials and their nanocomposites with metal oxides: Biosynthesis, electrochemical, photocatalytic and antimicrobial applications.* 2020. **83**: p. 79–116.

[30] Minale, M., et al., *Application of graphene-based materials for removal of tetracyclines using adsorption and photocatalytic-degradation: A review.* 2020. **276**: p. 111310.

[31] Madima, N., et al., *Carbon-based nanomaterials for remediation of organic and inorganic pollutants from wastewater: A review.* 2020. **18**(4): p. 1169–1191.

[32] Safajou, H., et al., *Green synthesis and characterization of RGO/Cu nanocomposites as photocatalytic degradation of organic pollutants in waste-water.* 2021. **46**(39): p. 20534–20546.

[33] Zhang, X., et al., *Powerful combination of 2D g-C3N4 and 2D nanomaterials for photocatalysis: Recent advances.* 2020. **390**: p. 124475.

[34] Hasani, A., et al., *Graphene-based catalysts for electrochemical carbon dioxide reduction.* 2020. **2**(2): p. 158–175.

[35] Velasco-Hernández, A., et al., *Synthesis and characterization of graphene oxide-TiO2 thin films by sol-gel for photocatalytic applications.* 2020. **114**: p. 105082.

[36] Madhusudan, P., et al., *Graphene-ZnO. 5Cd0. 5S nanocomposite with enhanced visible-light photocatalytic CO2 reduction activity.* 2020. **506**: p. 144683.

[37] Ni, Y., et al., *Graphitic carbon nitride (g-C3N4)-based nanostructured materials for photodynamic inactivation: Synthesis, efficacy and mechanism.* 2021. **404**: p. 126528.

[38] Singh, M., A. Kumar, and V.J.M.A. Krishnan, *Influence of different bismuth oxyhalides on the photocatalytic activity of graphitic carbon nitride: A comparative study under natural sunlight.* 2020. **1**(5): p. 1262–1272.

[39] Saleem, Z., et al., *Two-dimensional materials and composites as potential water splitting photocatalysts: A review.* 2020. **10**(4): p. 464.

[40] Wang, L., et al., *Graphitic carbon nitride (g-C3N4)-based nanosized heteroarrays: promising materials for photoelectrochemical water splitting.* 2020. **2**(2): p. 223–250.

[41] Cheng, C., et al., *Facile preparation of nanosized MoP as cocatalyst coupled with g-C3N4 by surface bonding state for enhanced photocatalytic hydrogen production.* 2020. **265**: p. 118620.

[42] Güy, N.J.A.S.S., *Directional transfer of photocarriers on CdS/g-C3N4 heterojunction modified with Pd as a cocatalyst for synergistically enhanced photocatalytic hydrogen production.* 2020. **522**: p. 146442.

[43] Xu, H., et al., *In situ construction of protonated g-C3N4/Ti3C2 MXene Schottky heterojunctions for efficient photocatalytic hydrogen production.* 2021. **42**(1): p. 107–114.

[44] Cao, Q., et al., *Graphitic carbon nitride and polymers: A mutual combination for advanced properties.* 2020. **7**(3): p. 762–786.

[45] Xing, Y., et al., *Recent advances in the improvement of g-C3N4 based photocatalytic materials.* 2021. **32**(1): p. 13–20.

[46] Han, C., et al., *Defective ultra-thin two-dimensional g-C3N4 photocatalyst for enhanced photocatalytic H2 evolution activity.* 2021. **581**: p. 159–166.

[47] Wen, J., et al., *Constructing multifunctional metallic Ni interface layers in the g-C3N4 nanosheets/amorphous NiS heterojunctions for efficient photocatalytic H2 generation.* 2017. **9**(16): p. 14031–14042.

[48] Yan, T., et al., *g-C3N4/α-Fe2O3 supported zero-dimensional Co3S4 nanoparticles form S-scheme heterojunction photocatalyst for efficient hydrogen production.* 2020. **35**(1): p. 856–867.

[49] Ghosh, U., A. Majumdar, and A.J.J.O.E.C.E. Pal, *Photocatalytic CO2 reduction over g-C3N4 based heterostructures: Recent progress and prospects.* 2021. **9**(1): p. 104631.

[50] Li, F., D. Zhang, and Q.J.C.C. Xiang, *Nanosheet-assembled hierarchical flower-like gC 3 N 4 for enhanced photocatalytic CO 2 reduction activity.* 2020. **56**(16): p. 2443–2446.

[51] Jiang, H., et al., *Recent advances in heteroatom doped graphitic carbon nitride (g-C3N4) and g-C3N4/metal oxide composite photocatalysts.* 2020. **24**(6): p. 673–693.

[52] Alharthi, F.A., et al., *Photocatalytic degradation of the light sensitive organic dyes: Metanofi blue and rose bengal by using urea derived g-C3N4/ZnO nanocomposites.* 2020. **10**(12): p. 1457.

[53] Zhang, J., J. Li, and X.J.O.M. Liu, *Ternary nanocomposite ZnO-g—C3N4—Go for enhanced photocatalytic degradation of RhB.* 2021. **119**: p. 111351.

[54] Sun, Z., et al., *g-C3N4 based composite photocatalysts for photocatalytic CO2 reduction.* 2018. **300**: p. 160–172.

[55] Beenish, T., T. Muhammad, and M.G.M.J.C.E.T. Nawawi, *Synthesis of WO3/g-C3N4 nanocomposite for photocatalytic CO2 reduction under visible light.* 2021. **83**: p. 247–252.

[56] Pirhashemi, M., et al., *Review on the criteria anticipated for the fabrication of highly efficient ZnO-based visible-light-driven photocatalysts.* 2018. **62**: p. 1–25.

[57] Patial, S., et al., *Boosting light-driven CO2 reduction towards solar fuels: Mainstream avenues for engineering ZnO-based photocatalysts.* 2021: p. 111134.

[58] Di Mauro, A., et al., *Synthesis of ZnO/PMMA nanocomposite by low-temperature atomic layer deposition for possible photocatalysis applications.* 2020. **118**: p. 105214.

[59] Cai, X., et al., *Tuning photocatalytic performance of multilayer ZnO for water splitting by biaxial strain composites.* 2020. **10**(10): p. 1208.

[60] Galdámez-Martínez, A., et al., *Photocatalytic hydrogen production performance of 1-D ZnO nanostructures: Role of structural properties.* 2020. **45**(56): p. 31942–31951.

[61] Mohammadzadeh, A., et al., *Synergetic photocatalytic effect of high purity ZnO pod shaped nanostructures with H2O2 on methylene blue dye degradation.* 2020. **845**: p. 156333.

[62] Alharthi, F.A., et al., *Low temperature ionothermal synthesis of TiO2 nanomaterials for efficient photocatalytic H2 production, dye degradation and photoluminescence studies.* 2020. **44**(11): p. 8362–8371.

[63] Salem, K.E., et al., *Ge-doped ZnO nanorods grown on FTO for photoelectrochemical water splitting with exceptional photoconversion efficiency.* 2021. **46**(1): p. 209–220.

[64] Sharma, M.D., C. Mahala, and M.J.I.J.O.H.E. Basu, *Sensitization of vertically grown ZnO 2D thin sheets by MoSx for efficient charge separation process towards photoelectrochemical water splitting reaction.* 2020. **45**(22): p. 12272–12282.

[65] Wang, J., et al., *Insight into charge carrier separation and solar-light utilization: rGO decorated 3D ZnO hollow microspheres for enhanced photocatalytic hydrogen evolution.* 2020. **564**: p. 322–332.

[66] Yendrapati, T.P., et al., *Formation of ZnO@ CuS nanorods for efficient photocatalytic hydrogen generation.* 2020. **196**: p. 540–548.

[67] Yang, X., et al., *Preparation of flower-like ZnO@ ZnS core-shell structure enhances photocatalytic hydrogen production.* 2020. **45**(51): p. 26967–26978.

[68] Liu, X., et al., *Enhanced visible-light-driven photocatalytic hydrogen evolution and NO photo-oxidation capacity of ZnO/g-C3N4 with N dopant.* 2020. **599**: p. 124869.

[69] Yoo, H., et al., *Z-scheme assisted ZnO/Cu2O-CuO photocatalysts to increase photoactive electrons in hydrogen evolution by water splitting.* 2020. **204**: p. 110211.

[70] Chavoshan, S., et al., *Photocatalytic degradation of penicillin G from simulated wastewater using the UV/ZnO process: isotherm and kinetic study.* 2020. **18**(1): p. 107.

[71] Islam, M.T., et al., *Development of photocatalytic paint based on TiO2 and photopolymer resin for the degradation of organic pollutants in water.* 2020. **704**: p. 135406.

[72] Chenab, K.K., et al., *Water treatment: Functional nanomaterials and applications from adsorption to photodegradation.* 2020. **16**: p. 100262.

[73] Demirci, S., et al., *A study of heating rate effect on the photocatalytic performances of ZnO powders prepared by sol-gel route: Their kinetic and thermodynamic studies.* 2020. **507**: p. 145083.

[74] Yaqoob, A.A., et al., *Advances and challenges in developing efficient graphene oxide-based ZnO photocatalysts for dye photo-oxidation.* 2020. **10**(5): p. 932.

[75] Lin, Y., et al., *Green synthesis of ZnO-GO composites for the photocatalytic degradation of methylene blue.* 2020. **2020**.

[76] Wu, Z., et al., *A ternary magnetic recyclable ZnO/Fe 3 O 4/gC 3 N 4 composite photocatalyst for efficient photodegradation of monoazo dye.* 2019. **14**(1): p. 1–14.

[77] Pant, B., et al., *Synthesis and characterization of ZnO-TiO2/carbon fiber composite with enhanced photocatalytic properties.* 2020. **10**(10): p. 1960.

[78] Deng, H., et al., *S-scheme heterojunction based on p-type ZnMn2O4 and n-type ZnO with improved photocatalytic CO2 reduction activity.* 2021. **409**: p. 127377.

[79] Zhang, F., et al., *Boosting the activity and stability of Ag-Cu2O/ZnO nanorods for photocatalytic CO2 reduction.* 2020. **268**: p. 118380.

[80] Hegazy, I., et al., *Influence of oxygen vacancies on the performance of ZnO nanoparticles towards CO2 photoreduction in different aqueous solutions.* 2020. **8**(4): p. 103887.

[81] Kim, D., and K.J.A.C.B.E. Yong, *Boron doping induced charge transfer switching of a C3N4/ZnO photocatalyst from Z-scheme to type II to enhance photocatalytic hydrogen production.* 2021. **282**: p. 119538.

[82] Li, M., et al., *Construction of highly active and selective polydopamine modified hollow ZnO/Co3O4 pn heterojunction catalyst for photocatalytic CO2 reduction.* 2020. **8**(30): p. 11465–11476.

[83] Perović, K., et al., *Recent achievements in development of TiO2-based composite photocatalytic materials for solar driven water purification and water splitting.* 2020. **13**(6): p. 1338.

[84] Peiris, S., et al., *Recent development and future prospects of TiO2 photocatalysis.* 2021. **68**(5): p. 738–769.

[85] Dontsova, T.A., et al., *Enhanced photocatalytic activity of TiO2/SnO2 binary nanocomposites.* 2020. **2020**.

[86] Chen, D., et al., *Photocatalytic degradation of organic pollutants using TiO2-based photocatalysts: A review.* 2020. **268**: p. 121725.

[87] Zhang, Y., and X.J.A.O. Xu, *Machine learning band gaps of doped-TiO2 photocatalysts from structural and morphological parameters.* 2020. **5**(25): p. 15344–15352.

[88] Karthik, K., et al., *Multifunctional properties of microwave assisted CdO—NiO—ZnO mixed metal oxide nanocomposite: Enhanced photocatalytic and antibacterial activities.* 2018. **29**(7): p. 5459–5471.

[89] Díaz-Sánchez, M., et al., *Synergistic effect of cu, F-codoping of titanium dioxide for multifunctional catalytic and photocatalytic studies.* 2021: p. 2000298.

[90] Uddin, M.T., M.E. Hoque, and M.C.J.R.A. Bhoumick, *Facile one-pot synthesis of heterostructure SnO 2/ZnO photocatalyst for enhanced photocatalytic degradation of organic dye.* 2020. **10**(40): p. 23554–23565.

[91] Bashiri, R., et al., *Influence of growth time on photoelectrical characteristics and photocatalytic hydrogen production of decorated Fe2O3 on TiO2 nanorod in photoelectrochemical cell.* 2020. **510**: p. 145482.

[92] Muscetta, M., et al., *Hydrogen production through photoreforming processes over Cu2O/TiO2 composite materials: A mini-review.* 2020. **45**(53): p. 28531–28552.

[93] Corredor, J., M.J. Rivero, and I.J.I.J.O.H.E. Ortiz, *New insights in the performance and reuse of rGO/TiO2 composites for the photocatalytic hydrogen production.* 2021. **46**(33): p. 17500–17506.

[94] Ibrahim, N.S., et al., *A critical review of metal-doped TiO2 and its structure-physical properties-photocatalytic activity relationship in hydrogen production.* 2020. **45**(53): p. 28553–28565.

[95] Padmanabhan, N., et al., *Morphology engineered spatial charge separation in superhydrophilic TiO2/graphene hybrids for hydrogen production.* 2020. **17**: p. 100447.

[96] Li, H., et al., *Atomic defects in ultra-thin mesoporous TiO2 enhance photocatalytic hydrogen evolution from water splitting.* 2020. **513**: p. 145723.

[97] Rusinque, B., S. Escobedo, and H.J.C. de Lasa, *Photoreduction of a Pd-doped mesoporous TiO2 photocatalyst for hydrogen production under visible light.* 2020. **10**(1): p. 74.

[98] Zhao, X., et al., *Salt templated synthesis of NiO/TiO2 supported carbon nanosheets for photocatalytic hydrogen production.* 2020. **587**: p. 124365.

[99] Qin, Z., et al., *TiO2/BiYO3 composites for enhanced photocatalytic hydrogen production.* 2020. **836**: p. 155428.

[100] Verma, P., D.J. Stewart, and R.J.C. Raja, *Recent advances in photocatalytic CO2 utilisation over multifunctional metal: Organic frameworks.* 2020. **10**(10): p. 1176.

[101] Meng, A., et al., *TiO2/polydopamine S-scheme heterojunction photocatalyst with enhanced CO2-reduction selectivity.* 2021. **289**: p. 120039.

[102] He, F., et al., *2D/2D/0D TiO2/C3N4/Ti3C2 MXene composite S-scheme photocatalyst with enhanced CO2 reduction activity.* 2020. **272**: p. 119006.

[103] Feng, S., et al., *Facile synthesis of Mo-doped TiO2 for selective photocatalytic CO2 reduction to methane: Promoted H2O dissociation by Mo doping.* 2020. **38**: p. 1–9.

[104] Wang, L., et al., *In situ irradiated XPS investigation on S-scheme TiO2@ ZnIn2S4 photocatalyst for efficient photocatalytic CO2 reduction.* 2021. **17**(41): p. 2103447.

[105] Wang, Z., et al., *Step-scheme CdS/TiO2 nanocomposite hollow microsphere with enhanced photocatalytic CO2 reduction activity.* 2020. **56**: p. 143–150.

[106] Dehkordi, A.B., et al., *Preparation of hierarchical g-C3N4@ TiO2 hollow spheres for enhanced visible-light induced catalytic CO2 reduction.* 2020. **205**: p. 465–473.

[107] Li, G., et al., *Ag quantum dots modified hierarchically porous and defective TiO2 nanoparticles for improved photocatalytic CO2 reduction.* 2021. **410**: p. 128397.

7 Functional Nanocomposites for Electrochemical Applications

Mehmood Shahid, Ahmed Usman, Waqar Ahmed,
Suresh Sagadevan, Chariya Kaewsaneha,
and Pakorn Opaprakasit

CONTENTS

7.1   INTRODUCTION

Nanomaterials range from 1–100 nm in size. The size of a nanomaterial is 10^{-9}, a one-billionth part of a meter. Owing to higher surface-to-volume ratios, nanomaterials have remained a hot topic for various applications among researchers. Nanosized material brought exceptional contributions to research and development bearing outstanding properties. The field whereby nanomaterials employ in the matter at the molecular level and at the atomic scale is known as nanotechnology [1, 2]. Figure 7.1 shows the contribution of nanotechnology in various fields [2].

Nanomaterials can be classified based on their structures; in the cases of nanowires and nanorods, one dimension should be less than 100 nm, structures consisting fiber plates and tubes should have two dimensions less than 100 nm and spherical nanomaterials with three dimensions should each be less than 100 nm [3, 4]. Nanomaterials can be divided into different categories: single-phase solids, matrix composites, or multiple composites based on their phase composition [5]. The materials at nanoscale

DOI: 10.1201/9781003263852-7

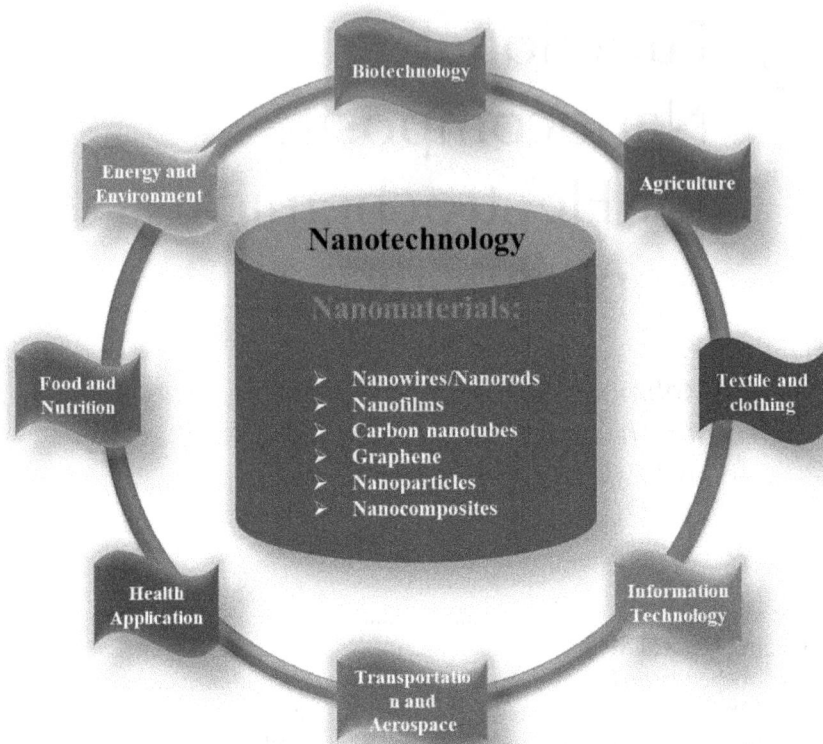

FIGURE 7.1 Applications of nanotechnology.

exhibit extraordinary optical, electrical, mechanical, and thermal properties when compared to their bulk counterparts [6]. Nanomaterials are classified into different categories according to their dimensionality—such as nanoparticles, nanowires, or nanotubes, nanoplates, or nanoribbons—and also depending upon their chemical composition as organic or inorganic nanomaterials [7]. The boom in advancement of nanotechnology and associated synthetic techniques have drawn considerable interest to synthesize nanomaterials utilizing materials science methods [8]. Different parameters of nanomaterials can be exceptionally controlled by tuning the molecules' composition, size, morphology, and surface chemistry for the desired applications. The controllability of various functionalities of nanomaterials makes them promising candidates for many potential technological commercial applications. For instance, nanomaterials—due to their hierarchical nanostructures—appear to be the potential building block in the construction of multicomponent nanomaterials as electrode materials, wearable electrochemical electronics, and energy storage devices [8]. These nanomaterials were also used for various other green energy conversion and storage applications [9] and also as nanocomposite in conjugate with other materials (graphene, CNTs, polymers, etc.) [10, 11]. The nanomaterials and nanostructures have drawn greater attention from the researchers in electrochemical applications. Their unique physical and chemical properties make them suitable candidates at nanoscale to construct the higher performance devices for electrochemical applications. Their high surface-to-volume ratio also plays a major role in analytical sensitivity [12].

Decades ago, carbon only existed in two allotropes: diamond and graphite. In 1985, a third allotrope of carbon was discovered in the form of fullerene, and single- and multi-wall carbon nanotubes (SWCNTs and MWCNTs) were later discovered [13, 14]. Later on, graphene was discovered by mechanical cleavage of graphite, which contains a honeycomb sp^2 hybridized lattice structure [15–17]. In the past few years, biochar (BC)—a type of porous carbon material [18]—has gained significant interest in electrochemical applications, as well; it is a solid carbon-rich residue and normally obtained from the combustion of biomass in the absence or partial presence of oxygen atmosphere [19, 20]. BC is an easily prepared, abundantly available, cheaper alternative raw material which has the potential to be employed as functional nanocomposite material with metals/metal oxides in electrochemical applications [21]. Similarly, conducting polymers also have the great potential to be used as hybrid material with nanomaterials for energy applications [22]. Another emerging two-dimensional (2D) material, MXene, has shown promising results for various electrochemical applications [12].

Moreover, the surface of material can be functionalized either directly during the modification or post modification could also be done [23]. With the passage of time and advances in synthesis techniques, nowadays it is possible to control the design shape, size, composition, and arrangements of materials [24]. In this chapter, we will focus on "nanomaterials" and "nanostructures" used with functionalized materials (graphene, CNTs, MXene, polymers, biochar, etc.) as nanocomposite materials for electrochemical applications.

7.2 ELECTROCHEMICAL METHODS

Various analytical methods have been used so far for the evaluation and characterization of a substance. These analytical methods consist of either spectroscopic methods or electrochemical methods. The spectroscopic methods consist of high-pressure liquid chromatography (HPLC), UV-visible spectrophotometry, gas chromatography, fluorescence spectroscopy, etc. [25, 26]. These methods are known as classical methods in the field of sensors, water contamination detection, and various other applications [27–30]. However, these methods are not promising due to great inadequacies and usage of many solvents, longer analysis time, pre-treatment, and expensive apparatus use. For these reasons, researchers have turned to find alternative methods for testing and found electrochemical methods to be promising for various technological applications. The electrochemical techniques are quick, cost-effective, eco-friendly, and reliable, and they can be used without any pre-treatment for analysis in the fields of physiological molecules detection, biosensors, water contamination detection, energy storage/conversion, water splitting, etc. [31–37], as summarized in Figure 7.2.

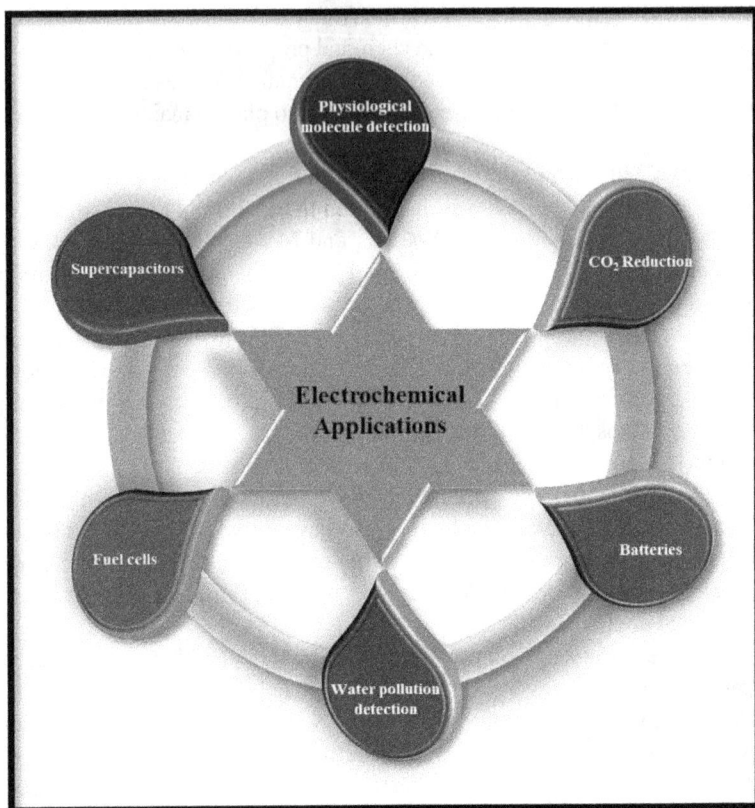

FIGURE 7.2   Schematic diagram of various important electrochemical applications.

7.3 ELECTROCHEMISTRY

Electrochemistry as a branch of science deals with the physical and chemical changes occurring due to the interaction of material under electrical influences (current, voltage, and electron charge). The term electroanalytical chemistry particularly deals with the analyte-containing solutions and results in quantifying the measured electrical signal as a result of electrochemical methods. In this system, electrodes are immersed in solutions/electrolytes which contain the analyte and this set-up consisting of electrodes, analytes, and solution/electrolyte is called an electrochemical cell. In this method, the current is recorded as a result of oxidation or reduction in the chemical reaction [38].

The upcoming sections of this chapter will cover various electrochemical applications using functional nanocomposite material as electrode material in the electrochemical cell containing solution/electrolyte and/or analytes. Furthermore, it will also explain the suitability of functional nanocomposites for a particular electrochemical application. The collection of functional nanocomposites for the most suitable electrochemical application will give insight to the reader to use and further develop the electrode material with the same characteristics and features for particular electrochemical applications. The following section will give an understanding to use the functionalized nanocomposite materials for various electrochemical applications.

7.4 CNTs-BASED NANOCOMPOSITES

A carbon nanotube (CNT) is a tube made of carbon with diameters typically measured in nanometers. The research on CNT has tremendously increased in the past few years owing to their potential in different fields, including photoelectricity, biomedicine, and electrochemistry [39]. CNTs possess unique electronic, structural, and chemical properties such as good conductivity, low charge transfer resistance, unique tubular nanostructure, excellent biocompatibility, modifiable sidewall, large surface area, and so on [40]. The overall activity of these carbon nanotubes depends on different factors, including the protocol opted to develop nanotubes, the surface properties of the final structure, and the applications related to the electrochemical sensing, water purification, energy conversion and energy storage, etc. Therefore, it is required to have significant prior knowledge to prepare carbon nanotubes [41].

In a report, He and Li in 2017 [42] synthesized the MWCNTs and loaded the Fe_2O_3 on the surface of these tubes by an atomic layer deposition method. They analyzed the effect of metal oxide on the overall properties of CNTs and synthesized the nanocomposite of Fe_2O_3 deposited MWCNTs by the calcination process with an extremely small size of the final product, i.e., 10 nm. After their successful synthesis, the material was assessed as an electrode in a high-performance supercapacitor as a potential candidate with a discharge current density of 1 A g^{-1}, and a higher specific capacitance was reported i.e., 787 F g^{-1} (Figure 7.3a). The charge transfer resistance of the electrode examined by EIS spectra is illustrated in Figure 7.3c; further, the equivalent circuit and structural morphology of the materials is investigated by transmission electron microscopy (TEM) are summarized in Figure 7.3d–f. the study reveals that Fe_2O_3-coated MWCNTs nanocomposite has shown higher electrocatalytic activity due to the presence of Fe_2O_3 nanoparticles providing a synergistic effect. The synthesized electrocatalytic material for the supercapacitor exhibited a great rate performance with a cyclic retention of 91.6%, even after running 5,000 cycles continuously (Figure 7.3b). The Fe_2O_3 nanoparticles significantly contributed to high theatrical capacitance, and MWCNTs played their part in providing high surface area along with higher conductivity. This report presents an overall high performance based on nanocomposite as a suitable candidate with demonstrated performance for supercapacitors applications.

Fan et al. [43] have synthesized the CNTs and coated them on Al_2O_3 substrate to develop a membrane structure for water contamination removal. They have assisted in the removal of the contaminants by using a membrane via an electrochemical process. The addition of electrochemistry with membrane filtration has resulted in an increased flux with greatly enhanced selectivity in removing the unfavorable contaminants. The prepared structure has excellent pore size, strong electroconductivity, and greater mechanical stability. Furthermore, with and without electrochemical assistance, membrane efficiency for water contaminants removal was studied and it was

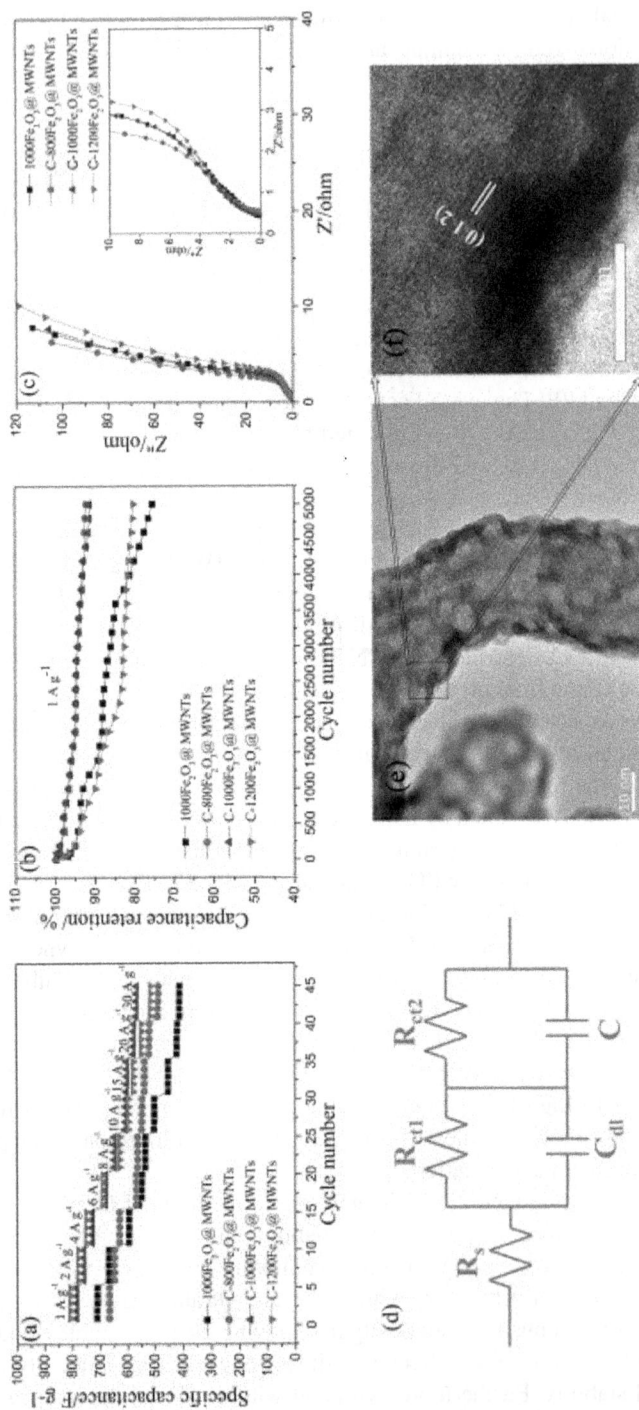

FIGURE 7.3 (a) The specific capacitances of C-1000Fe$_2$O$_3$@MWNTs against current densities; (b) Cycle retention of the C-1000Fe$_2$O$_3$@MWNTs at a current density of 1 A g^{-1}; (c) EIS of the 1000Fe$_2$O$_3$@MWNTs, C-800Fe$_2$O$_3$@MWNTs, C-1000Fe$_2$O$_3$@MWNTs, and C-1200Fe$_2$O$_3$@MWNTs; (d) The equivalent circuit fitting; and (e–f), TEM and HRTEM images of Fe$_2$O$_3$ nanoparticles and crystalline Fe$_2$O$_3$.

Source: (Reprinted with permission from [42])

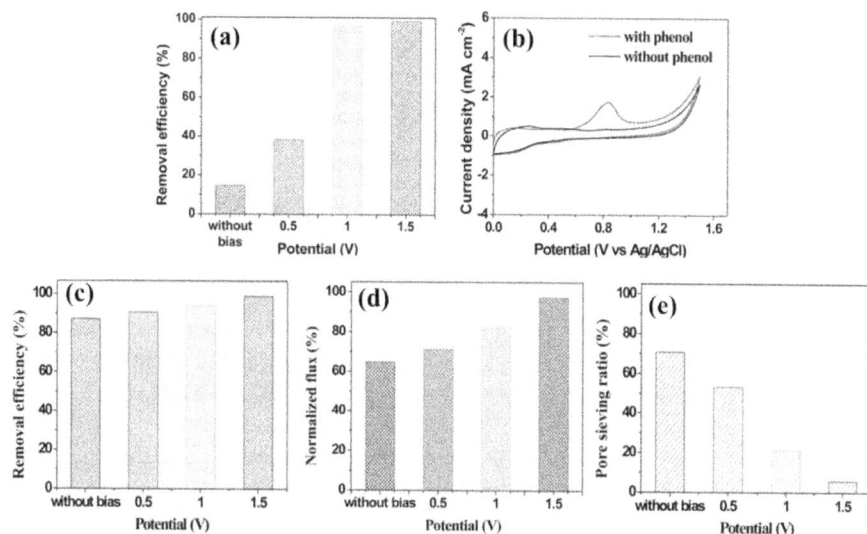

FIGURE 7.4 (a) CNTs/Al$_2$O$_3$ membrane efficiency for phenol removal in the presence and absence of electrochemical assistance; (b) CNTs/Al$_2$O$_3$ membrane based cyclic voltammetry with and without the addition of phenol; (c) CNTs/Al$_2$O$_3$ membrane efficiency for silica-sphere removal; (d) CNTs/Al$_2$O$_3$ membrane for the silica-sphere removal with normalized flux; (e) of silica spheres retention ratio on the membrane surface (the operation time for contamination removal was 30 min each).

Source: (Reprinted with permission from [43])

found that with the assistance of the electrochemical method CNTs/Al$_2$O$_3$ has exhibited a 1.6-fold higher permeate flux and an improvement of three folds in removing natural organic matter (NOM). Figure 7.4 illustrates the removal efficiency of NOM under the assistance of the electrochemical method and without its assistance; it depicts that the constructed membrane showed the capability of removing NOM, including silica-sphere, phenol, and total organic carbon. These results state that the addition of electrochemistry in the materials makes them permeable to selective substances. Another important finding was the recovery of the flux of fouled membrane through the backwashing approach which was again mediated by the electrochemical assistance.

7.5 POLYMERS NANOCOMPOSITES FOR ELECTROCHEMICAL APPLICATIONS

Exhibiting the unique electrochemical properties of conducting polymers including polyaniline (PANI), polypyrrole (PPy), and poly(3,4-ethylenedioxythiophene) polystyrene sulfonate (PEDOT: PSS) is the most promising conducting polymers which have been extensively explored by researchers. The energy storage and energy conversion properties of the conducting polymers have been highly investigated by researchers in the past few decades. Zhang, Fan, et al. (2012) [44] synthesized a hybrid three-layered structure consisting of carbon-based polyaniline, TiO_2 nanoparticles, and graphene nanosheets (called PTG). The introduction of polymers and graphene nanosheets with TiO_2 played a very important role in preventing the agglomeration of TiO_2 nanoparticles and hence participated in enhanced device performance. The schematic protocol used for the synthesis of PTG is expressed in Figure 7.5. The synthesized PTD material is employed as anode material in lithium-ion batteries (LIB). They analyzed the fast charging and discharging property of the PTG electrode having a high reversible performance of Coulombic efficiency of 99.19% at a current density of 1,000 mA/g after 100 cycles and discharge capacity of 149.8 mAh/g at 100 cycles [44].

The electrochemical performance of PMo_{12}/PANI/MWNTs based nanocomposite was investigated by Hu et al. in 2017 [45]. The structural and chemical composition of the designed material was examined by SEM/EDX, TEM, FTIR, XPS, and Raman spectroscopy. The material is used as the anode material in LIB, and the initial discharge and attained discharge value attained by the electrode was 1572 mAh/g and 1,000 mAh/g as sustained discharge capacity for 100 cycles, which is an extraordinarily stable performance exhibited by the anode material at a current density of 0.5 mA/cm² (Figure 7.6). They reported that the exponential electrochemical characteristics of the electrode were due to the mutual reinforcement effect of uniform dispersal of PMo_{12} cluster, establishing a layer of metal oxide on the surface of PANI/MWCNTs nanocomposite. PANI/MWCNTs offered a good platform for the diffusion of PMo_{12}, and also allowed a pathway for the efficient flow of electrons.

In comparison to other PV devices, DSSCs are considered superior because of their cost-effectiveness, simple method of assembly, and sustainable nature. The

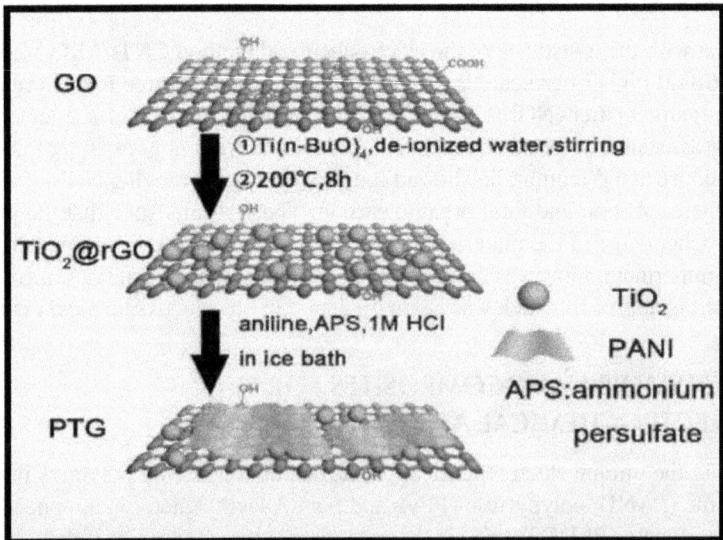

FIGURE 7.5   Mechanism for synthesis of PANI-TiO_2-rGO (PTG) nanocomposites.

Source: (Reprinted with permission from [44])

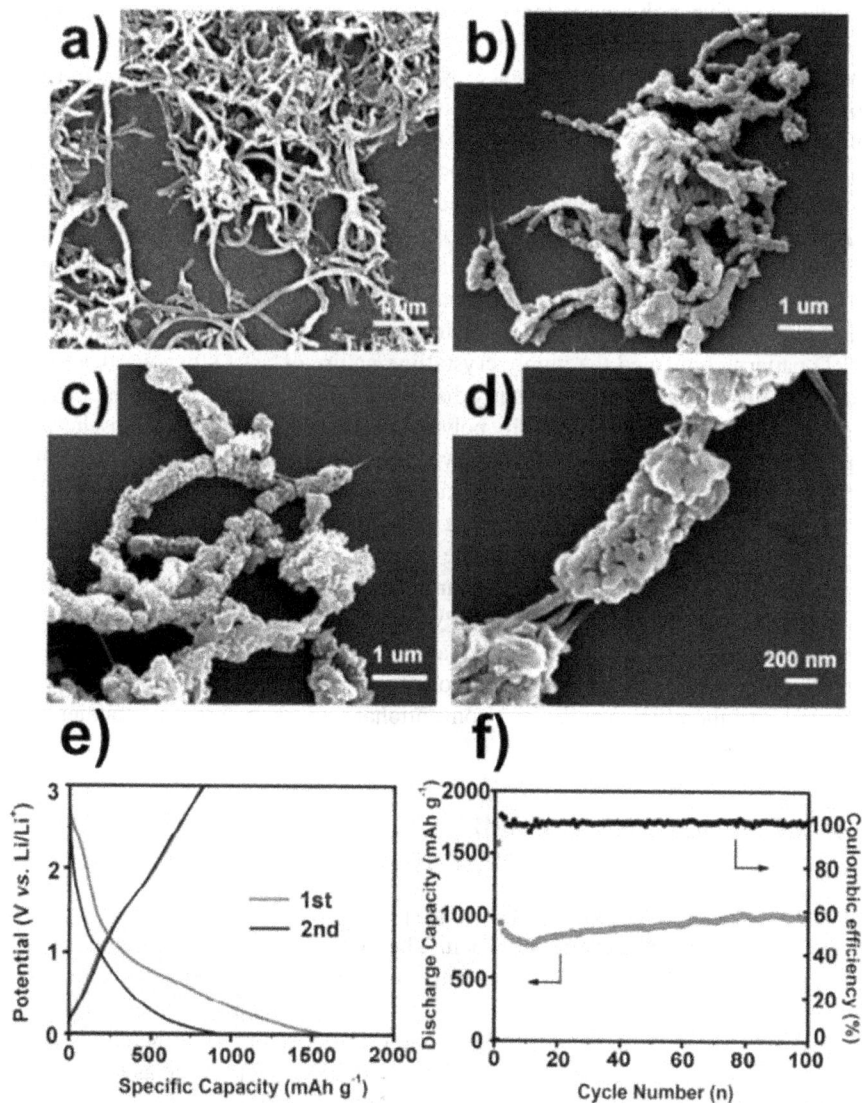

FIGURE 7.6 SEM images of: (a) MWNTs; (b) PANI/MWNTs; (c–d) PMo_{12}/PANI/MWNTs; (e) the charging and discharging curves of the lithium rechargeable battery at PMo_{12}/PANI/MWNTs I = 0.5 mA/cm²; (f) PMo_{12}/PANI/MWNTs discharge capacity and Coulombic efficiency vs cycle number; I = 0.5 mA/cm².

Source: (Reprinted with the permission from [45])

final performance of DSSCs is influenced by several factors, most importantly the composition of electrolytes and the nature of sensitizing dyes, photoanodes, and counter electrodes. In this research work, we have examined the performance of different concentrations of polyaniline (PANI) polymers being synthesized through simple oxidative polymerization. This PANI was later employed as a counter electrode in DSSC. PANI was first coated on FTO counter electrodes followed by examination of their performance by changing the film thickness with a doctor blade and spin coating approach. It is reported that the 20 mg concentration n of PANI-3 depicted an increased redox couple reaction which eventually resulted in the enhanced redox activity of the polymer [46]. Furthermore, in a report, the performance of PPy was also investigated to optimize the best conducting polymer-based counter electrode for DSSC [22]. It was observed that PPy provides a good catalytic performance to PANI. To enhance the efficiency of the PPy electrodes, SrTiO$_3$ nanoparticles were also incorporated into the PPy counter electrode. Finally, the synthesized materials were characterized through different techniques. The XRD pattern confirmed the highly crystalline nature of SrTiO$_3$ nanoparticles. This validated the formation of consistently dispersed nanoparticle–polymer interactions. Results of FESEM depicted that PPy polymer and SrTiO$_3$ represent a nanosphere and nanocubes like morphology, respectively (Figure 7.7). FTIR and UV-VIS spectroscopy confirmed the synthesis of prepared composites. The incorporation of SrTiO$_3$ was also found to increase and decrease the particle's surface area and low charge transfer resistance, respectively [22]. The synthesized nanocomposite of PPy-SrTiO$_3$ exhibited higher performance as compared to unaided PPy as a counter electrode in DSSCs. Different compositions of PPy-SrTiO$_3$ nanocomposite were prepared by varying the PPy concentration, and greater performance were shown by the PPy-SrTiO$_3$-50% counter electrode with a power conversion efficiency of 2.52% when compared to the efficiency (2.17%) conventionally used sputtered Pt counter electrodes (Figure 7.7d).

7.6 RGO-BASED NANOCOMPOSITES FOR ELECTROCHEMICAL APPLICATIONS

Recently in 2022, V. Shanmugam et al. [47] synthesized a novel nanocomposite of NiO/rGO via a hydrothermal approach. The synthesized material was employed as a photoanode for photocatalytic activity, which exhibited high performance toward the removal of color, showing the photocatalytic degradation toward methyl orange dye i.e., 99.9% in the initial 50 minutes. Furthermore, the efficiency of NiO/rGO composite was higher than the efficiency of rGO photoanode as alone which was about 95% removal in 90 minutes. The NiO/rGO photoanode also presented outstanding electrochemical properties due to the uniform distribution of NiO nanoparticles on rGO nanolayers, which provides high surface area for the flow of electrons. Moreover, the performance of the fabricated material was also evaluated for supercapacitor application. It was found that the electrode showed the capacitance of 192 F g^{-1} at the current density of 0.2 A g^{-1}. The high capacitance retention rate of 72.1% after 2,000 cycles determines the superior rate capability. In another study, the photocatalytic degradation of MB dye by reduced graphene oxide/silver nanowires was examined by Singh and Dhaliwal [48]. The rGO was deposited onto the surface of stainless steel through electroplating method, the Ag nanowires were uniformly distributed on the surface of rGO, and the structural and chemical composition of the electrode FESEM, FTIR, UV-vis, XRD, DSC/TGA, EDX, and Raman spectroscopy was determined. The electrochemical performance of rGO/

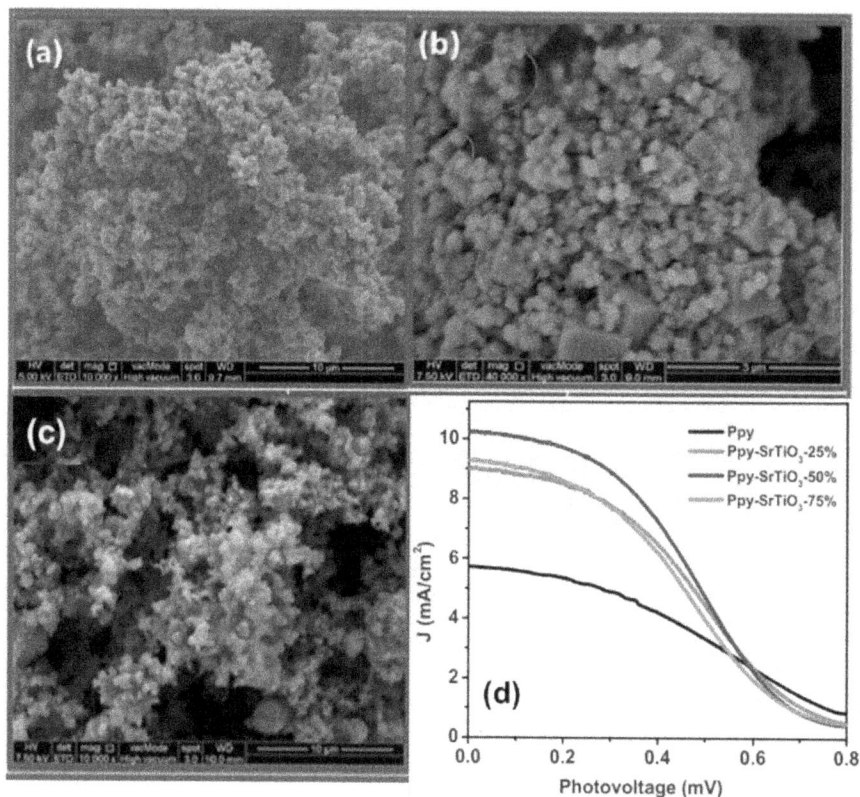

FIGURE 7.7 FESEM images of (a) nanospheres of PPy, (b) nanocubes of SrTiO$_3$ (c) PPy-SrTiO$_3$-50%, and (d) the *J-V* curves of the counter electrode of DSSC with different concentration of PPy (25%, 50%, and 75%) in the nanocomposite of PPy-SrTiO$_3$.

Source: (Reprinted with permission from [22])

AgNWs electrode as a photocatalyst for photocatalytic degradation was also investigated. It is found that with the different wt.% composition of Ag, the performance of the electrode varies. The highest photocatalytic degradation of almost 98.36 and electrochemical performance of 99% is exhibited by rGO/AgNWs having 9.5 wt.%. It was observed that the performance of rGO was enhanced with the incorporation of Ag nanowires. The complete mechanism of the experiment is illustrated in Figure 7.8.

The electrochemical sensing of an inorganic compound, hydrazine, was carried out using hydrothermally synthesized rGO-Co$_3$O$_4$@Au nanocomposite [49]. The functionalized graphene with the presence of oxygen functional group allowed to Co$_3$O$_4$ to interact on its surface by providing the active sites. The nanocomposite, along with other controlled electrodes, was characterized sufficiently to confirm the successful formation of nanocomposites. The FESEM images of rGO-Co$_3$O$_4$@ Au nanocomposite shown in Figure 7.9a shows the ternary nanocomposite in which Co$_3$O$_4$ nanocubes are finally decorated with the Au (8 mM) nanoparticles, these Au deposited Co$_3$O$_4$ nanocubes can be seen under the layers of graphene as blur image.

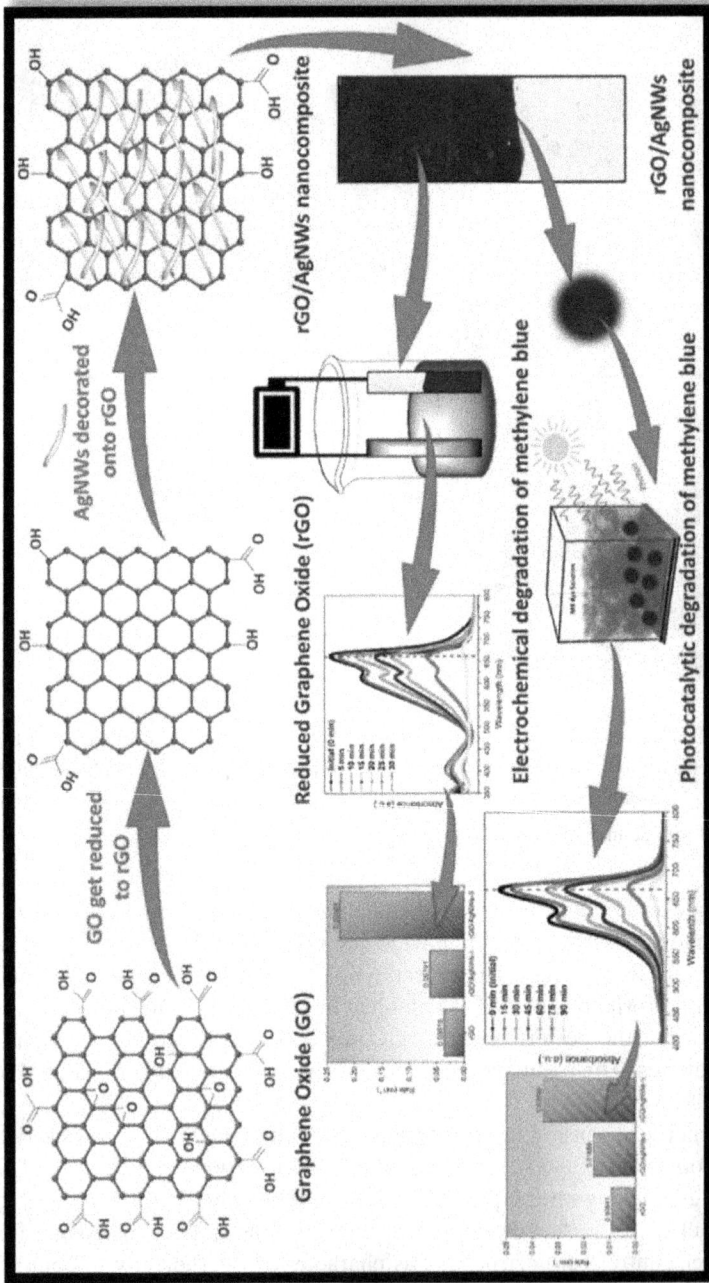

FIGURE 7.8 Synthesis mechanism of rGO/AgNWs nanocomposite and its use as catalyst for electrochemical and photocatalytic degradation of methylene blue dye.

Source: (Reprinted with permission from [48])

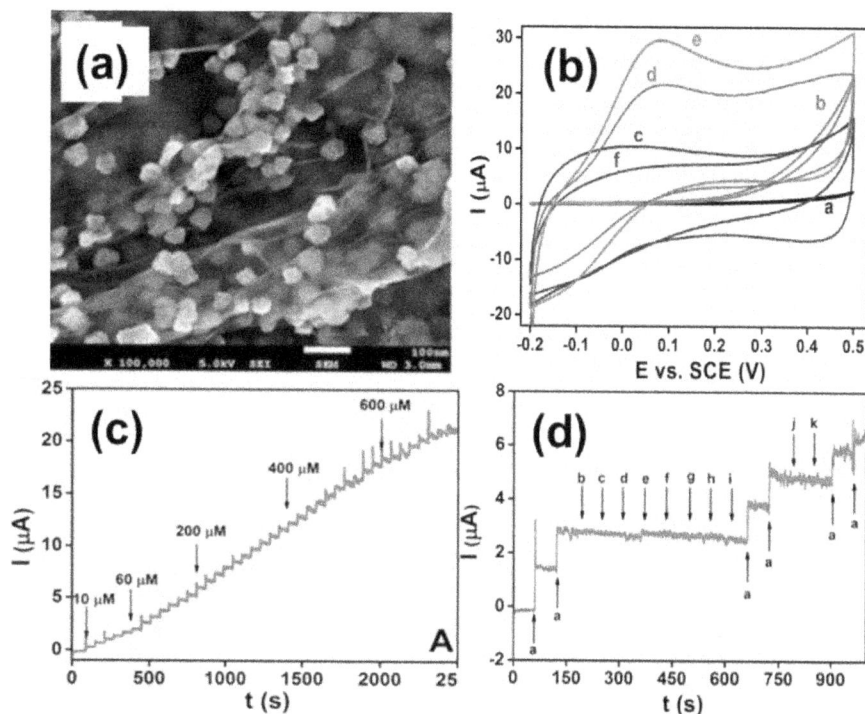

FIGURE 7.9 (a) FESEM images of rGO-Co_3O_4@Au nanocomposite; (b) cyclic voltammograms obtained at rGO-Co_3O_4@Au along with other controlled electrodes at a scan rate of 50 mV s$^-$; (c) amperometric detection (i–t) curve using rGO-Co_3O_4@Au nanocomposite (8 mM) for the successive addition of 10 µM of hydrazine; and (d) various interference molecules with 0.5 mM concentration were added to test the selectivity for the sensors are NO_3^-, SO_4^{2-}, Cl$^-$, Ag$^+$, Na$^+$, K$^+$, ethanol, 4-nitrophenol, ascorbic acid and glucose.

Source: (Reprinted with permission from [49])

The FESEM images further confirm the intercalation of Au deposited Co_3O_4 nanocubes into various layers of reduced graphene oxide sheets, and hence provided the greater electrocatalytic activity in the detection of hydrazine as compared to another controlled electrode as shown by cyclic voltammetry in Figure 7.9b. The detection of hydrazine was carried out by glassy carbon electrode modified by rGO-Co_3O_4@Au (8 mM) by employing chronoamperometric technique and it was found that rGO-Co_3O_4@Au nanocomposite can detect lower limit of hydrazine concentration i.e., be 0.443 µM and sensitivity of 0.58304 ± 0.00466 µA µM-1 (Figure 7.9c). The selectivity of the sensor was also tested by adding various co-existing molecules using CA (catechol) technique and it was observed that there is no interference caused by any external molecules added, which proves that the rGO-Co_3O_4@Au nanocomposite is highly selective toward the detection of hydrazine, as can be seen through interference study in Figure 7.9d.

7.7   BIOCHAR WITH METALS OR METAL OXIDES
FOR ELECTROCHEMICAL APPLICATIONS

The word "biocharcoal", formally known as biochar, is the combination of two different terminologies, i.e., biomass and charcoal. Biocharcoal includes the deposition of charcoal in the biomass composite of either vegetable or animal origin through the process of carbonization or pyrolysis in an oxygen-deficient environment [50]. The properties of final developed biochar entirely depend on the origin of initially used biomass. The thermal degradation at different steps, anywhere from 300–1,000°C, is also done to solidify the composite; however, this kind of degradation also results in the production of certain gaseous and liquid products.

The incorporation of metals and their oxides/sulfides enhances the electrochemical performance of biochar, and in the past few years, these coated biochar have been used extensively in various applications of electrochemistry; for example, energy storage, detection of substances in environmental wastes, and in the development of fuel cells. Xiang et al. in 2018 [21] designed a novel scalable, facile, and cost-effective approach to attach the gold nanoparticles on the surface of biochar. The final developed material exhibited rich pores, large surface area, and very good electric conductivity. Studies determined that these Au-NPs/BC nanocomposites are capable of detecting hydroquinone (HQ) and catechol (CA) even at their nanomolar range with a lower limit of detection as 3.4 nM for HQ and the sensitivities values of 45.1 and 13.9 $AM^{-1}{\cdot}cm^{-2}$. Similarly, the limit of detection for CA was recorded as 9.0 nM and the sensitivities values of 38.8 and 13.1 $AM^{-1}{\cdot}cm^{-2}$ (Figure 7.10). This shows that the he designed a biochar-based sensor that has the potential to detect the target analytes with a lower limit of detection and higher sensitivities. Material has a very

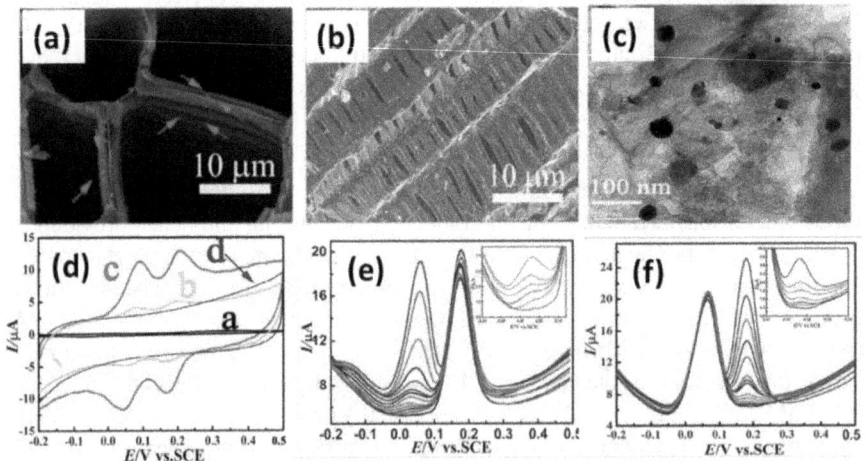

FIGURE 7.10   (a–b) SEM images of as Ithesized Au-NPs/BC-850; (c) TEM image of Au-NPs/BC-850; (d) cyclic voltammograms of 10 μMI and 10 μM CA in pH 6 PBS; (e) DPV curves of various concentration of HQ in the presence of 10 μM CA; and (f) vice versa.

Source: (Reprinted with permission from [21])

low limit of detection with high sensitivity. Both CA and HQ are pollutants which are extremely hazardous for the human body. They are the prevalent components of environmental wastes, including wastewater; therefore, the Au-NPs/BC developed in this study possess great potential to detect these pollutants in environmental samples.

In another experiment conducted by Thomas et al. in 2020 [51], the metal oxide–coated biochar has been exploited for application in energy storage. The banana stem from the agricultural wastes was taken as the biomass which was then employed for the development of two distinct composites; namely, magnetic biochar–polyaniline (PANI) composite (BPC) and magnetic biochar (MB). The magnetic behavior in this biochar was induced by loading iron oxide on them and this magnetism resulted in the a significant increase in surface area from 7.97 m^2g^{-1} to 283 m^2g^{-1}. Both BPC and MB displayed good energy storage ability with the capacitance value of 315.7 Fg^{-1} and 234.8 Fg^{-1}, respectively, which is much larger than the capacitance values of previously reported biochar; therefore, this study confirmed that the addition of metal oxides in the biocharcoals makes them good supercapacitors.

Xie et al., in 2021 [52] prepared the electrodes coated with biochar particles and evaluated their properties in the identification and degradation of 4-chlorophenol, a small organic compound which upon high exposure results in severe damage to the human kidney and liver. This compound has been reported in different environmental wastes, including wastewater. In the study conducted by Xie et al., biochar-based electrodes were prepared by using KOH as an activator of the reaction and $MnCl_2 \cdot 4H_2O$, $SnCl_4 \cdot 5H_2O$, and $SbCl_3$ as the reaction modifiers. In the in situ conditions, the designed electrodes significantly degraded 4-chlorophenol in the wastewater with an efficiency rate of 93%. it was reported that after the optimal degradation of 4-chlorophenol, there is the generation of an intermediate as shown by the rapid decrease in the concentration of 4-chlorophenol after the start of electrolysis, while an increase in the concentration of intermediates like catechol, oxalic acid, hydroquinone, maleic acid, p-benzoquinone, and fumaric acid. Besides that, the highest removal rates of metal ions were depicted by the 150 Mn/AC, 150 Sn/AC and 200 Sb/AC electrodes. It was found that 36.71% of 4-chlorophenol was removed within 2.5 hours of electrolysis. Moreover, the final product was reported to be easily degradable and less toxic in nature.

7.8 MXENE WITH METALS OR METAL OXIDES FOR ELECTROCHEMICAL APPLICATIONS

MXene materials are considered as ceramics but it is not necessary for ceramics to be MXene. The 2D unique sheet structure of carbide and nitrides of metals enhances the electronic and electrical characteristics of MXene. The various ions and molecules can easily disperse into MXene layers by the process called intercalation which provides a high surface area for the flow of ions and electrons into the layers. The predicted stable phase of MXene is considered by many researchers with the incorporation of different transition metals such as vanadium, molybdenum, titanium, and vanadium with nitrogen and carbon [53–55].

The chemical composition of MXene is $M_{n+1}X_nT_x$, where M represents the transition metal, Tx is the surface transmission of the ions (including O, F, and OH), and X donates the nitrogen or carbon ions, depending on carbide or nitride composition. To enhance the performance of the electrode material, the doubled-layered transition metals are separated by the layers of 2D carbides. The high metallic characteristic of MXene is due to the mobility of free electrons of the nitrides of carbides. The surface chemistry behind the selection of the grouping of transition metals and X composite helps to control the electrochemical performance of MXene electrodes [56].

Transition metal oxides (TMOs) are the most promising materials exhibiting the high electronically stable transport property, superior specific capacitance, and eco-friendly nature. In combination with $Ti_3C_2T_x$, the researchers reported improvement in the poor electrical conductivity of MXene; 1D MnO_2 in composited with $Ti_3C_2T_x$ as an ink like material was fabricated by Zhou et al. [57]. The reported negatively charged surface of both $Ti_3C_2T_x$ and MnO_2 repel each other and form regular colloid. The synergetic effect in the material is due to the structural composition. The MnO_2 nanowires incorporation into the $Ti_3C_2T_x$ film tends to increase the high surface area, and on the same side, 1D and 2D hybrid structures drastically strengthen the interlayers, spacing which helps in the acceleration of ion transport into the surface of the material. In another work, the positively charged MnO_2 and negatively charged $Ti_3C_2T_x$ develop a composite electrostatically stable film with enhanced conductivity of the electrons at the interface due to the high electrostatic attraction of oppositely charged MnO_2 and $Ti_3C_2T_x$ [58].

Additionally, many other metal oxides have also been loaded, including TiO_2 [59, 60], WO_3 [61], NiO [62], Fe_2O_3 [63], and MoO_3 [64]. In a research work conducted by Wang et al. [65], the approach of hydrothermal method was opted for use to attach the oxides of molybdenum and nickel metals in one single TMO form on the $Ti_3C_2T_x$ to prepare a composite, which resulted in the production of a unique flower-like nanostructure. The addition of transition metal oxides not only resulted in increasing the size of composite and accelerating the electrolyte, but also removed the stacking related issues which were previously reported in other 2D materials. The surface area of the final nanocomposite was increased from 24.15–152.3 $m^2\ g^{-1}$. The increased electrolyte is due to the reduction in sheet thickness which eventually enlarged the pitch exposure region toward the active sites. This data suggests that the addition of metal oxides increases the efficiency of composites [56].

7.9 CONCLUSION

The current chapter enlightened the use of metal oxide nanoparticles as nanocomposite with the family of carbon and conducting polymers as potential functional nanocomposite for electrochemical applications. The basic knowledge about the importance of metal oxides at nanoscale and their functional nanocomposites with reduced graphene oxide, CNTs, biochar, and conducting polymer was the focus of this chapter. Furthermore, in this chapter, we presented the use of functional nanocomposites as versatile materials for various analytical applications, i.e., electrochemical applications. The focused electrochemical applications were fuel cells, sensors, wastewater treatment, supercapacitors, batteries, and CO_2 reduction, etc.

This study suggests employing these functional nanocomposites in various other fields such as life science, medicine, biomedical engineering, chemical engineering, water treatments, etc.

7.10   ACKNOWLEDGMENT

This study was supported by Thammasat Postdoctoral Fellowship. The authors gratefully acknowledged support from the Center of Excellence in Materials and Plasma Technology (CoE M@P Tech), Thammasat University, Thailand.

REFERENCES

[1] Li, Z., et al., *Carbon-based functional nanomaterials: Preparation, properties and applications.* Composites Science and Technology, 2019. **179**: p. 10–40.

[2] Hassan, T., et al., *Functional nanocomposites and their potential applications: A review.* Journal of Polymer Research, 2021. **28**(2): p. 1–22.

[3] Fattakhova-Rohlfing, D., A. Zaleska, and T. Bein, *Three-dimensional titanium dioxide nanomaterials.* Chemical Reviews, 2014. **114**(19): p. 9487–9558.

[4] Chenot, C.C., R.L. Robiette, and S. Collin, *First evidence of the cysteine and glutathione conjugates of 3-sulfanylpentan-1-ol in hop (Humulus lupulus L.).* Journal of Agricultural and Food Chemistry, 2019. **67**(14): p. 4002–4010.

[5] Pavlovic, M., J. Mayfield, and B. Balint, *Nanotechnology and its application in medicine*, in *Handbook of Medical and Healthcare Technologies.* 2013, Springer. p. 181–205.

[6] Buzea, C., I.I. Pacheco, and K. Robbie, *Nanomaterials and nanoparticles: Sources and toxicity.* Biointerphases, 2007. **2**(4): p. MR17–MR71.

[7] Klaessig, F., M. Marrapese, and S. Abe, *Current perspectives in nanotechnology terminology and nomenclature*, in *Nanotechnology standards.* 2011, Springer. p. 21–52.

[8] Yin, Y., and D. Talapin, *The chemistry of functional nanomaterials.* Chemical Society Reviews, 2013. **42**(7): p. 2484–2487.

[9] Schaudel, B., et al., *Spirooxazine-and spiropyran-doped hybrid organic: Inorganic matrices with very fast photochromic responses.* Journal of Materials Chemistry, 1997. **7**(1): p. 61–65.

[10] Shahid, M.M., et al., *Enhanced electrocatalytic performance of cobalt oxide nanocubes incorporating reduced graphene oxide as a modified platinum electrode for methanol oxidation.* RSC Advances, 2014. **4**(107): p. 62793–62801.

[11] Rafique, S., et al., *Significantly improved photovoltaic performance in polymer bulk heterojunction solar cells with graphene oxide/PEDOT: PSS double decked hole transport layer.* Scientific Reports, 2017. **7**(1): p. 1–10.

[12] Zhang, A., et al., *MXene-based nanocomposites for energy conversion and storage applications.* Chemistry: A European Journal, 2020. **26**(29): p. 6342–6359.

[13] Feng, Y., et al., *A mechanically strong, flexible and conductive film based on bacterial cellulose/graphene nanocomposite.* Carbohydrate Polymers, 2012. **87**(1): p. 644–649.

[14] Zhai, F., et al., *Graphene-based chiral liquid crystal materials for optical applications.* Journal of Materials Chemistry C, 2019. **7**(8): p. 2146–2171.

[15] Shahid, M.M., et al., *An electrochemical sensing platform based on a reduced graphene oxide: Cobalt oxide nanocube@ platinum nanocomposite for nitric oxide detection.* Journal of Materials Chemistry A, 2015. **3**(27): p. 14458–14468.

[16] Zhao, L., et al., *Carbon nanotubes grown on electrospun polyacrylonitrile-based carbon nanofibers via chemical vapor deposition.* Applied Physics A, 2012. **106**(4): p. 863–869.

[17] Lv, P., et al., *Increasing the interfacial strength in carbon fiber/epoxy composites by controlling the orientation and length of carbon nanotubes grown on the fibers.* Carbon, 2011. **49**(14): p. 4665–4673.

[18] Vithanage, M., et al., *Interaction of arsenic with biochar in soil and water: A critical review.* Carbon, 2017. **113**: p. 219–230.

[19] Pandolfo, A.G., and A.F. Hollenkamp, *Carbon properties and their role in supercapacitors.* Journal of Power Sources, 2006. **157**(1): p. 11–27.

[20] Yuan, H., et al., *Oxygen reduction reaction catalysts used in microbial fuel cells for energy-efficient wastewater treatment: A review.* Materials Horizons, 2016. **3**(5): p. 382–401.

[21] Xiang, Y., et al., *Biochar decorated with gold nanoparticles for electrochemical sensing application.* Electrochimica Acta, 2018. **261**: p. 464–473.

[22] Ahmed, U., et al., *An efficient platform based on strontium titanate nanocubes interleaved polypyrrole nanohybrid as counter electrode for dye-sensitized solar cell.* Journal of Alloys and Compounds, 2021. **860**: p. 158228.

[23] Niemeyer, C.M., *Nanoparticles, proteins, and nucleic acids: Biotechnology meets materials science.* Angewandte Chemie International Edition, 2001. **40**(22): p. 4128–4158.

[24] Zhang, S., et al., *Synthesis, assembly, and applications of hybrid nanostructures for biosensing.* Chemical Reviews, 2017. **117**(20): p. 12942–13038.

[25] Karovičová, J., and P. Šimko, *Determination of synthetic phenolic antioxidants in food by high-performance liquid chromatography.* Journal of Chromatography A, 2000. **882**(1–2): p. 271–281.

[26] Delgado-Zamarreno, M., et al., *Analysis of synthetic phenolic antioxidants in edible oils by micellar electrokinetic capillary chromatography.* Food Chemistry, 2007. **100**(4): p. 1722–1727.

[27] Rohaizad, A., et al., *Green synthesis of silver nanoparticles from Catharanthus roseus dried bark extract deposited on graphene oxide for effective adsorption of methylene blue dye.* Journal of Environmental Chemical Engineering, 2020. **8**(4): p. 103955.

[28] Katugampalage, T.R., et al., *A smart magnetically separable MIL-53 (Al) MOF-coated nano-adsorbent for antibiotic pollutant removal with rapid and non-contact inductive heat regeneration.* Chemical Engineering Journal Advances, 2021. **8**: p. 100160.

[29] Shahabuddin, S., et al., *Synthesis of chitosan grafted-polyaniline/Co3O4 nanocube nanocomposites and their photocatalytic activity toward methylene blue dye degradation.* RSC Advances, 2015. **5**(102): p. 83857–83867.

[30] Muthukumaran, M., et al., *Green synthesis of cuprous oxide nanoparticles for environmental remediation and enhanced visible-light photocatalytic activity.* Optik, 2020. **214**: p. 164849.

[31] Sagadevan, S., et al., *Reduced graphene/nanostructured cobalt oxide nanocomposite for enhanced electrochemical performance of supercapacitor applications.* Journal of Colloid and Interface Science, 2020. **558**: p. 68–77.

[32] Yusoff, N., et al., *Amperometric detection of nitric oxide using a glassy carbon electrode modified with gold nanoparticles incorporated into a nanohybrid composed of reduced graphene oxide and Nafion.* Microchimica Acta, 2017. **184**(9): p. 3291–3299.

[33] Yusoff, N., et al., *A facile preparation of titanium dioxide-iron oxide@silicon dioxide incorporated reduced graphene oxide nanohybrid for electrooxidation of methanol in alkaline medium.* Electrochimica Acta, 2016. **192**: p. 167–176.

[34] Shahid, M.M., P. Rameshkumar, and N.M. Huang, *Morphology dependent electrocatalytic properties of hydrothermally synthesized cobalt oxide nanostructures.* Ceramics International, 2015. **41**(10, Part A): p. 13210–13217.

[35] Shahid, M.M., et al., *An electrochemical sensing platform of cobalt oxide@gold nanocubes interleaved reduced graphene oxide for the selective determination of hydrazine.* Electrochimica Acta, 2018. **259**: p. 606–616.

[36] Shahid, M.M., et al., *Cobalt oxide nanocubes interleaved reduced graphene oxide as an efficient electrocatalyst for oxygen reduction reaction in alkaline medium.* Electrochimica Acta, 2017. **237**: p. 61–68.

[37] Anuar, N.S., et al., *Fabrication of platinum nitrogen-doped graphene nanocomposite modified electrode for the electrochemical detection of acetaminophen.* Sensors and Actuators B: Chemical, 2018. **266**: p. 375–383.

[38] Demir, E., H. Silah, and N. Aydogdu, *Electrochemical applications for the antioxidant sensing in food samples such as citrus and its derivatives, soft drinks, supplementary food and nutrients,* in *Citrus.* 2021, IntechOpen.

[39] Gong, K., et al., *Electrochemistry and electroanalytical applications of carbon nanotubes: A review.* Analytical Sciences, 2005. **21**(12): p. 1383–1393.

[40] Pumera, M., *The electrochemistry of carbon nanotubes: Fundamentals and applications.* Chemistry: A European Journal, 2009. **15**(20): p. 4970–4978.

[41] Campbell, J.K., L. Sun, and R.M. Crooks, *Electrochemistry using single carbon nanotubes.* Journal of the American Chemical Society, 1999. **121**(15): p. 3779–3780.

[42] Li, M. and H. He, *Study on electrochemical performance of multi-wall carbon nanotubes coated by iron oxide nanoparticles as advanced electrode materials for supercapacitors.* Vacuum, 2017. **143**: p. 371–379.

[43] Fan, X., et al., *Enhanced permeability, selectivity, and antifouling ability of CNTs/Al2O3 membrane under electrochemical assistance.* Environmental Science & Technology, 2015. **49**(4): p. 2293–2300.

[44] Zhang, F., et al., *Enhanced anode performances of polyaniline—TiO2—reduced graphene oxide nanocomposites for lithium ion batteries.* Inorganic Chemistry, 2012. **51**(17): p. 9544–9551.

[45] Hu, J., F. Jia, and Y.-F. Song, *Engineering high-performance polyoxometalate/PANI/MWNTs nanocomposite anode materials for lithium ion batteries.* Chemical Engineering Journal, 2017. **326**: p. 273–280.

[46] Ahmed, U., et al., *Influence of concentration of polyaniline (PANI) as counter electrode in dye sensitized solar cell,* 5th IET International Conference on Clean Energy and Technology (CEAT2018), 2018, p. 1–5.

[47] Shanmugam, V., et al., *Hydrothermal development of a novel NiO/rGO nanocomposites for dual supercapacitor and photocatalytic applications.* Materials Chemistry and Physics, 2022: p. 126425.

[48] Singh, J., and A. Dhaliwal, *Electrochemical and photocatalytic degradation of methylene blue by using rGO/AgNWs nanocomposite synthesized by electroplating on stainless steel.* Journal of Physics and Chemistry of Solids, 2022. **160**: p. 110358.

[49] Shahid, M.M., et al., *An electrochemical sensing platform of cobalt oxide@ gold nanocubes interleaved reduced graphene oxide for the selective determination of hydrazine.* Electrochimica Acta, 2018. **259**: p. 606–616.

[50] Dai, L., et al., *Integrated process of lignocellulosic biomass torrefaction and pyrolysis for upgrading bio-oil production: A state-of-the-art review.* Renewable and Sustainable Energy Reviews, 2019. **107**: p. 20–36.

[51] Thomas, D., et al., *Iron oxide loaded biochar/polyaniline nanocomposite: Synthesis, characterization and electrochemical analysis.* Inorganic Chemistry Communications, 2020. **119**: p. 108097.

[52] Xie, S., et al., *In-situ preparation of biochar-loaded particle electrode and its application in the electrochemical degradation of 4-chlorophenol in wastewater.* Chemosphere, 2021. **273**: p. 128506.

[53] Zhang, C., et al., *Achieving quick charge/discharge rate of 3.0 V s− 1 by 2D titanium carbide (MXene) via N-doped carbon intercalation.* Materials Letters, 2019. **234**: p. 21–25.

[54] Hassan, F.M., et al., *Pyrrolic-structure enriched nitrogen doped graphene for highly efficient next generation supercapacitors*. Journal of Materials Chemistry A, 2013. **1**(8): p. 2904–2912.

[55] Kresse, G., and J. Furthmüller, *Efficient iterative schemes for ab initio total-energy calculations using a plane-wave basis set*. Physical Review B, 1996. **54**(16): p. 11169.

[56] Yang, J., et al., *MXene-based composites: Synthesis and applications in rechargeable batteries and supercapacitors*. Advanced Materials Interfaces, 2019. **6**(8): p. 1802004.

[57] Zhou, J., et al., *A conductive and highly deformable all-pseudocapacitive composite paper as supercapacitor electrode with improved areal and volumetric capacitance*. Small, 2018. **14**(51): p. 1803786.

[58] Yan, J., et al., *Polypyrrole—MXene coated textile-based flexible energy storage device*. RSC Advances, 2018. **8**(69): p. 39742–39748.

[59] Zhu, J., et al., *Composites of TiO2 nanoparticles deposited on Ti3C2 MXene nanosheets with enhanced electrochemical performance*. Journal of the Electrochemical Society, 2016. **163**(5): p. A785.

[60] Zheng, W., et al., *Microwave-assisted synthesis of SnO2-Ti3C2 nanocomposite for enhanced supercapacitive performance*. Materials Letters, 2017. **209**: p. 122–125.

[61] Ambade, S.B., et al., *2D Ti3C2 MXene/WO3 hybrid architectures for high-rate supercapacitors*. Advanced Materials Interfaces, 2018. **5**(24): p. 1801361.

[62] Zhang, K., et al., *Three-dimensional porous Ti3C2Tx-NiO composite electrodes with enhanced electrochemical performance for supercapacitors*. Materials, 2019. **12**(1): p. 188.

[63] zou, R., et al., *Self-assembled MXene (Ti3C2Tx)/α-Fe2O3 nanocomposite as negative electrode material for supercapacitors*. Electrochimica Acta, 2018. **292**: p. 31–38.

[64] Zhu, J., X. Lu, and L. Wang, *Synthesis of a MoO 3/Ti 3 C 2 T x composite with enhanced capacitive performance for supercapacitors*. RSC Advances, 2016. **6**(100): p. 98506–98513.

[65] Wang, Y., et al., *2D/2D heterostructures of nickel molybdate and MXene with strong coupled synergistic effect towards enhanced supercapacitor performance*. Journal of Power Sources, 2019. **414**: p. 540–546.

8 Functional Nanomaterials for Chemical Sensors

Ab Rahman Marlinda

CONTENTS

8.1 INTRODUCTION

A nanomaterial is a material with dimensions ranging from 1–100 nanometers (nm) or at least one dimension in the nanometer range [1,2]. Generally, nanomaterials can be categorized based on their basic properties, including dimension, composition, morphology, homogeneity and distribution. Because of their unique physical and chemical qualities, such as with a vast interfacial ratio and high mechanical strength, they are key components of nanotechnology because they provide fundamentals for the creation of numerous devices and modified materials. Nanomaterials have been widely employed in a variety of sectors due to their unique qualities, including opto-electronic devices, data processing, catalysis, biological research, earth sciences, energy production and storage, sophisticated military systems, and many more [3]. There are a few types of nanomaterials, such as inorganic-based (including metals and metal oxides), carbon-based, organic-based, and composite-based nanomaterials.

DOI: 10.1201/9781003263852-8

149

Sensors are devices that receive a physical, chemical, or biological stimuli response which later can be converted into an output signal by a transducer. Meanwhile, chemical sensors are tools used to track and detect the trace of chemical targets such as toxins and metal ions that can be detrimental to human health and also the environment. Chemical sensors are very useful in safety, health, hygiene, and anti-terrorism applications due to their specific interactions between a modified electrode and target analytes. Chemical sensors contain two basic functional units: a receptor and a transducer. Some sensors may also include a separator, which is a membrane. Chemical sensors are categorized according to the transducer's working mechanism. Chemical sensors utilize electrochemical, optical, electrical, and mass-sensitive operating principle classes [4,5]. The definition of the operating principle classes is defined in Figure 8.1.

This chapter gives readers a thorough knowledge of the principles before looking into particular types of nanomaterials and their functionality. The essential background will be covered, including an introduction to some fundamental characteristics of nanomaterials and their synthesis or chemical functionalization methods, and their advantages. Following that, we will discuss the many types of nanomaterials and how they may be functionally modified through surface alterations, molecular adjustments, and optimal multi-dimensional ordered or hybrid nanostructures. The modification overview on electrochemical, optical, electrical, and mass-sensitive sensors will be discussed. Examples of functional nanomaterials with their detection, absorption, and degradation features should be mentioned in Section 8.4. Their corresponding applications in chemical sensors are also discussed. The challenges and prospects of remarkable enhanced functional properties over unmodified

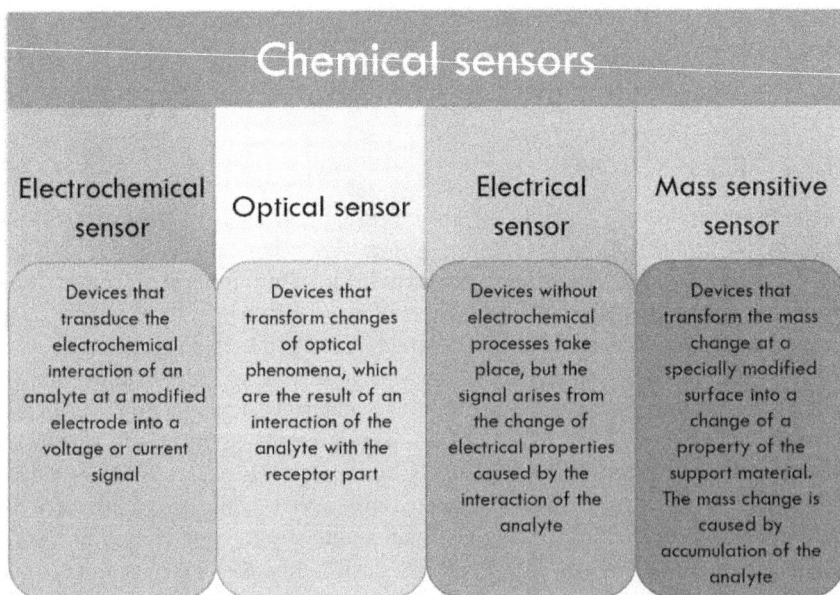

FIGURE 8.1 Classification of chemical sensor working principles.

nanomaterials open up new avenues for developing and applying chemical sensors. We wish that this chapter will be a valuable resource for future research related to functional nanomaterials. Thus, it provides critical possibilities for the needful multidisciplinary collaboration of scientists in various fields, including materials science, chemistry, physics, biology, nanotechnology, and engineering.

8.2 PROPERTIES OF FUNCTIONAL NANOMATERIALS

8.2.1 Unique Features of Functional Nanomaterials

The discovery of functional nanomaterials has sparked global scientific and technological interest, intending to revolutionize many frontiers in nanotechnology. The fascinating properties of functional materials were expected to offer up more opportunities in the material universe, particularly for chemical sensors and functional materials-based nanocomposites. Nanotechnology can alter functional materials by utilizing nanotechnology to create materials with extremely tiny sizes and enhanced characteristics. Then, it is able to improve the prospects of material science by creating a new class of materials with multifunctional properties for application to specific chemical sensors. Nanomaterials are materials with one or more nanoscale exterior dimensions or internal structures. Materials with nanoscale structures frequently differ from bulk equivalents in optical, thermal conductivity, and mechanical or electrical characteristics [6]. Nanomaterials are categorized as shown in Figure 8.2.

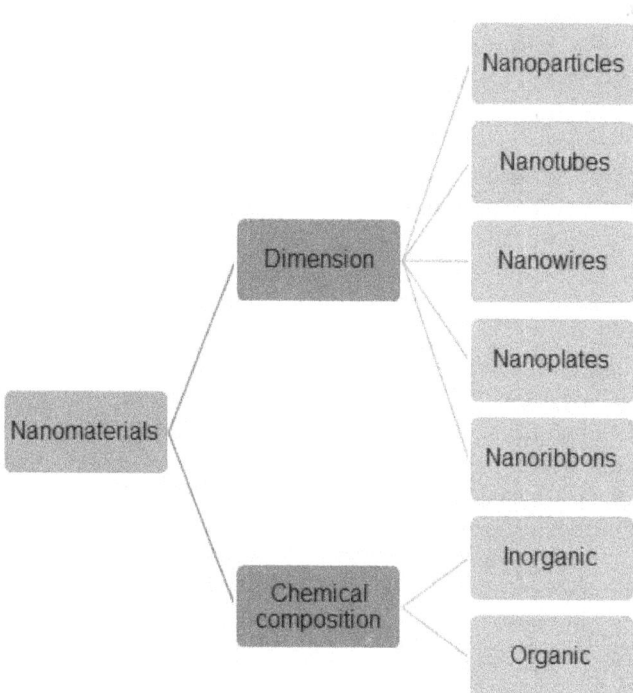

FIGURE 8.2   Categories of nanomaterials for possible commercial uses.

Their dimensionality [7] and chemical composition were strongly associated with various synthesis techniques based on controllable molecular structure, size, composition, and surface chemistry of nanomaterials to alter their properties for possible commercial uses [8].

For example, nanoparticles (NPs) and ultrafine particles are particles sized from 1–100 nm [9]. These NPs are easily dispersed in various gaseous, liquid, or solid media. The nanowires are similar to nanorods. Their structures are dominated by wire-like tendrils with a diameter range of several tens of nanometers to a micrometer and length in the range of several tens to several hundred micrometers. Meanwhile, nanotubes possess wire-like nanostructures composed of hollow cores that can be observed via transmission electron microscopy (TEM).

Nanomaterials have appeared as possible key components for developing multifunctional fabrics with superhydrophobic characteristics, thermochromism, photochromism, and antibacterial activity [10,11]. Their unique structures offer several opportunities to investigate their sensing behavior by developing new nanosensing devices. Meanwhile, advances in the development of highly sensitive electrodes and large capacity of the electrochemical energy storage devices have been made possible by their multicomponent nanomaterials with hierarchical nanostructures [12,13]. Nanomaterial hybridization results in fast photochromic reactions and great photostability, making them ideal for capturing and using green solar energy [14,15].

Additionally, carbon-based nanomaterials are unique nanomaterials that have been generated in various dimensionalities throughout the previous decades, including zero-dimensional (0D) fullerenes, one-dimensional (1D) carbon nanotubes (CNTs), and two-dimensional (2D) graphene. CNTs, for example, have demonstrated theoretical and practical findings with very high elastic moduli larger than 1 TPa, and strengths 10–100 times stronger than steel [16]. The unique properties of CNTs–which have superior thermal and electrical properties that stable up to $2800°C$ in a vacuum condition—have an electrical conductivity of ~103 S cm^{-1} with 1,000 times greater carrying capacity than diamond, and thermal conductivity of ~1900 W m^{-1} K^{-1}, which is about two times higher than diamond [17,18]. The physical and catalytic features of CNTs make them excellent for chemical sensors. They may also be used to build nanoscale electronics and create multifunctional composite materials.

Aside from CNTs, there is another carbon-based nanomaterial known as carbonaceous materials, including carbon black, graphite, graphene, amorphous porous carbon, and carbon nanofiber [19–21]. Carbonaceous materials' inert qualities are commonly employed as counter electrodes due to their superior and unique electrical properties (CE). Chemical stability, strong thermal conductivity, good mechanical qualities, superelectroconductivity, and increased optical properties are among the unique features of functional carbon nanomaterials of nanoscale dimensions. Since their discovery, carbonaceous materials have piqued the curiosity of several researchers. Furthermore, the unique properties of functional carbon nanostructures hold immense promise for many applications such as material processing, environmental research, energy storage, biology, and medicine.

8.2.2 ADVANTAGES OF THE FUNCTIONAL NANOMATERIALS FOR CHEMICAL SENSORS

These nanomaterials have received a lot of interest in recent years since they are regarded current sophisticated architectures [22,23]. Nanomaterials differ from bulk equivalents in the following ways, which is why they are employed in electrode modification [24–26]. Among the main advantages of having functional nanomaterials for the chemical sensor is the speedup of the electrochemical reaction. This may be accomplished by increasing the specific surface area of functional nanomaterials with different morphologies. The morphologies and sizes of nanomaterials such as nanozymes play significant role in defining enzyme-like activity since the smaller sizes of the nanozymes offer greater specific areas, resulting in enhanced enzyme-like activity. As a result, the nanozymes' specific surface areas will boost the catalytic effectiveness of nanocatalysts. This is explained by regulating the density of their active crystallographic facets to increase the surface reactivity on dangling bonds of the crystal facets [27]. Yang et al. explain that by adjusting the density of their active crystallographic facets, the peroxidase mimetic activity of CeO_2 nanomaterials is increased [28].

Besides morphologies and sizes of nanomaterials, the supportive active sites of nanomaterials can contribute to excellent chemical sensor activity. The surface modification of nanomaterials with functional groups such as carboxyl and hydroxyl strongly influence the chemical activity. Song and colleagues, for example, initially revealed that carboxyl-modified graphene oxide (GO-COOH) displayed inherent peroxidase-like activity. The GO-COOH may catalyze the oxidation of a typical peroxidase substrate of 3,30,5,50-tetramethylbenzidine (TMB) with the help of hydrogen peroxide, promoting a better affinity for a TMB substrate than HRP (horseradish peroxidase) [29]. As a result of the alteration of carboxyl and hydroxyl groups, GO-COOH demonstrates exceptional catalytic efficiency by allowing quicker electron transfer with peroxidase substrates H_2O_2 and TMB. Another example is Li et al., who created carboxyl-modified C_{60} ($C_{60}[C(COOH)_2]_2$) to boost peroxidase-like activity [30]. $C_{60}[C(COOH)_2]_2$ has more peroxidase-like activity than water-dispersible C_{60}. The sizes are also found to alter the reaction activity in this work, with the lower hydrodynamic diameter for $C_{60}[C(COOH)_2]_2$ (50 nm) than the size of water-dispersible C_{60} (68 nm), potentially increasing the specific surface area and amount of active sites. Thus, it is directly promoting the excellent catalytic activity of chemical reactions.

8.3 SYNTHESIS AND CHEMICAL FUNCTIONALIZATION OF NANOMATERIALS

Two different approaches generally synthesize nanomaterials: a 'top-down' physical approach and a 'bottom-up'—primarily a chemical—approach (Figure 8.3). The top-down approach is defined as the synthesis of nanomaterials from bulk materials. It can be synthesized in high yields with simple and low-cost processing methods. Nevertheless, this approach can obtain nanomaterials with heterogeneous chemical compositions. Meanwhile, the bottom-up approach is a method to build

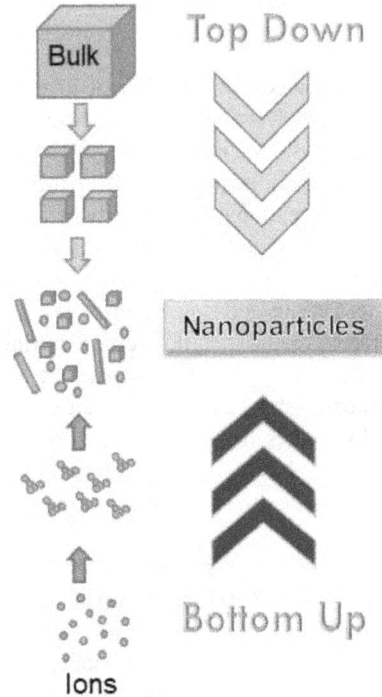

FIGURE 8.3 Synthesis flow of nanomaterials via top-down and bottom-up approaches.

up nanomaterials from atoms or molecules with a more uniform size distribution. However, this approach is more expensive than the top-down approach because it adopts methods that require a piece of expensive equipment for the synthesis processes. Somehow, it also produced more minimal yields than the top-down approach.

8.3.1 TOP-DOWN

Several physical methods have been identified for the top-down approach, including mechanical milling (also known as ball milling), thermal ablation/laser ablation, explosion process, chemical etching, and sputtering techniques. Most of the top-down approaches are utilized by the industry since they offer a high yield of products and are easy to upscale. Among listed methods, mechanical milling offers a lower processing cost and is also the most straightforward method. Mechanical milling or ball milling can produce nanomaterials using mechanical attrition of grinding medium to reduce the size of a material [31,32]. Meanwhile, a well-known lithographic method approach requires higher energy and expensive equipment, as well as facilities [33]. For the most part, this approach can manufacture micron-sized nanomaterial features. For numerous decades, the lithography technology has been utilized to print circuits and computer boards.

8.3.2   Bottom-Up

Meanwhile, several chemical methods have been identified for the bottom-up approach, including electrochemical precipitation, atomic/molecular condensation, vapor deposition, solvothermal/hydrothermal, sol-gel process, microemulsion, spray pyrolysis, aerosol process, and biochemical reductions [34–37]. The bottom-up technique is appealing to obtain nanomaterials because various forms—such as nanoparticles, nanorods, and nanowires—may be generated utilizing a variety of reaction conditions and surfactants or solvents [38–40]. The sol-gel technique is a flexible soft chemical process that is commonly used to synthesize metal oxides, ceramics, and glass materials [33]. For example, synthesis of SiO_2 and Al_2O_3 nanomaterials is primarily obtained via the sol-gel method, while TiO_2 and $BaTiO_3$ nanomaterials can be synthesis using hydrothermal route [41]. The sol-gel process has several advantages, including generating highly stable materials with superior purity and uniform composition of the yields at low temperatures. Furthermore, this mechanism changes particle shape throughout the chemical transition of the molecular precursor to the final oxidic network. Many researchers have utilized the sol-gel method to studies on the synthesis of iron oxide and alumina nanomaterials for dielectric and electronic applications [41–44].

8.4   OVERVIEW OF THE DEVELOPMENT OF FUNCTIONAL NANOMATERIALS FOR CHEMICAL SENSORS

8.4.1   Functional Nanomaterials for Electrochemical Sensors

Electrochemical sensing is the most superior technique to study and sensitively identify a variety of biological and chemical analytes. Among the different electrochemical techniques, potentiometric and amperometric techniques are mostly reported for electroanalytical transduction. Meanwhile, carbon-based nanomaterials are the main functional nanomaterials that have received intensive attention among researchers worldwide for electrochemical sensors. There are numerous forms of functional carbon-based nanomaterials, such as carbon nanospheres, carbon nanofibers, and carbon nanotubes. These carbon nanomaterials produce superior properties that enhance electroanalytical activities for the electrochemical sensor. Carbon nanotubes (CNTs), for example, are among the most intensively investigated electrochemical sensors due to their unique characteristics including the ability to enhance the electrochemical reactivity of biomolecules and promote electron transfer reactions due to their surface conductivity. Hence, it enables use in nanoscale sensors with higher sensitivity. For example, Hyun et al. immobilized GOx on CNTs to detect glucose [45]. A large content of CNT has contributed to enhancing the amperometric response of glucose with sensitivity value of 53.5 $\mu A\ mM^{-1}\ cm^{-2}$ and a high rate of electron transfer constant (1.14 s^{-1}). The investigations also found that GOx/CNT had long-term stability, with 86 percent of the glucose activity remaining after two weeks. Liu group has reported earlier on fundamental studies of electron transfer of GOx that is immobilized into a single-walled carbon nanotube–poly(ethylenimine) (SWCNT–PEI) matrix [46]. CNTs that have been concurrently treated with GOx exhibit electron transfer kinetics similar to those seen in isolated flavine adenine dinucleotide (FAD), allowing for increased sensing sensitivity.

However, following those years, there has been a surge in interest in the electrochemical functionalization of CNTs with metallic nanoparticles, notably in applications of sensing and catalysis, with excellent analytical performance; for example, the functionalization of CNTs with $Ni_{12}P_5$ via one-pot hot-solution synthesis method [47]. Wang et al. reported the fabrication of $Ni_{12}P_5$/CNTs nanohybrid structure with current densities of 2 and 10 mA cm^{-2} over applied potentials of 65 mV and 129 mV when evaluating hydrogen evolution reaction (HER) catalyst that operated in acidic electrolytes. Moreover, the $Ni_{12}P_5$/CNTs nanohybrid as an anode material also exhibits enhanced electrochemical performance for lithium-ion batteries with high capacity, excellent cycling stability, and good rate performance. Similarly, Pham et al. synthesized palladium (Pd) nanostructures modified SWCNTs electrode for an electrochemical nitrite sensor [48]. The Pd with urchin-like nanostructures were coated on pre-patterned SWCNTs exhibited current response of 1.5 times higher than the spherical Pd particles for detecting nitrite with a lower LOD value of 0.25 µM ($S/N = 3$) and resulted in two linear ranges (2–238 µM and 283–1230 µM) under the controlled condition, as shown in Figure 8.4.

8.4.2 FUNCTIONAL NANOMATERIALS FOR OPTICAL SENSORS

Optical sensors include the interaction of a chemically selective layer with a species of interest to create changes in the optical characteristics of the sensor [49]. The optical sensor provides a simple, quick, and low-cost method for detecting diverse analyte targets. An optical sensor, in general, contains an identification unit that may link with the desired target of the analyte and transducer element, signaling the binding event [50]. The combination of functional nanomaterials assisting target response can improve sensitivity for the sensing platform. The abilities of functional nanomaterials to specifically recognize, trigger optical signals, and directly amplify the detection signal make them frequently employed to develop the optical sensor. Basically, utilizing nanomaterial-based electroanalyticals can give us some ideas and concepts for the diagnosis of cancer cells and bone disease. A novel functional nanomaterial such as gold nanoparticles (AuNPs) is extensively synthesized in the creation of perfect sensors due to its distinct physical and optical properties. Liu et al. fabricated rhodamine B-functionalized with gold nanoparticles for developing a highly sensitive probe for pesticide sensor [51]. A strong bonding interaction between the active thiol group (thiocholine) with AuNPs prevents the production of thiocholine and results in the recovery of fluorescence of RB, as depicted in Figure 8.5. The study reported detection of carbaryl and malathion with good sensitivity at lower LODs of 0.1 and 0.3 µg L^{-1}, respectively.

Besides the example given, various additional fluorescence nanomaterials—such as QDs [52,53], upconversion nanoparticles [54], carbon-based [55], and fluorescent dye [56,57]—have been used to create AuNPs-based enzyme platforms.

Other nanomaterials include nanozymes, metal oxides, and organic fluorescent molecules–based nanomaterials are also used as elements in optical sensors for determining contaminants, quality, and authenticity of food for food safety. For example,

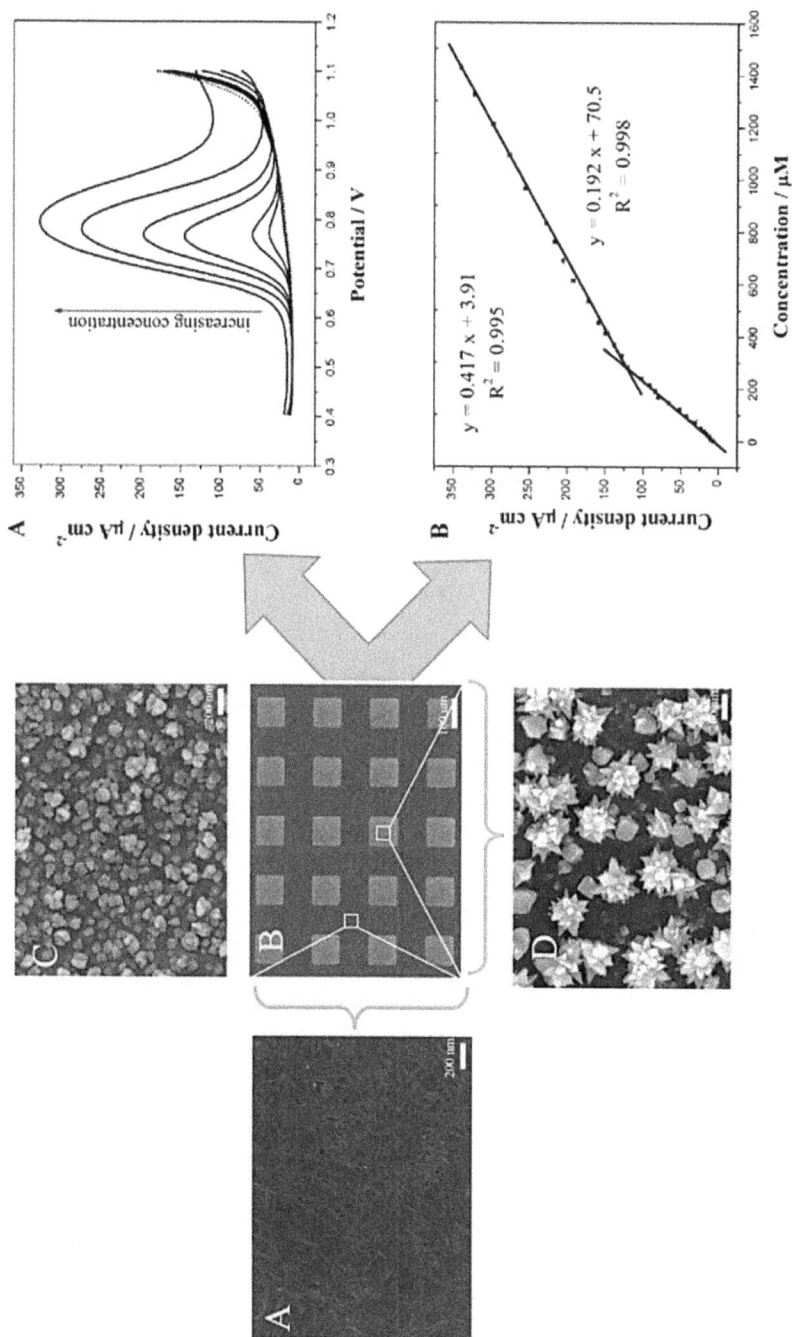

FIGURE 8.4 FESEM images of (A) pristine SWCNT film; (B) Pd/SWCNT; (C) Spherical Pd NPs; and (D) urchin-like Pd/SWCNT thin-film electrodes that obtained (A) differential pulse voltammograms and (B) calibration plots for determination of nitrite at concentrations of 0, 10, 48, 91, 304, 500, 880, and 1231 μM.

Source: (Images reproduced from [48] with permission)

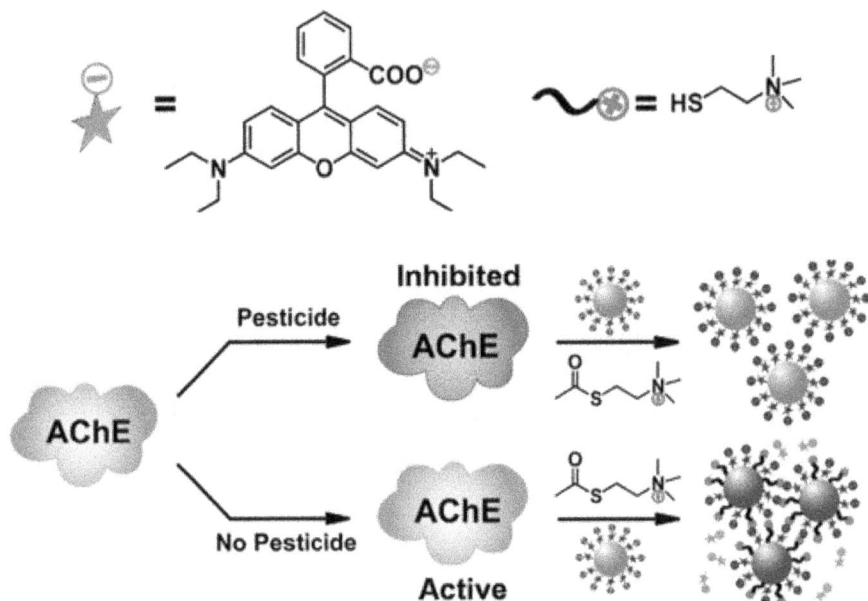

FIGURE 8.5   A design for pesticide test with two readouts (colorimetric and fluorometric).

Source: (Image adapted from [50] with permission)

nanozymes can change color's spectrum after an analytical chemical reaction by using fluorescence [58]. Thus, its catalytic reaction can amplify the detection signal. The nanozymes can be classified as oxidoreductases (for example, oxidase, peroxidase, catalase, and superoxide dismutase) or hydrolases (for example, nuclease and protease) based on their catalytic processes [59]. Hundreds of nanomaterials have been discovered to operate as biocatalysts since the discovery of the first nanozyme (Fe_3O_4 NPs). According to Yan et al., the peroxidase enzyme-like activity of ferromagnetic nanoparticles (Fe_3O_4 NPs) is an intrinsic enzyme mimetic activity similar to that of natural peroxidase, which is widely used to oxidize organic substrates in the treatment of wastewater, which could immediately trigger and boost the oxidation of peroxidase substrates in the presence of hydrogen peroxide (H_2O_2) [60]. Nanozyme-based biosensors (NBs) have been widely used in a variety of applications, including food quality and safety evaluation, clinical illness diagnosis, biological metabolite measurement, and environmental pollutant monitoring. Metal oxide nanoparticles are sensitive to surface chemistry, as well as having enzyme-like catalytic activity. As a result, they can be used for visual detection if they have a larger percentage of surface metal and surface peroxide species; for example, MnO_2 nanosheets synthesized by Yao et al. for fluorescence sensing of glutathione (GSH) detection [61]. According to Liu and his colleagues, the MnO_2 nanosheets can react with glutathione and convert it into Mn^{2+} ions that effectively oxidize o-phenylenediamine (OPDA). As a result, the MnO_2-OPDA sensing platform showed a compassionate response to GSH at 10 nM–40 µM with a low detection limit of 10 nM. Therefore, it can be concluded that metal oxide nanoparticles are frequently utilized to detect

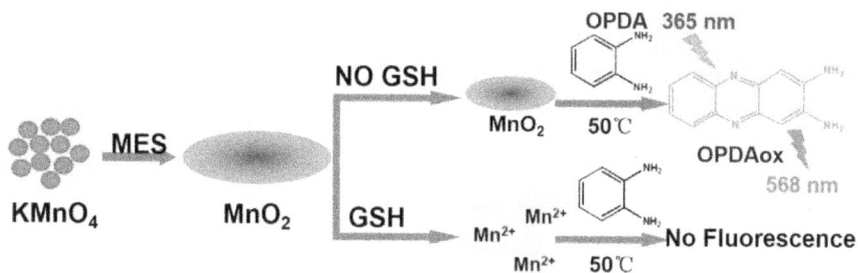

FIGURE 8.6 The MnO_2 nanosheet-based fluorescence assay for GSH sensing is depicted schematically.

Source: (Image adapted from [61] with permission)

specific targets by adsorbing or reacting with ligands [62]. In organic fluorescent molecule-based nanomaterials, the spectrum characteristics, tunability, and biocompatibility of specific molecular organic fluorescent dyes may be coupled with the intensity, chemical stability, and colloidal stability of inorganic compounds [63]. The organic fluorescent molecule can be built with excellent selectivity for detecting targets based on diverse targets.

8.4.3 FUNCTIONAL NANOMATERIALS FOR ELECTRICAL SENSORS

A novel functional nanomaterial with diverse, unique properties has offered outstanding progress for broad applications in sensing devices. Biomedical applications are among the potential applications of the advanced functional nanomaterials that are much utilized in analytical diagnostics, point-of-care devices, and clinical and medical industries. This is because nanobiosensors based on functional nanomaterials can create high sensitivity and ease miniaturization that contributes much to develop a new analytical design for clinical and field-deployable instruments [64]. Because of their unique thermal, electrical, and mechanical characteristics, carbon nanomaterials—particularly graphene—are now an appealing possibility for sensor development. Graphene is a two-dimensional honeycomb lattice structure in which carbon atoms share sp^2 electrons with three carbon atoms from nearby six-membered rings [65]. Graphene has high electrical strength because of its peculiar molecular structure, which is created by the interference of electrons on its surface. Furthermore, graphene has 60 times the electrical conductivity of SWNTs [66]. For example, Qu *et al.* developed a distinct 3D-linked graphene framework that is homogeneously coupled by a few-layer graphene (FLG) sheet. The 3DG/FLG serves as a binder with a unique pore structure for high-performance lithium-ion batteries (LIB). This hybrid structure has a high specific capacity of 770 mA h/g and an energy density of 388 Wh/kg [67]. A large surface area and flexible graphene nanosheets have attracted many researchers to work on graphene-based hybrids as anode and cathode materials. For instance, Zhou et al. fabricated graphene nanosheets (GNSs) with Fe_3O_4 particles via in situ reduction technique of iron hydroxide and GNSs [68]. The modified GNS/Fe$_3$O$_4$ nanocomposite possesses a high specific capacity of 1,026 mA h g^{-1}, and after 30 cycles at 35 mA g^{-1} and 580 mAh g^{-1}, respectively. Meanwhile, the specific capacity show increment after 100 cycles at 700 mA g^{-1} (Figure 8.7). This study reveals the great performance for lithium storage capacity, as well as improvement in cyclic stability and excellent rate capability.

In addition to graphene hybrid metal oxide–based nanocomposites, different metals such as Pt, Au, Ag, Cu, Fe, and others are blended with graphene to produce nanocomposites appropriate for biomedical sensing applications. Li et al. developed a salbutamol sensor based on Ag-NPs and N-doped reduced graphene oxide (Ag-N-rGO) using a simple and environmentally friendly chemical process [69]. The proposed sensor exhibited an active catalytic performance for salbutamol detection in human urine and pig samples. For detecting salbutamol, the improved Ag-N-rGO composite electrode achieved a reduced LOD of 7.0 nM with a linear range of 0.03–20.00 μM. The manufactured Ag-N-rGO composite also has great stability and repeatability. Marlinda et al. developed an amperometric sensor detecting β-nicotinamide adenine dinucleotide (NADH) using rGO-decorated gold nanorods (Au NR@rGO) [70]. Figure 8.8 shows the developed modified electrode of AuNR@rGO/GCE demonstrating better electrocatalytic activity toward NADH detection. This sensor showed current response increased linearly with NADH concentration throughout the 1–31 μM range, with a low LOD value of 0.22 μM (S/N = 3). In addition, the developed sensor displayed high stability in the presence of various interferents such as glucose, uric acid, ascorbic acid, and dopamine. Thus, it devises encouragement for biomedical applications.

8.4.4 FUNCTIONAL NANOMATERIALS FOR MASS-SENSITIVE SENSORS

A higher sensitivity of chemical sensing using real samples is often more demanding than the standard laboratory tests. For technological reasons, more industries are interested to get greater sensor responses with less material [71]. Thus, well-known devices based on the operation concept of quartz crystal microbalance (QCM) and silicon microcantilevers are attractive possibilities for increasing the sensitivity of mass-sensitive sensors [71,72]. These device systems can increase the fundamental device frequency sensitive to different external parameters, such as mass perturbations, and are highly reported for volatile organic compounds (VOCs) or bacteria detection [73–75]. Besides, the QCM system is very sensitive. It also detects the change in the mass of a molecule at room temperature, making the QCM widely utilized as a transducer for environmental monitoring applications [72]. This effort can be achieved by coating a sensing material layer such as nanoparticles on the active electrode of the QCM [76,77]. The selective nanoparticle layers, with nanoparticle sizes ranging from 1–100 nm, can overcome sensitivity constraints in thin-film sensor layers by boosting accessibility and hence the effectivity of interaction sites [71].

Metallic and metal oxide nanoparticles have been used often to change the surfaces of mass-sensitive crystals, enhancing mass transport, effective sensor surface area, and sensitivity. Lieberzeit et al. introduced Molybdenum disulfide (MoS$_2$) (particle size 200–300 nm) coated on the QCM electrode for two analytical applications that interact with thiols in the air as well as engine oil degradation measurements

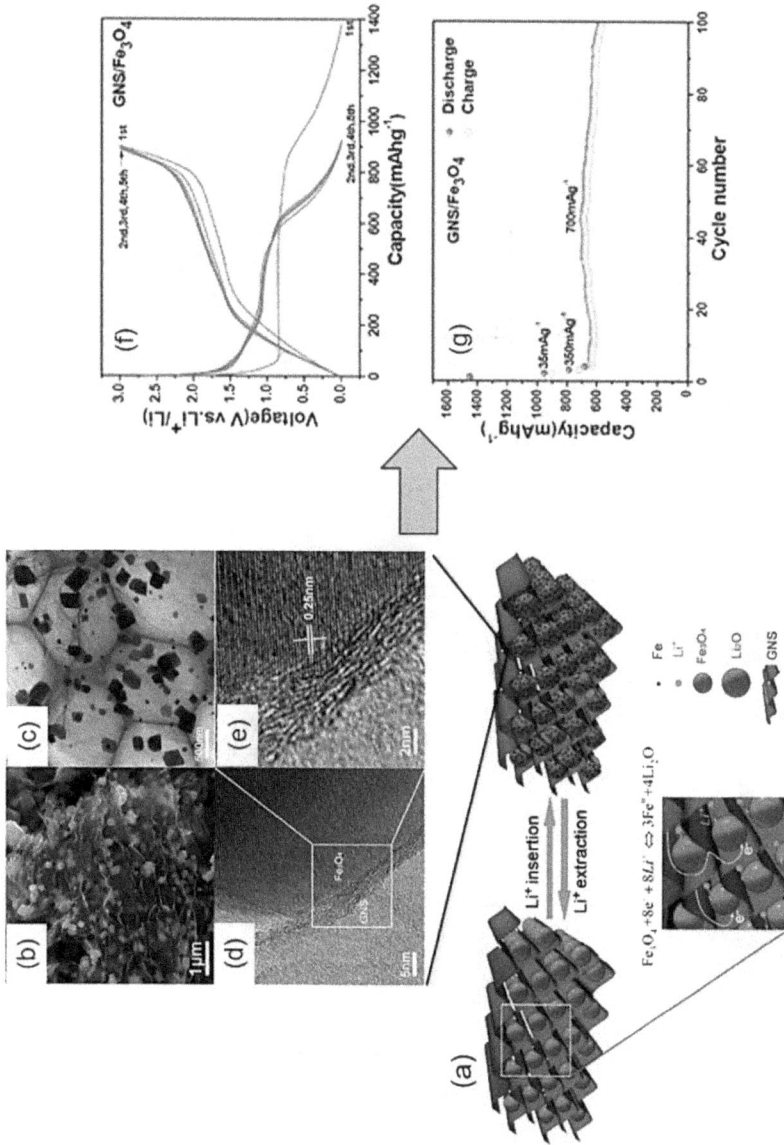

FIGURE 8.7 (a) illustrates a schematic of a flexible interleaved structure of GNSs/Fe$_3$O$_4$ particles; (b–c) SEM images of the cross-section of GNS/Fe$_3$O$_4$ composite; (d–e) are high-resolution TEM images of GNS/Fe$_3$O$_4$ composite of the square area in (d); (f) discharge/charge profiles of the GNS/Fe$_3$O$_4$ composite, and (g) cycling performance of the GNS/Fe$_3$O$_4$ composite at a current density of 700 mA g^{-1} for 100 cycles.

Source: (Images adapted from [68] with permission)

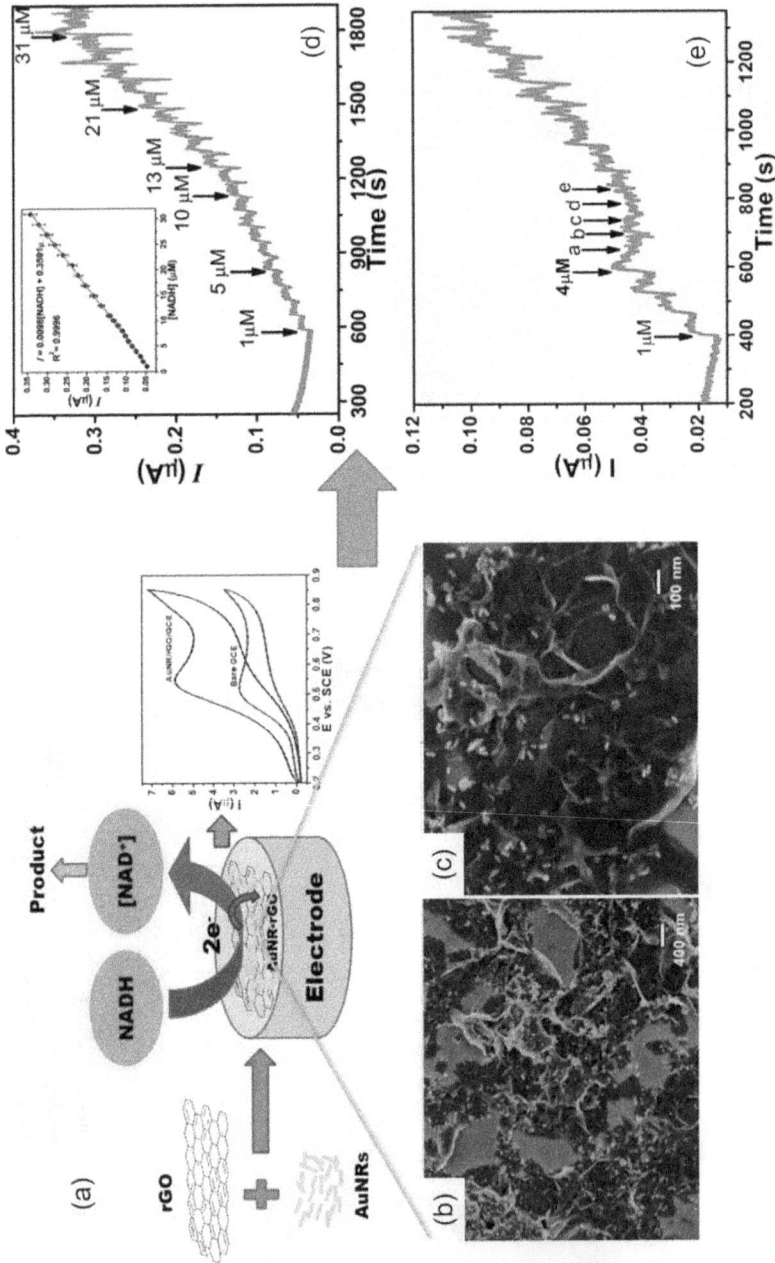

FIGURE 8.8 (a) A schematic representation of the electrocatalytic oxidation of NADH at the AuNR@rGO/GCE-modified electrode; (b–c) FESEM images of AuNR@rGO at two different magnifications; (c) amperometric i–t curve achieved for NADH at AuNR@rGO/GCE-modified electrode with injection of 10 M of interferences.

Source: (Images adapted from ref [70] with permission)

[71]. It was discovered that MoS_2 was a very suitable material for interacting with organic thiols, resulting in a tenfold improvement in sensor sensitivity compared to a pure gold layer represented in Figure 8.9. Furthermore, the nanoparticle layers exhibit fully reversible sensor signals. The synthesis particles were also integrated with the molecular imprinting approach: a precipitation procedure demonstrated that molecularly imprinted titanate sol-gel (TiO_2) particles with a size range of 200–300 nm could be generated for engine oil degradation measurements.

Unfortunately, electrical instability, high-cost production, the necessity for simple manufacturing techniques, and toxicity of nanoparticles are significantly limiting the use of metal and metal oxide for sensor applications. Therefore, melanin nanoparticles have sparked significant interest in biomedical disciplines in recent years because of their biocompatibility, antioxidant activity, free radical scavenging, metal ion chelation, strong near infrared absorption, and high photothermal

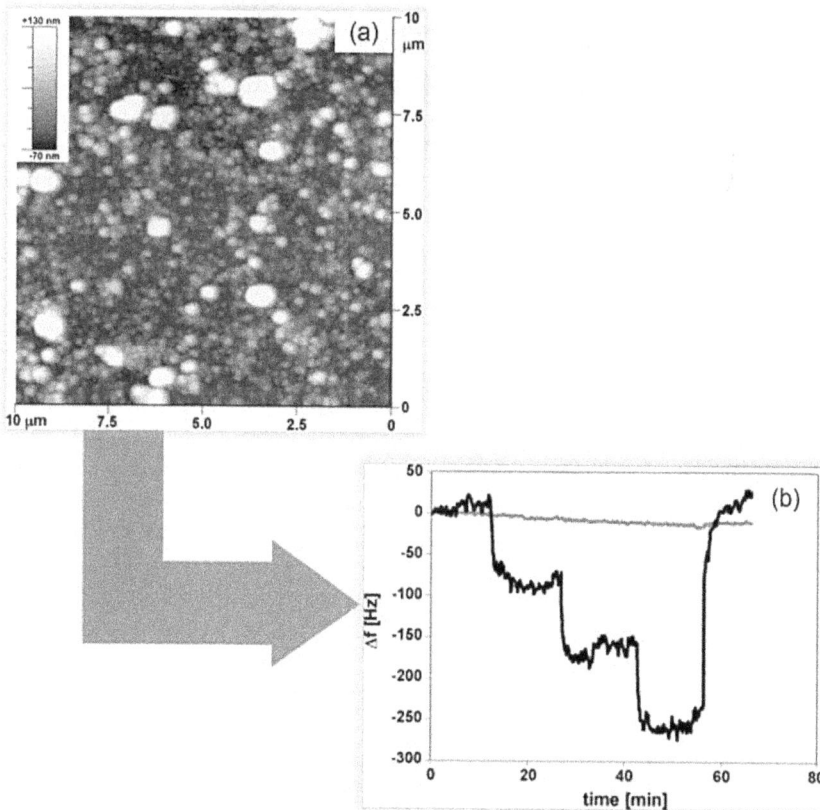

FIGURE 8.9 (a) AFM image of a MoS_2 nanoparticle layer on a sensor substrate; and (b) QCM sensor responses to varying butane thiol concentrations for a MoS_2 nanoparticle layer where black line represents layer height 3 kHz and a gray line represents reference channel.

Source: (Images adapted from [71] with permission)

conversion efficiency [78–80]. Melanin is a brownish-yellow pigment generated by melanosomes that is extensively dispersed in living creatures. Meanwhile, bacteria, fungus, black tea leaves, chestnut shells, catfish, and cuttlefish ink are being employed as natural sources of melanin [78]. Demir et al. utilized melanin nanoparticles to functionalize a Quartz tuning fork (QTF) to create a recognition layer for a specific mass-sensitive biosensor [81]. Demir and her team have used dip-coating and electroplating methods under different parameters. The efficiency of the surface modification technique was evaluated with the change in resonance frequency of QTF as a function of coating parameters. The results showed that deposition of melanin nanoparticles at 110°C using 30 μL of melanin nanoparticle solution creates the most significant decrement in resonance frequencies within dip-coating conditions with relatively low operating costs and convenience of usage.

8.5 CHALLENGES AND PROSPECTS OF THE FUNCTIONAL NANOMATERIALS FIELD FOR CHEMICAL SENSOR APPLICATIONS

The advancement of nanotechnology is linked to the advent of various functionalized nanomaterials such as 2D conducting carbon hybrid materials, organic metal frameworks, conducting polymers, metals, and metal oxides/hydroxides nanoparticles. Because of their chemical resilience, precision, high sensitivity, and specificity, nanomaterials have emerged as a possible choice for measuring diverse biomolecules [82]. Furthermore, the integration of nanomaterials has led to the shrinking of biosensors in general. Rapid advancements in the science field and challenging renewable technologies in creating functional nanomaterials may offer a plethora of new opportunities potentially leading to entirely new applications and improvements in existing technologies. In this chapter, functional nanomaterials for developing chemical sensors have been demonstrated with more promising and effective methods. The combination of carbon nanomaterials can facilitate an effective phonon transport and rapid ion and electron motion along a particular direction in structural electrodes to enhance sensing performance for practical applications. Tailoring nanomaterial architectures such as molecular composition, size, structure, and surface chemistry of nanomaterials can control their functionality for great potential in commercial applications [8]. Some issues and obstacles, however, remain unresolved. interfacial relationships among carbon nanomaterials and their fillers—for example, polymers or active inorganic materials—remain crucial. This factor can have a considerable impact on the distribution of fillers, changing the characteristics of composites and their application performance. Meanwhile, the weak linkages reduce composite stability, impede high-efficiency ion and charge transfer, and reduce the lifespan and performance of extensive light-driven actuators [83,84], thermal dissipation materials [85–87], and electrochemical storage devices [88–90]. Furthermore, typical processing techniques for these materials are time-consuming, costly, environmentally hazardous, and challenging to establish at the industrial level [8]. Despite the fact that numerous carbon-based nanoparticles were already used in actuation, heat conduction, extensive light conversion, and energy storage applications, a vast spectrum of carbon-based and functional composite materials remains unexplored.

The intelligent integration of pseudocapacitive or optimal architectures of the active anode materials for sophisticated wearable and flexible electronic devices with superior thermal dissipation qualities is a highly interesting research field [91–96]. Thus, adopting sensing technology in those areas to increase the sensitivity and selectivity for practical applications, with the appropriate design of an appropriate mimic of the natural enzyme, is critical. For example, sensing platforms based on surface-enhanced Raman spectroscopy (SERS), fluorescence, and electrochemistry must be further focused. It is predicted, in particular, that ultrasensitive detection of objects will be possible with the help of SERS technology. Furthermore, the sensitive identification of particular compounds based on nanozymes should be expanded.

8.6 CONCLUSION

The increased study of nanotechnology, particularly nanomaterials, is in accordance with contemporary demands for the speed and efficiency of a new tool in upgrading technology utilization. At the same time, it may assist consumers from all kinds of backgrounds and ages. Following that, the summary of this chapter is meant to bring some insight and serve as a guide for scholars interested in furthering their studies in the field of nanomaterials. It can generally guide promoting nanomaterial self-assembly, developing novel nanomaterials, and exploring the potential applications to meaningful multidisciplinary cooperation in various subjects such as materials science, biomedical engineering, tissue engineering, analytical science, and energy.

REFERENCES

[1] Gogotsi Y. *Nanomaterials Handbook* (1st Ed.). Boca Raton: CRC Press; 2006. https://doi.org/10.1201/9781420004014
[2] Kreyling WG, Semmler-Behnke M, Chaudhry Q. A complementary definition of nanomaterial. *Nano Today*. 2010;5(3):165–168. https://doi.org/10.1016/j.nantod.2010.03.004
[3] Yin Y, Talapin D. The chemistry of functional nanomaterials. *Chem Soc Rev*. 2013;42(7):2484–2487. doi:10.1039/C3CS90011H
[4] Hulanicki A, Glab S, Ingman F. Chemical sensors: Definitions and classification. *Pure Appl Chem*. 1991;63(9):1247–1250.
[5] Kassal P, Steinberg MD, Steinberg IM. Wireless chemical sensors and biosensors: A review. *Sensors Actuators B Chem*. 2018;266:228–245.
[6] Buzea C, Pacheco II, Robbie K. Nanomaterials and nanoparticles: Sources and toxicity. *Biointerphases*. 2007;2(4):MR17–MR71. doi:10.1116/1.2815690
[7] Klaessig F, Marrapese M, Abe S. Current perspectives in nanotechnology terminology and nomenclature. In: Murashov V, Howard J, eds. *Nanotechnology Standards*. New York: Springer; 2011:21–52. doi:10.1007/978-1-4419-7853-0_2
[8] Li Z, Wang L, Li Y, Feng Y, Feng W. Carbon-based functional nanomaterials: Preparation, properties and applications. *Compos Sci Technol*. 2019;179:10–40.
[9] Ghosh SK, Pal T. Interparticle coupling effect on the surface plasmon resonance of gold nanoparticles: From theory to applications. *Chem Rev*. 2007;107(11):4797–4862. doi:10.1021/cr0680282
[10] Gao Y-N, Wang Y, Yue T-N, Weng Y-X, Wang M. Multifunctional cotton non-woven fabrics coated with silver nanoparticles and polymers for antibacterial, superhydrophobic

and high performance microwave shielding. *J Colloid Interface Sci.* 2021;582:112–123. https://doi.org/10.1016/j.jcis.2020.08.037

[11] Hong HR, Kim J, Park CH. Facile fabrication of multifunctional fabrics: Use of copper and silver nanoparticles for antibacterial, superhydrophobic, conductive fabrics. *RSC Adv.* 2018;8(73):41782–41794. doi:10.1039/C8RA08310J

[12] Yuan C, Wu H Bin, Xie Y, Lou XW (David). Mixed transition-metal oxides: Design, synthesis, and energy-related applications. *Angew Chemie Int Ed.* 2014;53(6):1488–1504. https://doi.org/10.1002/anie.201303971

[13] Hu J, Xu Z, Li X, et al. Partially graphitic hierarchical porous carbon nanofiber for high performance supercapacitors and lithium ion batteries. *J Power Sources.* 2020;462:228098. https://doi.org/10.1016/j.jpowsour.2020.228098

[14] Zhang X, Hou L, Samorì P. Coupling carbon nanomaterials with photochromic molecules for the generation of optically responsive materials. *Nat Commun.* 2016;7(1):11118. doi:10.1038/ncomms11118

[15] Zhao X, Hu X, Sun J, et al. VO2-based composite films with exemplary thermochromic and photochromic performance. *J Appl Phys.* 2020;128(18):185107. doi:10.1063/5.0015382

[16] Ahmadi M, Zabihi O, Masoomi M, Naebe M. Synergistic effect of MWCNTs functionalization on interfacial and mechanical properties of multi-scale UHMWPE fibre reinforced epoxy composites. *Compos Sci Technol.* 2016;134:1–11.

[17] Ibrahim KS. Carbon nanotubes-properties and applications: A review. *Carbon Lett.* 2013;14(3):131–144.

[18] Collins PG, Avouris P. Nanotubes for electronics. *Sci Am.* 2000;283(6):62–69.

[19] Ma-Hock L, Strauss V, Treumann S, et al. Comparative inhalation toxicity of multi-wall carbon nanotubes, graphene, graphite nanoplatelets and low surface carbon black. *Part Fibre Toxicol.* 2013;10(1):1–20.

[20] Ma C, Wu L, Zheng L, et al. Preparation and capacitive performance of modified carbon black-doped porous carbon nanofibers. *J Nanoparticle Res.* 2019;21(2):1–12.

[21] Borchardt L, Zhu Q-L, Casco ME, et al. Toward a molecular design of porous carbon materials. *Mater Today.* 2017;20(10):592–610.

[22] Asif M, Ajmal M, Ashraf G, et al. The role of biosensors in coronavirus disease-2019 outbreak. *Curr Opin Electrochem.* 2020;23:174–184. https://doi.org/10.1016/j.coelec.2020.08.011

[23] Baig N, Sajid M, Saleh TA. Recent trends in nanomaterial-modified electrodes for electroanalytical applications. *TrAC Trends Anal Chem.* 2019;111:47–61. https://doi.org/10.1016/j.trac.2018.11.044

[24] Ashraf G, Asif M, Aziz A, et al. Facet-energy inspired metal oxide extended hexapods decorated with graphene quantum dots: Sensitive detection of bisphenol A in live cells. *Nanoscale.* 2020;12(16):9014–9023.

[25] Asif M, Aziz A, Azeem M, et al. A review on electrochemical biosensing platform based on layered double hydroxides for small molecule biomarkers determination. *Adv Colloid Interface Sci.* 2018;262:21–38. https://doi.org/10.1016/j.cis.2018.11.001

[26] An'amt MN, Yusoff N, Sagadevan S, Wahab YA, Johan MR. Recent progress in nitrates and nitrites sensor with graphene-based nanocomposites as electrocatalysts. *Trends Environ Anal Chem.* Published online 2022:e00162.

[27] Tian N, Zhou Z-Y, Sun S-G, Ding Y, Wang ZL. Synthesis of tetrahexahedral platinum nanocrystals with high-index facets and high electro-oxidation activity. *Science (80-).* 2007;316(5825):732–735.

[28] Yang Y, Mao Z, Huang W, et al. Redox enzyme-mimicking activities of CeO2 nanostructures: Intrinsic influence of exposed facets. *Sci Rep.* 2016;6(1):1–7.

[29] Song Y, Qu K, Zhao C, Ren J, Qu X. Graphene oxide: Intrinsic peroxidase catalytic activity and its application to glucose detection. *Adv Mater.* 2010;22(19):2206–2210.

[30] Li R, Zhen M, Guan M, et al. A novel glucose colorimetric sensor based on intrinsic peroxidase-like activity of C60-carboxyfullerenes. *Biosens Bioelectron.* 2013;47:502–507.

[31] Abid N, Khan AM, Shujait S, et al. Synthesis of nanomaterials using various top-down and bottom-up approaches, influencing factors, advantages, and disadvantages: A review. *Adv Colloid Interface Sci.* 2022;300:102597. https://doi.org/10.1016/j.cis.2021.102597

[32] Huot J, Ravnsbæk DB, Zhang J, Cuevas F, Latroche M, Jensen TR. Mechanochemical synthesis of hydrogen storage materials. *Prog Mater Sci.* 2013;58(1):30–75.

[33] Arole VM, Munde SV. Fabrication of nanomaterials by top-down and bottom-up approaches-an overview. *J Mater Sci.* 2014;1:89–93.

[34] Hong K, Lee TH, Suh JM, Yoon S-H, Jang HW. Perspectives and challenges in multilayer ceramic capacitors for next generation electronics. *J Mater Chem C.* 2019;7(32):9782–9802. doi:10.1039/C9TC02921D

[35] Hayashi H, Noguchi T, Islam NM, Hakuta Y, Imai Y, Ueno N. Hydrothermal synthesis of BaTiO3 nanoparticles using a supercritical continuous flow reaction system. *J Cryst Growth.* 2010;312(12):1968–1972. https://doi.org/10.1016/j.jcrysgro.2010.03.034

[36] Ismail RA, Zaidan SA, Kadhim RM. Preparation and characterization of aluminum oxide nanoparticles by laser ablation in liquid as passivating and anti-reflection coating for silicon photodiodes. *Appl Nanosci.* 2017;7(7):477–487. doi:10.1007/s13204-017-0580-0

[37] Marlinda AR, Pandikumar A, Yusoff N, Huang NM, Lim HN. Electrochemical sensing of nitrite using a glassy carbon electrode modified with reduced functionalized graphene oxide decorated with flower-like zinc oxide. *Microchim Acta.* 2015;182(5): 1113–1122.

[38] Nikam A V, Prasad BLV, Kulkarni AA. Wet chemical synthesis of metal oxide nanoparticles: A review. *CrystEngComm.* 2018;20(35):5091–5107. doi:10.1039/C8CE00487K

[39] Park J, Joo J, Kwon SG, Jang Y, Hyeon T. Synthesis of monodisperse spherical nanocrystals. *Angew Chemie Int Ed.* 2007;46(25):4630–4660. https://doi.org/10.1002/anie.200603148

[40] Vaseghi Z, Nematollahzadeh A. Nanomaterials: Types, synthesis, and characterization. *Green Synth Nanomater Bioenergy Appl.* Published online 2020:23–82.

[41] Tawade BV, Apata IE, Singh M, et al. Recent developments in the synthesis of chemically modified nanomaterials for use in dielectric and electronics applications. *Nanotechnology.* 2021;32(14):142004.

[42] Mahdavi R, Talesh SSA. Sol-gel synthesis, structural and enhanced photocatalytic performance of Al doped ZnO nanoparticles. *Adv Powder Technol.* 2017;28(5):1418–1425.

[43] Yeganeh M, Shahtahmasebi N, Kompany A, et al. The magnetic characterization of Fe doped TiO2 semiconducting oxide nanoparticles synthesized by sol—gel method. *Phys B Condens Matter.* 2017;511:89–98.

[44] Kafshgari LA, Ghorbani M, Azizi A. Synthesis and characterization of manganese ferrite nanostructure by co-precipitation, sol-gel, and hydrothermal methods. *Part Sci Technol.* Published online 2018.

[45] Hyun K, Han SW, Koh W-G, Kwon Y. Direct electrochemistry of glucose oxidase immobilized on carbon nanotube for improving glucose sensing. *Int J Hydrogen Energy.* 2015;40(5):2199–2206.

[46] Liu Y, Dolidze TD, Singhal S, Khoshtariya DE, Wei J. New evidence for a quasi-simultaneous proton-coupled two-electron transfer and direct wiring for glucose oxidase captured by the carbon nanotube—polymer matrix. *J Phys Chem C.* 2015;119(27):14900–14910.

[47] Wang C, Ding T, Sun Y, Zhou X, Liu Y, Yang Q. Ni 12 P 5 nanoparticles decorated on carbon nanotubes with enhanced electrocatalytic and lithium storage properties. *Nanoscale*. 2015;7(45):19241–19249.

[48] Pham X-H, Li CA, Han KN, et al. Electrochemical detection of nitrite using urchin-like palladium nanostructures on carbon nanotube thin film electrodes. *Sensors Actuators B Chem*. 2014;193:815–822.

[49] Regan F. Sensors | Overview☆. In: Worsfold P, Poole C, Townshend A, Miró M, eds. *E of AS* (Third Ed.). Oxford: Academic Press; 2019:172–178. https://doi.org/10.1016/B978-0-12-409547-2.14540-8

[50] Yan X, Li H, Su X. Review of optical sensors for pesticides. *TrAC Trends Anal Chem*. 2018;103:1–20. https://doi.org/10.1016/j.trac.2018.03.004

[51] Liu D, Chen W, Wei J, Li X, Wang Z, Jiang X. A highly sensitive, dual-readout assay based on gold nanoparticles for organophosphorus and carbamate pesticides. *Anal Chem*. 2012;84(9):4185–4191.

[52] Guo J, Li H, Xue M, et al. Highly sensitive detection of organophosphorus pesticides represented by methamidophos via inner filter effect of Au nanoparticles on the fluorescence of CdTe quantum dots. *Food Anal Methods*. 2014;7(6):1247–1255.

[53] Nasrin F, Chowdhury AD, Takemura K, et al. Single-step detection of norovirus tuning localized surface plasmon resonance-induced optical signal between gold nanoparticles and quantum dots. *Biosens Bioelectron*. 2018;122:16–24. https://doi.org/10.1016/j.bios.2018.09.024

[54] Long Q, Li H, Zhang Y, Yao S. Upconversion nanoparticle-based fluorescence resonance energy transfer assay for organophosphorus pesticides. *Biosens Bioelectron*. 2015;68:168–174.

[55] Angiola M, Rutherglen C, Galatsis K, Martucci A. Transparent carbon nanotube film as sensitive material for surface plasmon resonance based optical sensors. *Sensors Actuators B Chem*. 2016;236:1098–1103.

[56] Fang C, Dharmarajan R, Megharaj M, Naidu R. Gold nanoparticle-based optical sensors for selected anionic contaminants. *TrAC Trends Anal Chem*. 2017;86:143–154. https://doi.org/10.1016/j.trac.2016.10.008

[57] Amourizi F, Dashtian K, Ghaedi M. Polyvinylalcohol-citrate-stabilized gold nanoparticles supported congo red indicator as an optical sensor for selective colorimetric determination of Cr(III) ion. *Polyhedron*. 2020;176:114278. https://doi.org/10.1016/j.poly.2019.114278

[58] Wang X, Qin L, Lin M, Xing H, Wei H. Fluorescent graphitic carbon nitride-based nanozymes with peroxidase-like activities for ratiometric biosensing. *Anal Chem*. 2019;91(16):10648–10656.

[59] Wang W, Gunasekaran S. Nanozymes-based biosensors for food quality and safety. *TrAC trends Anal Chem*. 2020;126:115841.

[60] Gao L, Zhuang J, Nie L, et al. Intrinsic peroxidase-like activity of ferromagnetic nanoparticles. *Nat Nanotechnol*. 2007;2(9):577–583.

[61] Yao C, Wang J, Zheng A, Wu L, Zhang X, Liu X. A fluorescence sensing platform with the MnO2 nanosheets as an effective oxidant for glutathione detection. *Sensors Actuators B Chem*. 2017;252:30–36. https://doi.org/10.1016/j.snb.2017.05.136

[62] Liu B, Liu J. Sensors and biosensors based on metal oxide nanomaterials. *TrAC Trends Anal Chem*. 2019;121:115690.

[63] Svechkarev D, Mohs AM. Organic fluorescent dye-based nanomaterials: Advances in the rational design for imaging and sensing applications. *Curr Med Chem*. 2019;26(21):4042–4064.

[64] Kneipp J. Interrogating cells, tissues, and live animals with new generations of sur-face-enhanced raman scattering probes and labels. *ACS Nano.* 2017;11(2):1136–1141. doi:10.1021/acsnano.7b00152

[65] Yang G, Li L, Lee WB, Ng MC. Structure of graphene and its disorders: A review. *Sci Technol Adv Mater.* 2018;19(1):613–648. doi:10.1080/14686996.2018.1494493

[66] Liu C, Alwarappan S, Chen Z, Kong X, Li C-Z. Membraneless enzymatic biofuel cells based on graphene nanosheets. *Biosens Bioelectron.* 2010;25(7):1829–1833.

[67] Ye M, Dong Z, Hu C, et al. Uniquely arranged graphene-on-graphene structure as a binder-free anode for high-performance lithium-ion batteries. *Small.* 2014;10(24):5035–5041.

[68] Zhou G, Wang D-W, Li F, et al. Graphene-wrapped Fe3O4 anode material with improved reversible capacity and cyclic stability for lithium ion batteries. *Chem Mater.* 2010;22(18):5306–5313.

[69] Li J, Xu Z, Liu M, et al. Ag/N-doped reduced graphene oxide incorporated with molecularly imprinted polymer: An advanced electrochemical sensing platform for salbutamol determination. *Biosens Bioelectron.* 2017;90:210–216. https://doi.org/10.1016/j.bios.2016.11.016

[70] Marlinda AR, Sagadevan S, Yusoff N, et al. Gold nanorods-coated reduced graphene oxide as a modified electrode for the electrochemical sensory detection of NADH. *J Alloys Compd.* 2020;847:156552. https://doi.org/10.1016/j.jallcom.2020.156552

[71] Lieberzeit PA, Afzal A, Rehman A, Dickert FL. Nanoparticles for detecting pollut-ants and degradation processes with mass-sensitive sensors. *Sensors Actuators B Chem.* 2007;127(1):132–136. https://doi.org/10.1016/j.snb.2007.07.020

[72] Mecea VM. Is quartz crystal microbalance really a mass sensor? *Sensors Actuators A Phys.* 2006;128(2):270–277.

[73] Saiz PG, Gandia D, Lasheras A, et al. Enhanced mass sensitivity in novel magnetoelas-tic resonators geometries for advanced detection systems. *Sensors Actuators B Chem.* 2019;296:126612. https://doi.org/10.1016/j.snb.2019.05.089

[74] Khoshaman AH, Bahreyni B. Application of metal organic framework crystals for sens-ing of volatile organic gases. *Sensors Actuators B Chem.* 2012;162(1):114–119.

[75] Wang J, McIvor MJ, Elliott CT, Karoonuthaisiri N, Segatori L, Biswal SL. Rapid detec-tion of pathogenic bacteria and screening of phage-derived peptides using microcantile-vers. *Anal Chem.* 2014;86(3):1671–1678.

[76] Wang L, Cha X, Wu Y, et al. Superhydrophobic polymerized n-octadecylsilane surface for BTEX sensing and stable toluene/water selective detection based on QCM sensor. *ACS omega.* 2018;3(2):2437–2443.

[77] Chen JY, Penn LS, Xi J. Quartz crystal microbalance: Sensing cell-substrate adhesion and beyond. *Biosens Bioelectron.* 2018;99:593–602.

[78] Kaleli-Can G, Ozlu B, Özgüzar HF, et al. Natural melanin nanoparticle-decorated screen-printed carbon electrode: Performance test for amperometric determination of hexavalent chromium as model trace. *Electroanalysis.* 2020;32(8):1696–1706.

[79] Wang D, Chen C, Ke X, et al. Bioinspired near-infrared-excited sensing platform for in vitro antioxidant capacity assay based on upconversion nanoparticles and a dopamine: Melanin hybrid system. *ACS Appl Mater Interfaces.* 2015;7(5):3030–3040.

[80] Eom T, Woo K, Cho W, et al. Nanoarchitecturing of natural melanin nanospheres by layer-by-layer assembly: Macroscale anti-inflammatory conductive coatings with opto-electronic tunability. *Biomacromolecules.* 2017;18(6):1908–1917.

[81] Demir D, Gündoğdu S, Gundogdu S, Kılıç Ş, Kilic S, Kartallıoğlu T, Kartallioglu T, Alkan Y, Baysoy E, Kaleli Can GA. Comparison of different strategies for the modifica-tion of quartz tuning forks based mass sensitive sensors using natural melanin nanopar-ticles. *Akıllı Sist. ve Uygulamaları Derg.* 2021;4(2):128–132.

[82] Dhara K, Debiprosad RM. Review on nanomaterials-enabled electrochemical sensors for ascorbic acid detection. *Anal Biochem.* 2019;586:113415. https://doi.org/10.1016/j. ab.2019.113415

[83] Dong L, Feng Y, Wang L, Feng W. Azobenzene-based solar thermal fuels: Design, properties, and applications. *Chem Soc Rev.* 2018;47(19):7339–7368. doi:10.1039/ C8CS00470F

[84] Zhitomirsky D, Cho E, Grossman JC. Solid-state solar thermal fuels for heat release applications. *Adv Energy Mater.* 2016;6(6):1502006. https://doi.org/10.1002/ aenm.201502006

[85] Zhang Z, Qu J, Feng Y, Feng W. Assembly of graphene-aligned polymer composites for thermal conductive applications. *Compos Commun.* 2018;9:33–41. https://doi. org/10.1016/j.coco.2018.04.009

[86] Zhang F, Feng Y, Qin M, et al. Stress controllability in thermal and electrical conductivity of 3D elastic graphene-crosslinked carbon nanotube sponge/polyimide nanocomposite. *Adv Funct Mater.* 2019;29(25):1901383. https://doi.org/10.1002/adfm.201901383

[87] Chen S, Feng Y, Qin M, Ji T, Feng W. Improving thermal conductivity in the through-thickness direction of carbon fibre/SiC composites by growing vertically aligned carbon nanotubes. *Carbon N Y.* 2017;116:84–93. https://doi.org/10.1016/j.carbon.2017.01.103

[88] Yao F, Pham DT, Lee YH. Carbon-based materials for lithium-ion batteries, electrochemical capacitors, and their hybrid devices. *ChemSusChem.* 2015;8(14):2284–2311. https://doi.org/10.1002/cssc.201403490

[89] Bulbula ST, Lu Y, Dong Y, Yang X-Y. Hierarchically porous graphene for batteries and supercapacitors. *New J Chem.* 2018;42(8):5634–5655. doi:10.1039/C8NJ00652K

[90] Mukherjee S, Ren Z, Singh G. Beyond graphene anode materials for emerging metal ion batteries and supercapacitors. *Nano-Micro Lett.* 2018;10(4):1–27.

[91] Huang Z, Guo H, Zhang C. Assembly of 2D graphene sheets and 3D carbon nanospheres into flexible composite electrodes for high-performance supercapacitors. *Compos Commun.* 2019;12:117–122. https://doi.org/10.1016/j.coco.2019.01.010

[92] Zhao F, Li Y, Feng W. Recent advances in applying vulcanization/inverse vulcanization methods to achieve high-performance sulfur-containing polymer cathode materials for Li—S batteries. *Small Methods.* 2018;2(11):1800156. https://doi.org/10.1002/ smtd.201800156

[93] Liang X, Cheng Q. Synergistic reinforcing effect from graphene and carbon nanotubes. *Compos Commun.* 2018;10:122–128. https://doi.org/10.1016/j.coco.2018.09.002

[94] Sagadevan S, Johan MR, Marlinda AR, et al. Chapter one—Background of energy storage. In: Arshid N, Khalid M, Grace AN, eds. *A in S and S.* Singapore: Elsevier; 2021:1–26. https://doi.org/10.1016/B978-0-12-819897-1.00003-3

[95] Sagadevan S, Chowdhury ZZ, Johan MR Bin, Rafique RF, Aziz FA. One pot synthesis of hybrid ZnS—Graphene nanocomposite with enhanced photocatalytic activities using hydrothermal approach. *J Mater Sci Mater Electron.* 2018;29(11):9099–9107.

[96] Sagadevan S, Marlinda AR, Johan MR, et al. Reduced graphene/nanostructured cobalt oxide nanocomposite for enhanced electrochemical performance of supercapacitor applications. *J Colloid Interface Sci.* 2020;558:68–77.

9 Functional Nanomaterials for Gas Sensors

Sugandha Gupta, Ravish Kumar Uppadhayay, and Ajeet Singh

CONTENTS

9.1 INTRODUCTION

A sensor is an instrument that detects variations or changes in a measured quantity and outputs the result. A mercury-in-glass thermometer, for example, turns observed temperature into liquid expansion and contraction, which can then be read on a calibrated glass tube. Similarly, a gas sensor detects concentration of toxic gases and impurities in the environment which ultimately results in the form of optical and electrical signals as an output. The environment in which we breathe, the workplace in which we work, the home in which we live—all of them are surrounded by poisonous gases which need to be monitored or sensed for us to have sustained healthy lives. Inevitably, the occasional escape of hazardous gases still occurs, resulting in major threat to the lives of living beings and ultimately to the green environment. It is well known that infants and young children are more prone to health risks due to toxic gas pollution, as they have large surface area per unit body weight, and hence their inhalation rate is higher as compared to adults. Every year, the World Health Organization (WHO) releases its mortality rate due to toxic gas pollution. A toxic

DOI: 10.1201/9781003263852-9

gas can cause significant disease or death by harming living tissue and the central nervous system. Some hazardous gases aren't visible, can't be smelled, and/or don't have an immediate effect, but they can kill humans. As a result, relying solely on human sense without the use of sensing devices may not be appropriate [1].

According to WHO, regarding the COVID-19 pandemic situation, the global gas sensors business is predicted to increase at a CAGR of 7.0 percent from 2021–2026, from USD 1.12 billion in 2021 to USD 1.51 billion in 2026. Raised requirements for gas sensor devices in vital industries, design and execution of several health and protection guidelines globally, interconnection of gas sensor devices in HVAC (heating, ventilation, and air conditioning) systems and ambient air monitors, vastly increased pollution level of air hence the necessity for environmental monitoring in smart urban, and rising demands for gas sensor devices in healthcare manufacturing are all factors driving this market's growth.

9.2   NEED FOR GAS SENSORS

Gas sensors were first used in coal mines, where precise continuous monitoring of hazardous gases was necessary. Gas sensors began to arise in industrial applications, severe environmental measuring units, and healthcare of living beings shortly after that. The first gas sensor was commercialized in 1923. It was comprised of hot platinum wire working at several hundred degrees utilizing the catalyst to detect combustible gases in the atmosphere. The operating process is simply to measure resistance change resulting from temperature rise due to toxic gases. The signal output was only few millivolts at 1,000 ppm of isobutene. In 1960, widespread demand for gas sensors arose in Japan owing to gas bottle explosion of LPG (liquefied petroleum gas) which resulted in development of semiconductor gas sensors using Zinc oxide by Sieyama et al [2]. Later Taguchi invented gas sensor using stannic oxide and patented it as detector for toxic gases [3]. Earlier gas sensors were limited to professional users; however, with technological advancement and commercialization, gas sensors were implemented into many new domains and applications.

One of the most imperative applications of gas sensor (Figure 9.1) devices is the investigation of organic vapors such as toluene, methanol, and benzene for research laboratory and industrial safety [4] and breath analysis for traffic protection [5] and disease diagnosis [6] An automated nose, which is made on displays of gas sensors, is a relatively recent manifestation of gas sensing that has drawn curiosity due to its greater analytical capability [7]. Among other things, it's used to test the smell of food, perfumes, and synthetic aromas. The following subsections describe the real-time industrial applications contributing to the boon created by gas sensors.

9.2.1   CRITICAL INDUSTRIES

Gas sensors are used in critical industries (Figure 9.2). like oil and gas, chemicals, mining, and efficiency in detecting and monitoring the presence of flammable and toxic gases. These vital businesses emit a huge number of gases into the atmosphere, including carbon monoxide, carbon dioxide NH_4, ammonia, hydrogen sulfide (H_2S), and

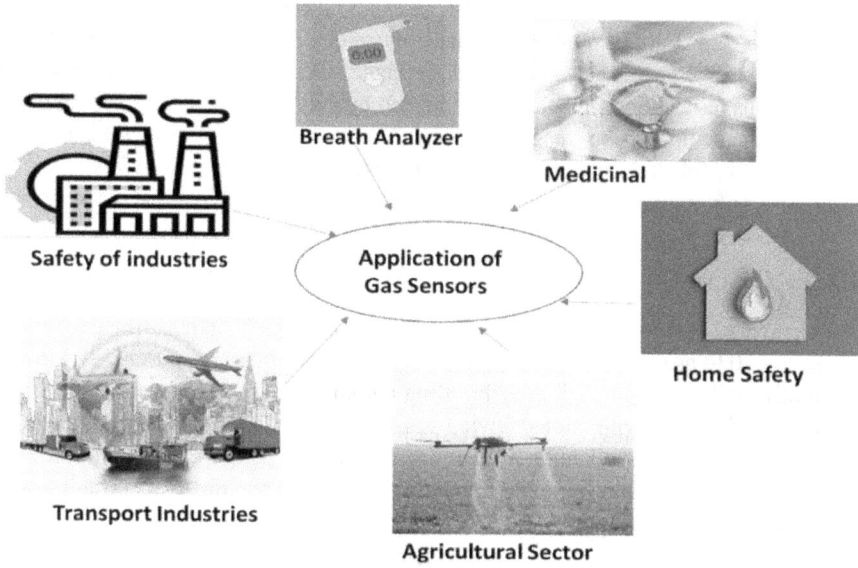

FIGURE 9.1 Various applications of gas sensors.

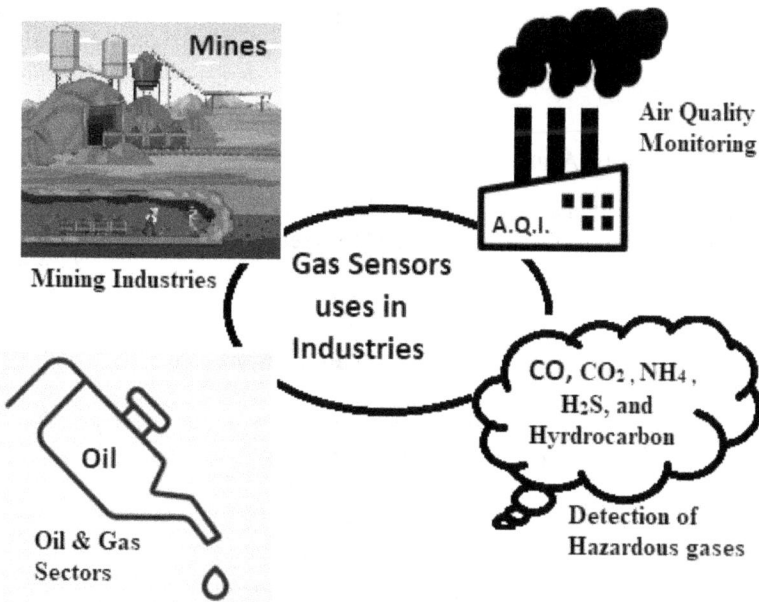

FIGURE 9.2 Various use of gas sensors in critical industries.

hydrocarbons. Substantial releases of such gases into the atmosphere can be hazardous to human health. Furthermore, these major industries and domestic equipment may produce explosive gases such as methane, propane, and butane, which could result in fires. Various restrictions are being implemented by many regulatory agencies in order to protect the ecology from dangerous gases. This necessitates the detection, sensing, and monitoring of these gases, as well as the implementation of remedial actions to ensure that only a small volume of these gases is released directly into the atmosphere.

9.2.2 Oxygen Gas Sensors

The oxygen gas sensor is expected to account for the largest share of the gas sensor market by 2026 as oxygen detectors, analyzers, and monitors are widely used in the automotive industry, smart cities & building automation, food & beverage industry, and other industrial applications. Owing to COVID-19 pandemic situation, medical equipment is another application area of oxygen sensors, as they are used in incubators, ventilators, hypoxic hypoxia situations, anesthesia monitors, respirators, and oxygen concentrators.

9.2.3 Automotive and Transport Industry

The gas sensor market is projected to continue to be dominated by the automotive and transportation sectors. Pollutant gases enter the automobile cabin via ventilation systems, and pollutants enter through window apertures, while harmful gases enter through misdirected exhaust fumes due to a lack of fresh air intake. Gas sensor systems must be included in all vehicle and transportation OEMs' portfolios due to stringent environmental and safety laws. The diagrammatic representation of the inter-relationships of the gas sensor ecosystem and its market dependency is shown in Figure 9.1.

9.2.4 Household Applications

Gas sensors are used in a variety of everyday tasks (Figure 9.3): CO_2 sensors, intelligent refrigerators or ovens, fire alarms, and CO_2 sensors are used to detect dangerous

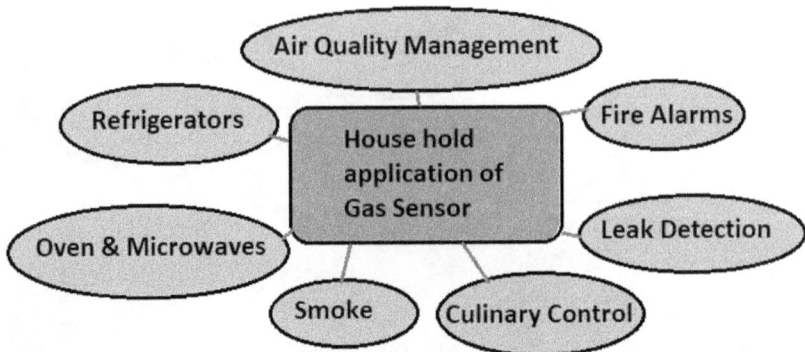

FIGURE 9.3 Various household applications of gas sensors.

TABLE 9.1
Classification of Conventional Gas Sensors

Category of Gas Sensors	Conducting Principles
Electro-chemical	Variations in current (I), voltage (v), capacitance, or impedance
Electrical	• Electrolytic conductivity • Organic conductivity • Metal oxide conductivity
Sensitive to mass	• Variations in weight amplitude, phase, frequency, size, and shape • Quartz crystal microbalance • Surface acoustic wave propagation • Cantilever
Magnetic	Variations in paramagnetic gas characteristics
Optical Devices	Variations in light intensity, color, emission spectra: • Absorbance • Luminescence • Reflectance • Refractive index • Light scattering (Raman scattering, plasmon resonance)
Thermometric	Heat effects of specific chemical reaction. Changes in temperature, heat flow, heat content: • Thermoelectric • Pyroelectric • Thermal conductivity

Source: (Hulanicki et al. [1991]. Published by International Union of Pure and Applied Chemistry)

gases or smoke in homes as a result of events such as fires and explosions (CO, CH_4, H_2 sensors); air quality management; air purifiers; culinary control; natural gas heating; and leak detection (humidity, odor and H_2 sensors) [8]. An array of gas sensors, in other words, are designated as electronic noses which are nowadays the most prospective candidates for a sustained and healthy lifestyle.

9.3 CATEGORIES OF GAS SENSORS

Conventionally, gas sensor devices were classified on the basis of different materials and operating principles [9]. Figaro, the world's leading company designing sensors since 1991, dominates the market in wide variety of sensors from explosive gas sensor applications to air quality monitoring detection. Table 9.1 illustrates the IUPAC Analytical Chemistry Division's classification of conventional gas sensors, which was proposed in 1991.

9.4 TECHNOLOGICAL DEVELOPMENT IN GAS SENSORS

Sensor technology is also one of the areas in which nanotechnology has made great impact. Nowadays, gas sensors are categorized in terms of surface chemistry because the gas detection process involves reactions with gas molecules at the surface. Hence, parameters such as reactivity, sensitivity, and selectivity are closely related to surface-to-volume ratio. Therefore, selectivity of sensors is significantly affected by the size of surface features. The most essential criteria in determining a gas sensor's performance are its limitation of detection (LOD) and selectivity. The gas sensor device having the lowest LOD detects the smallest amounts of analyte, while a highly selective gas sensor senses very small volumes of the target gas in a combination of numerous additional gases. Low power consumption, fast response and recovery time, stability (low drift rate), and sensitivity, i.e., the slope of the sensor output versus the analyte concentration, are other important sensor performance characteristics [10].

In an attempt to improve the sensitivity of gas sensors, the status quo of sensors has made a significant drift towards nanotechnology. Compared to conventional technology, nanotechnology-fabricated sensors offer a promising future in the arena of sensing technology. This is due to the fact that surface area expands exponentially as the size reduces to micro or nano range. According to the Shottky effect, as the particle size is reduced to smaller than a certain dimension extent, sensitivity increases beyond the boundary effects. Heat losses from the heater owing to convection, conduction, or radiation should be minimized, and the power intake of an ideal gas sensor design should be as low as possible. To minimize the energy consumption, the capacity of both the heater and the sensor device must be reduced, and the design must be optimized for heating uniformity. Miniaturization has an impact on both the heater and the sensor material.

Nanosized gas sensor elements may be faster, use less power, have a lower detection limit, function at lower temperatures, do not require costly catalysts, are extra resistant to heat shock, and may surprisingly be less expensive than their macro-counterparts.

There have been significant advancements in two crucial areas during the last two decades that may make this potential a certainty. The initial step is to develop a range of high-performance nanostructured gas sensors (e.g., nanometal oxide semiconductors [NMOSs]), which are the most often utilized as gas sensing resources; and the second step is the development of very low power loss miniature heater elements. Due to a wide range of potential applications, advanced nano- or micro–nano gas sensor devices have received much consideration. This is due to the fact that microstructure is highly defined in the nano domain, as grain boundaries fade away to produce more stable neck-like structures. Because the adsorbed oxygen has the potential to lock up all of the electrons from these narrow necks, the most sensitive areas of sensing materials are often found here, resulting in entirely electron-depleted highly resistive regions. It's vital to remember that porosity, grain size, and film thickness all affect the LOD and sensitivity of gas sensors [11]. This is because the high aspect ratio of nanostructured materials enormously strengthens the chemical reactivity of sensing material.

9.5 TYPES OF NANOSTRUCTURED GAS SENSORS

As previously stated, the exponential increase in the creation of nanostructured gas sensor devices has fueled the introduction of well-known materials such as

CNTs and graphene into the gas sensors field, with the expectation of accompanying a new age of next-generation autonomous sensing technology, because carbon nanotubes combine outstanding finding sensitivity with fascinating transduction characteristics in a single layer [12]. Carbon nanomaterials with high-quality crystal lattices, such as CNTs or graphene, exhibit excellent carrier motion (e.g., ballistic charge carriage) and little noise. While the first two properties are critical for transmission, the facet chemistry of low-dimensional, high-quality crystal formations is theoretically considerably simpler and manageable than that of polycrystalline materials. Sole crystalline nanostructures are thus ideal perfect ingredients for computational chemistry research, allowing scientists to understand more about their gas sensing operations. Carbon nanoparticles' sensitivity and selectivity can be improved by creating flaws and grafting functional groups to their surfaces in a regulated manner using a variety of methods. Carbon nanomaterials, which can be made in a variety of ways, are frequently amenable to producing devices using traditional processes, like lithography, which are cost effective. Furthermore, because of their mechanical qualities, they are well suited for use in flexible electrical devices. They have a high sensitivity-to-cost ratio even when operated at ambient temperature. Because of their low power requirements, they are ideal candidates for remote control. Electronic signal transduction of chemical environmental analytes (e.g., a resistance change) has several advantages over optical approaches, including reduced cost, enhanced equipment easiness, advanced sample data, and better portability [13].

9.5.1 Gas Sensors Made of Carbon Black

Carbon black particles are placed in insulating organic polymers to create individual sensor elements. The electrical conductivity of the sheets is provided by the carbon black, while the various organic polymers provide chemical variety among sensor parts. When exposed to a gas or vapor, the polymer swells, increasing the film's resistance and giving a very easy technique for knowing the existence of a gas or vapor [14–17]. Lewis and his colleagues developed and defined a set of components that react to a variety of vapors (or compound mixes of vapors) in a distinct, recognizable manner [18,19]. On each sensor element, they used carbon black that was readily available and varied polymer compositions. The electrical resistance signals output by the 17-element array were further processed using standard chemometrics. The goal was to determine the existence of various organic solvent vapors. This method may readily be implemented into software or hardware classification processors, enabling the fusion of sensing and analytical capabilities together into small, moderate-power, low-cost, and convenient vapor analyzers. The aging of the polymer matrix and the rearranging of carbon black particles within the polymer, resulting in variations in the percolation routes, can cause baseline and response drift in carbon black—polymer composite sensors. These rearrangements develop over time as a result of the polymer matrix's multiple swelling/shrinking processes as a result of frequent finding and repossession cycles [20].

9.5.2 CARBON NANOFIBER GAS SENSORS

Zhang and his colleagues suggested a method for dispersing carbon nanofibers inside a polymer matrix [21,22] as a solution to address the unpredictability of polymer composites made up of carbon black. When the composite absorbs vapor, carbon black nanoparticles have a tendency to reaggregate, lowering the matrix viscosity and increasing the composite's volume. Once vapor is absorbed and desorbed, however, by reducing movement within the polymer composites, distributing carbon nanofibers throughout the polymer improves vapor sensing stability. As a result, after absorbed vapor has been desorbed from the matrix, the original electrical percolation routes in these composites are retained. Metal oxide nanonodules on the facet of carbon nano fibers improve facet area and attraction for DMMP (Dimethyl methylphosphonate) vapors, resulting in high sensitivity of carbon nanofiber gas sensors [23].

9.5.3 CARBON NANOTUBE GAS SENSORS

The electrical characteristics of carbon nanotubes have been discovered to be particularly sensitive to the chemical environment in which they are situated. Because of their chemical sensitivity, they are great candidates for use in chemical sensor design. The first carbon nanotube gas sensors were published in the history of science by the Dai [44] and Zett [45], who used semiconductor single walled CNTs on SiO_2 substrates. In the presence of nitrogen dioxide and ammonia vapors, the conductivity of this sensor was measured. Later that year, Zett and Llobet reported on SWNT's exceptional oxygen sensitivity [24,25]. When defects and residual contaminants were present, Goldoni et al. investigated the gas sensing characteristics of SWNT [26]. In the presence of oxygen, nitrogen, carbon monoxide, moisture, nitrogen dioxide, and sulfur oxide, the photoemission spectra of SWNT mats were studied. Before and after an ultra-high vacuum heat treatment, the ammonia levels were altered, according to this study. Much research has shown that by functionalizing the sidewalls of carbon nanotubes, it is possible to develop a better chemical interaction between a certain chemical species and the nanotube, and that the selectivity of the adsorption process may be improved [27,28].

9.5.4 GRAPHENE GAS SENSORS

Graphene constituents have sparked a lot of curiosity in chemical species sensing. The current breakthrough in the area of graphene-based gas sensor devices was studied, highlighting current developments and exploring future difficulties and opportunities in this fascinating subject of graphene-based gas sensors [29]. In 2007, Novoselov's group published the first research of graphene's use in gas sensing [30], which established the electrical detection of gas molecules adsorbed on multi-terminal Hall bars. These were made from single-layer or few-layer graphene that had been mechanically split from graphite using traditional lithographic procedures. The devices showed concentration-dependent variations in resistivity after adsorption of gases. while heating at 150°C in a vacuum, the baseline might be restored. For different gases, the amount of the changes in resistivity varied, and whether the gas was an electron acceptor was indicated by the symbol of the variation (e.g., nitrogen dioxide, moisture) or an electron donor (e.g., oxygen, carbon monoxide, ethanol, and ammonia). Thanks to the low sound levels in their graphene equipment, on exceedingly dilute nitrogen dioxide samples, Novoselov and colleagues were able to conduct long-term testing, finding step-like variations in resistivity during adsorption and desorption. They took the results of statistical analyses of these quantized data as confirmation that individual gas molecule adsorption or desorption had been discovered. Variations in carrier density, mobility, or both appear to be the origin of the reported experimental results, given that conductivity is proportional to the product of carrier density and mobility. However, the magnitude of these two elements' contributions is still being studied. Soon after the development of graphene gas sensors, many computational chemistry studies were carried out to theoretically investigate the adsorption of various molecules on graphene (moisture, nitrogen dioxide, nitric oxide, ammonia, carbon monoxide, carbon dioxide, oxygen, and nitrogen) [31,32]. Johnson and colleagues [33] revealed that making gas detecting devices with traditional nanolithography leaves resist remains on the graphene facet. This pollution chemically dopes graphene, increases carrier dispersion, and functions as an absorbent layer that concentrates gas molecules on graphene's surface, improving gas responsiveness. Reduced graphene oxide has also been suggested as a potential material for gas sensor development. Graphene oxide is significantly easier to work with than graphene, and it allows one to customize the number of functional groups by adjusting the degree of reduction. Reduced graphene oxide was demonstrated as the active material for high-sensitivity gas sensors by Robinson et al. [34].

9.5.5 0-D, 1-D, 2-D NANOMATERIALS GAS SENSORS: GENERAL ASPECTS

Low-dimensional (0D, 1D, and 2D) nanomaterials (Figure 9.4) [35–45] exhibit higher gas sensing abilities to bulk material due to their high surface-to-volume ratio (S/V), ample active locations, and quantum effect in specific proportions. While none of the low-dimensional nanomaterials had a significant advantage in any of the three decisive parameters for gas sensing act, Because 0D nanomaterials have more active sites and a greater S/V, they are better at receptor function and utility, but their transducer function is ineffective [46]. In contrast, 1D or 2D nanomaterials have a high competence in electrical signal transduction and a significant benefit in micro- or nano-devices due to their comparatively large scale in a specific dimension, which can bridge two electrodes of resistive and FET devices [47–50]. One may take advantage of the benefit of all of the features of nanocomposites by functionalizing 0D nanomaterials on 1D or 2D backbones [51–53]. Furthermore, aggregation of 0D nanoparticles can be avoided through the help of 1D or 2D nanomaterials [54,55]. Some 0D nanomaterials or molecules, such as noble metal NPs and iron porphyrin [56,57], are not ideal for sensor fabrication, but they can afford an active site to selectively catalyze or absorb specific gas molecules. Selectivity can be achieved by placing them on 1D or 2D semiconductors [58–59]. Finally, heterojunction [60–64] between two material systems improves gas sensing capability.

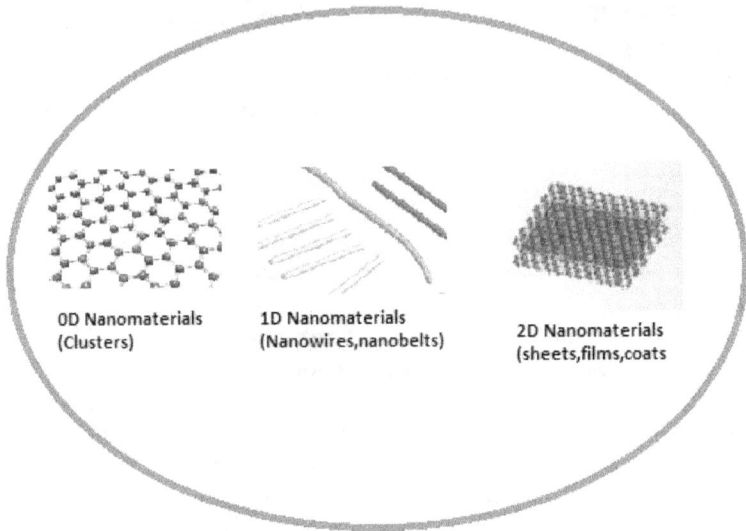

0D Nanomaterials   1D Nanomaterials   2D Nanomaterials
(Clusters)   (Nanowires,nanobelts)   (sheets,films,coats

FIGURE 9.4   Representation of 0D, 1D, and 2D nanomaterials.

9.6   CONCLUSION

The sensing performance of materials is highly dependent on their microstructures, necessitating the development of materials processing procedures to get the appropriate microstructures and morphologies. Understanding how materials react at the nanoscale while working and how controllably they are created is critical for future technological applications. Understanding how materials react at the nanoscale during working and how controllably they are created will be critical in future technological applications. Any physical changes inside the structure are inextricably linked to intrinsic chemical changes. As a result, researchers are constantly developing new fabrication procedures. Selectivity, reversibility, rapid responsibility, high sensitivity, and durability are just a few of the requirements and characteristics of an excellent sensor surface. It must also be non-contaminating and non-poisoning, have a simple procedure, be small enough to be portable, have a simple fabrication process with low noise and low temperature sensitivity, and most importantly, have low production costs.

Metal oxide semiconductor sensors, specifically synthesized from ZnO, have been the preferred candidate for gas sensors for many years. However, it is well known that traditional sensors have some advantages as well as disadvantages, such as low selectivity, poor sensitivity, and extra limitation of detection (LOD). The new class of sensors, Nanoscale materials-based sensors, are predicted to have several advantages over traditional sensors. Carbon nanomaterials, for example, have a number of features that make them suitable for use as sensors. They can be employed alone or in combination with other materials. As a result, one of the future directions for sensing research will be toward those working with carbon nanomaterials. The reason for this could be that these nanomaterials are extremely sensitive to variations in their chemical environment. This could be due to the electrical formation of the molecules that interact. Because of their sensitivity, they are good materials for gas sensor effectiveness. The use of carbon nanotubes as gas sensor device is a specific example (described previously in the Section 9.5.3) of the unique application of carbon nanomaterials in gas sensing. Adsorbed oxidative gas molecules, such as NOx, are said to induce the Fermi level of a nanotubes made up of carbon as a gas sensor to shift closer to the valence band. Adsorbed oxidative gas molecules, such as NOx, cause the Fermi level to shift closer to the nanotube's valence band, according to the theory. Hole transport in nanotubes improves as a result of this migration. As a result, the resistance of nanotubes has been reduced. Carbon nanomaterials are becoming more popular as sensors, according to this study, not only because of their electrical, optical, and mechanical capabilities, but also because of their high sensitivity, low power consumption, and low cost. However, due to difficulties in mass producing CNTs with desirable properties, commercial manufacture of CNTs and their construction as gas sensors remains a challenge. Several research projects are being conducted to identify cost-effective, scalable manufacturing procedures that preserve the key properties of such resources, ensuring a future with a healthy and sustainable lifestyle.

REFERENCES

[1] Singh, T. & Bonne, U. (2017). Gas sensors. In *Reference Module in Materials Science and Materials Engineering*, Elsevier, Amsterdam, The Netherlands, pp. 1–54.

[2] Sieyama, T. & Kato, A. (1962). Temperature effects on gas sensing properties of electrodeposited chlorine doped and undoped n-type cuprous oxide thin films. *Anal Chem.*, 34, 1504.

[3] Taguchi, N. (1962, Oct.). Published patent applications in Japan S37–47677.

[4] Wang, X., Carey, W. P. & Yee, S. S. (1995). Monolithic thin film metal oxide gas sensor arrays with application to monitoring of organic vapors. *Sens. Actuat. B Chem.*, 28, 63–70, doi: 10.1016/0925–4005(94)01531-L

[5] Mitsubayashi, K., Matsunaga, H., Nishio, G., Ogawa, M. & Saito, H. (2004). Biochemical gassensor (Bio-sniffer) for breath analysis after drinking. In SICE Annual Conf, Sapporo, Japan, 4–6.

[6] McEntegart, C. M., Penrose, W. R., Strathmann, S. & Stetter, J. R. (2000). Detection and discrimination of coliform bacteria with gas sensor arrays. *Sens. Actuat. B Chem.*, 70, 170–176, doi: 10.1016/S0925–4005(00)00561-X

[7] Arshak, K., Moore, E., Lyons, G. M., Harris, J. & Clifford, S. (2004). A review of gas sensors employed in electronic nose applications. *Sens. Rev.*, 24, 181–198, doi: 10.1108/02602280410525977

[8] Korotcenkov, G. (2013). *Handbook of Gas Sensor Materials: Properties, Advantage and Shortcomings for Applications*, Vol. 1, Springer, Berlin, Germany.

[9] Korotcenkov, G. (2014). *Handbook of Gas Sensor Materials: Properties, Advantage and Shortcomings for Applications: New Trends in Technologies*, Vol. 2, Springer, Berlin, Germany.

[10] Sharma, S. & Madou, M. (2012). A new approach to gas sensing with nanotechnology. *Phil. Trans. R. Soc. A*, 370, 2448–2473, doi: 10.1098/rsta.2011.0506.

[11] Lee, S. W., Tsai, P. P. & Chen, H. (2000). Comparison study of SnO2 thin and thick film gas sensors. *Sens. Actuat. B Chem.*, 67, 122–127, doi: 10.1016/S0925–4005(00)00390–7

[12] Roberts, M. E., LeMieux, M. C. & Bao, Z. (2009). Sorted and aligned single-walled carbon nanotube networks for transistor-based aqueous chemical sensors. *ACS Nano*, 3, 3287–3293.

[13] Llobet, E. (2012). Gas sensors using carbon nanomaterials: A review. *Sens. Actuators B: Chem.*, http://dx.doi.org/10.1016/j.snb.2012.11.014

[14] Norman, R.H. (1970). *Conductive Rubbers and Plastics*, Elsevier, Amsterdam.

[15] Lundberg, B. & Sundqvist, B. (1986). Resistivity of a composite conducting polymer as a function of temperature, pressure, and environment: Applications as a pressure and gas concentration transducer. *Journal of Applied Physics*, 60, 1074.

[16] Ruschau, G. R., Newnham, R. E., Runt, J. & Smith, B. E. (1989). 0–3 ceramic/polymer composite chemical sensors. *Sensors and Actuators*, 20, 269–275.

[17] Talik, P., Zabkowskawaclawek, M. & Waclawek, W. (1992). Sensing properties of the CB—PCV composites for chlorinated hydrocarbon vapours. *Journal of Materials Science*, 27, 6807.

[18] Lonergan, M. C., Severin, E. J., Doleman, B. J., Beaber, S. A., Grubbs, R. H. & Lewis, N. S. (1996). Array-based vapor sensing using chemically sensitive, carbon black-polymer resistors. *Chemistry of Materials*, 8, 2298–2312.

[19] Doleman, B. J., Lonergan, M. C., Severin, E. J., Vaid, T. P. & Lewis, N. S. (1998). Carbonblack-polymer composite vapor detectors. *Analytical Chemistry*, 70, 4177–4190.

[20] Llobet, E. (2012). Gas sensors using carbon nanomaterials: A review. *Sens. Actuators B: Chem.*, http://dx.doi.org/10.1016/j.snb.2012.11.014

[21] Llobet, E. (2012). Gas sensors using carbon nanomaterials: A review. *Sens. Actuators B: Chem.*, http://dx.doi.org/10.1016/j.snb.2012.11.014

[22] Zhang, B., Fu, R., Zhang, M., Dong, X., Wang, L. & Pittman, C. U. (2006). Gas sensitive vapor grown carbon nanofiber/polystyrene sensors. *Materials Research Bulletin*, 41, 553–562.

[23] Lee, J. S., Kwon, O. S., Park, S. J., Park, E. Y., You, S. A., Yoon, H. & Jang, J. (2011). Fabrication of ultrafine metal-oxide-decorated carbon nanofibres for DMMP sensor application. *ACS Nano*, 10, 7992–8001.

[24] Llobet, E. (2012). Gas sensors using carbon nanomaterials: A review. *Sens. Actuators B: Chem.*, http://dx.doi.org/10.1016/j.snb.2012.11.014

[25] Zett, A., Collins, P. G., Bradley, K. & Ishigami, M. (2000). Extreme oxygen sensitivity of electronic properties of carbon nanotubes. *Science*, 287, 1801–1804.

[26] Goldoni, A., Larciprete, R., Petaccia, L. & Lizzit, S. (2003). Single-wall carbon nanotube interaction with gases: Sample contaminants and environmental monitoring. *Journal of the American Chemical Society*, 125, 11329–11333.

[27] Kauffman, D. R. & Star, A. (2008). Carbon nanotube gas and vapor sensors. *Angewandte Chemie International Edition*, 47, 6550–6570.

[28] Bondavalli, P., Legagneux, P. & Pribat, D. (2009). Carbon nanotubes based transistors as gas sensors: State of the art and critical review. *Sensors and Actuators B: Chemical*, 140, 304–318.

[29] Varghese, S. S., Lonkar, S., Singh, K. K., Swaminathan, S. & Abdala, A. (2015). The first study of graphene use in gas sensing was reported in 2007 by Novoselov's group. *Sens. Actuators, B*, 218, 160–183.

[30] Schedin, F., Geim, A. K., Morozov, S. V., Hill, E. W., Blake, P., Katsnelson, M. I. & Novoselov, K. S. (2007). Detection of individual gas molecules adsorbed on graphene. *Nature Materials*, 6, 652–655.

[31] Llobet, E. (2012). Gas sensors using carbon nanomaterials: A review. *Sens. Actuators B: Chem.*, http://dx.doi.org/10.1016/j.snb.2012.11.014

[32] Leenaerts, O., Partoens, B. & Peeters, F. M. (2008). Adsorption of H2O, NH3, CO, NO2 and NO on graphene: A first-principles study. *Physical Review B*, 77, 125416.

[33] Huang, B., Li, Z., Liu, Z., Zhou, G., Hao, S., Wu, J., Gu, B.-L. & Duan, W. (2008). Adsorption of gas molecules on graphene nanoribbons and its implication for nanoscale molecule sensor. *Journal of Physical Chemistry C*, 112, 13442–13446.

[34] Robinson, J. T., Perkins, F. K., Snow, E. S., Wei, Z. & Sheehan, P. E. (2008). Reduced graphene oxide molecular sensors. *Nano Letters*, 8, 3137–3140.

[35] Tingqiang, Y., Yueli, L., Huide, W., Wang, Y., Duo, B., Zhang, Y., Yanqi, G., Han, Z. & Wen, C. (2020). Recent advances in 0D nanostructure-functionalized low-dimensional nanomaterials for chemiresistive gas sensors. *J. Mater. Chem. C*, 8, 7272–7299.

[36] Tao, W., Ji, X., Xu, X., Islam, M. A., Li, Z., Chen, S., Saw, P. E., Zhang, H., Bharwani, Z., Guo, Z., Shi J. & Farokhzad, O. C. (2017). Antimonene quantum dots: Synthesis and application as near-infrared photothermal agents for effective cancer therapy. *Angewandte Chemie-International Edition*, 56, 11896–11900.

[37] Xu, Y., Wang, Z., Guo, Z., Huang, H., Xiao, Q., Zhang, H. & Yu, X.-F. (2016). Solvothermal synthesis and ultrafast photonics of black phosphorus quantum dots. *Advanced Optical Materials*, 4, 1223–1229.

[38] Sun, Z., Zhao, Y., Li, Z., Cui, H., Zhou, Y., Li, W., Tao, W., Zhang, H., Wang, H., Chu, P. K. & Yu, X.-F. (2017). TiL₄-coordinated black phosphorus quantum dots as an efficient contrast agent for in vivo photoacoustic imaging of cancer. *Small*, 13, 1602896.

[39] Li, J. & Zheng, G. (2017). One-dimensional earth-abundant nanomaterials for water-splitting electrocatalysts. *Advanced Science*, 4, 1600380.

[40] Gong, S. & Cheng, W. (2017). One-dimensional nanomaterials for soft electronics. *Advanced Electronic Materials*, 3, 1600314.

[41] Lu, J., Zhang, K., Liu, X. F., Zhang, H., Sum, T. C., Castro Neto, A. H. & Loh, K. P. (2013). Order-disorder transition in a two-dimensional boron-carbon-nitride alloy. *Nature Communications*, 4, 2681.

[42] Sang, D. K., Wang, H., Qiu, M., Cao, R., Guo, Z., Zhao, J., Li, Y., Xiao, Q., Fan, D. & Zhang, H. (2019). Two dimensional β-InSe with layer-dependent properties: Band alignment, work function and optical properties. *Nanomaterials*, 9, 82.

[43] Zhang, M., Wu, Q., Zhang, F., Chen, L., Jin, X., Hu, Y., Zheng, Z. & Zhang, H. (2019). 2D black phosphorus saturable absorbers for ultrafast photonics. *Advanced Optical Materials*, 7, 1800224.

[44] Yang, W., Gan, L., Li, H. & Zhai, T. (2016). Two-dimensional layered nanomaterials for gas-sensing applications. *Inorg. Chem. Front.*, 3, 4.

[45] Yoo, R., Cho, S., Song, M.-J. & Lee, W. (2015). Highly sensitive gas sensor based on Al-doped ZnO nanoparticles for detection of dimethyl methylphosphonate as a chemical warfare agent simulant. *Sensors Actuators B: Chem.*, 221, 217–223.

[46] Yuan, Z., Zhou, C., Tian, Y., Shu, Y., Messier, J., Wang, J. C., van de Burgt, L. J., Kountouriotis, K., Xin, Y., Holt, E., Schanze, K., Clark, R., Siegrist, T. & Ma, B. (2017). One-dimensional organic lead halide perovskites with efficient bluish white-light emission. *Nature Communications*, 8, 14051.

[47] Ge, Y., Chen, S., Xu, Y., He, Z., Liang, Z., Chen, Y., Song, Y., Fan, D., Zhang, K. & Zhang, H. (2017). Few-layer selenium-doped black phosphorus: Synthesis, nonlinear optical properties and ultrafast photonics applications. *Journal of Materials Chemistry C*, 5, 6129–6135.

[48] Wu, Q., Chen, S., Wang, Y., Wu, L., Jiang, X., Zhang, F., Jin, X., Jiang, Q., Zheng, Z., Li, J., Zhang, M. & Zhang, H. (2019). MZI-based all-optical modulator using MXene Ti₃C₂Tₓ (T = F, O, or OH) deposited microfiber. *Advanced Materials Technologies*, 4, 1800532.

[49] Xing, C., Jing, G., Liang, X., Qiu, M., Li, Z., Cao, R., Li, X., Fan, D. & Zhang, H. (2017). Graphene oxide/black phosphorus nanoflake aerogels with robust thermo-stability and significantly enhanced photothermal properties in air. *Nanoscale*, 9, 8096–8101.

[50] Concina, I., Comini, E., Kaciulis, S. & Sberveglieri, G. (2014). Quantum dots as mediators in gas sensing: A case study of CdS sensitized WO_3 sensing composites. *Appl. Surf. Sci.*, 290, 295–300.

[51] Liu, Y., Wang, H., Chen, K., Yang, T., Yang, S. & Chen, W. (2019). Acidic site-assisted ammonia sensing of novel $CuSbS_2$ quantum dots/reduced graphene oxide composites with an ultralow detection limit at room temperature. *ACS Applied Materials & Interfaces*, 11, 9573–9582.

[52] Guo, Z., Chen, S., Wang, Z., Yang, Z., Liu, F., Xu, Y., Wang, J., Yi, Y., Zhang, H., Liao, L., Chu, P. K. & Yu, X.-F. (2017). Metal-ion-modified black phosphorus with enhanced stability and transistor performance. *Adv. Mater.*, 29, 1703811.

[53] Yin, L., Chen, D., Hu, M., Shi, H., Yang, D., Fan, B., Shao, G., Zhang, R. & Shao, G. (2014). Microwave-assisted growth of In_2O_3 nanoparticles on WO_3 nanoplates to improve H_2S-sensing performance. *J. Mater. Chem. A*, 2, 18867–18874.

[54] Han, W., Zang, C., Huang, Z., Zhang, H., Ren, L., Qi, X. & Zhong, J. (2014). Enhanced photocatalytic activities of three-dimensional graphene-based aerogel embedding TiO_2 nanoparticles and loading MoS_2 nanosheets as co-catalyst. *Int. J. Hydrogen Energy*, 39, 19502–19512.

[55] Sharma, B. & Kim, J.-S. (2018). MEMS based highly sensitive dual FET gas sensor using graphene decorated Pd-Ag alloy nanoparticles for H_2 detection. *Scientific Reports*, 8, 5902.

[56] Savagatrup, S., Schroeder, V., He, X., Lin, S., He, M., Yassine, O., Salama, K. N., Zhang, X.-X. & Swager, T. M. (2017). Bio-inspired carbon monoxide sensors with voltage-activated sensitivity. *Angew. Chem. Int. Ed.*, 56, 14066–14070. doi: 10.1002/anie.201707491

[57] Kim, S.-J., Choi, S.-J., Jang, J.-S., Kim, N.-H., Hakim, M., Tuller, H. L. & Kim, I.-D. (2016). Mesoporous WO_3 nanofibers with protein-templated nanoscale catalysts for detection of trace biomarkers in exhaled breath. *ACS Nano.*, 10, 5891–5899. doi: 10.1021/acsnano.6b01196

[58] Li, J., Liu, H., Fu, H., Xu, L., Jin, H., Zhang, X., Wang, L. & Yu, K. (2019). Synthesis of 1D α-MoO_3/0D ZnO heterostructure nanobelts with enhanced gas sensing properties. *J. Alloys Compd.*, 788, 248–256.

[59] Park, S. (2017). Acetone gas detection using TiO_2 nanoparticles functionalized In_2O_3 nanowires for diagnosis of diabetes. *J. Alloys Compd.*, 696, 655–662.

[60] Wang, Y., Zhou, Y., Meng, C., Gao, Z., Cao, X., Li, X., Xu, L., Zhu, W., Peng, X., Zhang, B., Lin, Y. & Liu, L. (2016). A high-response ethanol gas sensor based on one-dimensional TiO_2/V_2O_5 branched nanoheterostructures. *Nanotechnology*, 27, 425503.

[61] Kolmakov, A., Klenov, D. O., Lilach, Y., Stemmer, S. & Moskovits, A. M. (2005). Enhanced gas sensing by individual SnO_2 nanowires and nanobelts functionalized with Pd catalyst particles. *Nano Lett.*, 5, 667–673.

[62] Zeng, W., Wang, H. & Li, Z. (2016). Nanomaterials for sensing applications. *Journal of Nanotechnology*, 1–2.

[63] Gong, H., Hu, J. Q., Wang, J. H., Ong, C. H. & Zhu, F. R. (2006). Nano-crystalline cu-doped ZnO thin film gas sensor for CO. *Sensors and Actuators B*, 115, 247–251.

[64] Zhou, T. & Zhang, T. (2021). Recent progress of nanostructured sensing materials from 0D to 3D: Overview of structure-property-application relationship for gas sensors. *Small Methods*, 2100515. doi: 10.1002/smtd.202100515.

10 Electronic Devices Including Nanomaterial-Based Sensors

Preeti Singh, Syed Wazed Ali, Ravinder Kale,
Suresh Sagadevan, and Himanshu Aggarwal

CONTENTS

10.1 INTRODUCTION

Electronic devices are equipment that governs the passage of electrical currents in order to process data and control systems. Transistors and diodes are two prominent examples. Electronic devices are typically tiny and can be packaged together into integrated circuits. Semiconductor materials are used in the majority of modern electronic gadgets. Since the advent of the transistor, these solids' unique features have ushered in a revolution in electronics [1]. Current electronic equipment, such

DOI: 10.1201/9781003263852-10

185

as harvesting energy and storage systems, light-emitting devices, and electronically driving instruments, are stiff, weighty, and cumbersome, making it difficult to meet the demands of flexible electronics. As a result, future trends in the creation of next-generation electrical appliances include bendable, ultralight, and shape-controllable electronics. A variety of different microelectronics devices rely on nanometer-scale phenomena to function [2]. In electronics, this allows for faster, smaller, and more portable devices. Nanoelectronics improve electrical device capabilities, boost memory chip density, and reduce power consumption and transistor size in integrated circuits. In electronics, nanotechnology allows for faster, smaller, and more portable devices. Nanoelectronics improves electrical device capabilities, boosts memory chip density, and reduces power consumption and transistor size in integrated circuits [3,4]. These nanomaterials are used in nanotechnology in various fields such as industry, medical, energy transportation, electronics, information technology, polymers, environmental science and many more [5]. With their minute or extremely small size, nanomaterials achieve specific properties, thus greatly extending the material science in various fields. The application has helped us to work smoothly in various field. It aids in the early diagnosis of the patient rather than live hazardous materials in it in medicine. Nanotechnology aids work in a variety of disciplines in electronics, and one can witness the significant development in it nowadays. One example is the handheld device, which has transformed a normal man's life from rural to urban. This is especially true in the sensor sector [6]. For the past ten years, various institutions have been researching nanosensors. A nanosensor is a sensor that is based on nanometer measurements and is built on an atomic scale. There have been numerous advancements in nanosensor investigations and its expansion for a variety of applications. The medical industry, national security, aircraft, integrated circuits, and others are only a few of the key uses. Nanosensors come in a variety of sizes, shapes, and varieties, as well as a variety of manufacturing methods. The creation of these nanosensors has a lot of problems at the moment, but when they are polished for consistent use, they will have immeasurable benefits over the sensors employed based on today's knowledge. The global market for electrical products based on nanosensors has grown dramatically during the last few decades [7]. The current chapter discusses the sensors materials and classified parameters of electronic device-based nanomaterials.

10.2   NANOSENSORS

Sensors are necessary in our daily lives in order for society to function properly. Cars at traffic lights are recognized by road sensors, which cause the flow through intersections to change accordingly [8]. When sensors in the shopping center detect your presence, doors open to welcome you in. Sensors keep an eye on the amount of water in your washing machine to make sure it does not overflow. Nanosensors work in a similar fashion, but they can detect minute particles or extremely small amounts of things. Sensors that operate on the scale of atoms and molecules are more compact and lighter, and they require less power. Although sensors have a long and distinguished history, nanosensors are a newer technology.

Between 1994 and 2005, a timeline of the development of several nanosensors was created. Nanosensors are sensing devices with a sensing dimension of less than 100 nanometers on at least one side [9]. In nanotechnology, nanosensors are used to: (a) monitor physical and chemical events in difficult-to-reach regions; (b) detect biochemicals in cellular organelles; and (c) quantify nanoscopic particles in industry and the environment. Technological advancements with nanosensor implants are naturally followed by advancements in electrical devices [10]. Nanosensors operate at the "nano" level. The term "nano" refers to a measuring unit of approximately 10^{-9} meters. A nanosensor is a tool that can provide information and data on the behavior and characteristics of microscopic particles to a macroscopic level. Nanosensors can be used to detect chemical or mechanical information on the nanoscale, such as chemical species and nanoparticles, or to monitor physical parameters such as temperature [11]. The construction and application of nanosensors can be used to classify them. Nanosensors are divided into two types based on their structure: optical nanosensors and electrochemical nanosensors. Chemical nanosensors, biosensors, electrometers, and deployable nanosensors are some of the types of nanosensors available [12]. The nanosensor has applications in the chemical, optical, medical, food, and electronics industries (Figure 10.1). Nanosensors, in particular, hold great promise for medical diagnostics, food and water quality detection, and other applications. Optical signal detection is one of several nanosensors' most promising applications. Many classes of nanosensors are possible based on their optical selectivity [13].

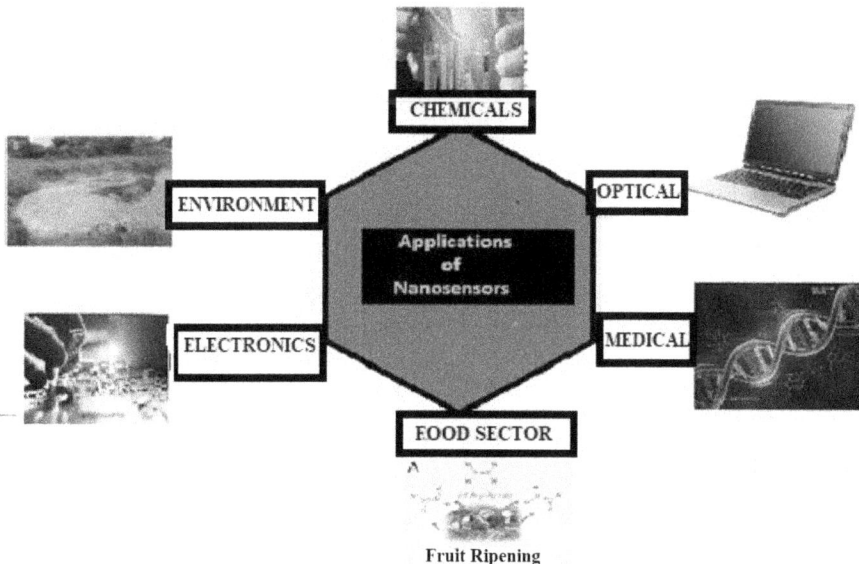

FIGURE 10.1 Represents the applications of nanosensors in different sectors.

10.3 ELECTRONIC DEVICES–BASED NANOSENSORS

A nanosensor is a nanoscale sensor whose primary aim is to acquire data at the microscopic level and convert it into facts that can be simply evaluated. "A chemical/physical sensor created employing nanoscale mechanisms, usually microscopic or sub-microscopic in size," according to another explanation given by Mousavi et al. [14]. These ultra-sensitive sensors may identify sole virus particles or uniform ultra-low amounts of a potentially dangerous material. Because so little is known about this skill, it is problematic to give a definitive explanation of what a nanosensor is. Nanosensors provide a number of advantages over traditional sensors, including their small size, low power consumption, higher sensitivity, and improved specificity, which makes them ideal for application in electronic devices. They are often used to monitor electrical changes in the sensor materials utilized in the devices [15]. Many people's daily lives are influenced by photovoltaics, batteries, and fuel cells. Although electrochemical sensors are less well-known, they are crucial to human health and well-being. Electrochemical glucose biosensors have without a doubt improved the quality of life for hundreds of millions of diabetics throughout the world. Electrochemical sensors are used in hospitals as part of automated clinical analyzers to determine blood level, DNA, electrolytes, etc. More distributed and decentralized electrochemical analysis is possible thanks to miniaturized electronics and wireless signal transfer. One promising example is the current development of wearable electrochemical sensors for non-invasive monitoring of biomarkers in sweat or interstitial fluid. Blood gases and electrolytes potentiometric pH sensors are common equipment in chemical labs [16–18].

10.4 APPLICATIONS OF NANOSENSORS IN ELECTRONIC DEVICES

Electronic devices implanted with nanosensors are very useful, economic and saves time. The following subsections describe some applications:

10.4.1 ELECTROMETRIC DEVICES

An electrometer is an electrical instrument that is used to measure electric charge or potential difference [19]. The electrometer is a nanometer-scale mechanical electrometer such as a pH detector (an electrometric device) as shown in Figure 10.2. Nanosensors simplify the determination in short span of time.

10.4.2 BIOMEDICAL DEVICES

One of the most well-funded areas of nanosensor research is biosensors. This is mostly owing to the technology's ability to aid in the early diagnosis of diseases such as cancer. Specific forms of DNA can also be detected using biosensors. They can be used to detect DNA and other biomaterials with the help of encoded antibodies [20].

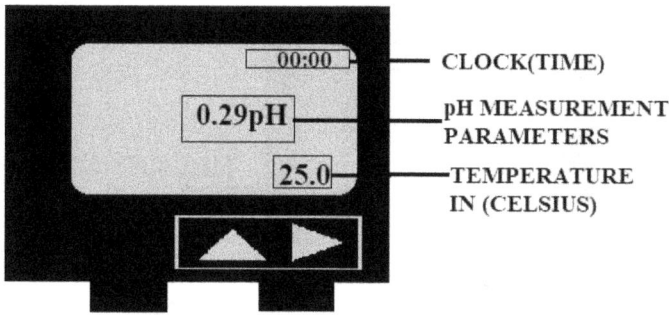

FIGURE 10.2   A pH (electrometric) meter device.

10.4.3   AERONAUTIC ELECTRONIC DEVICES

Another biosensor currently being developed has the potential to be used in the field of aviation. These nanosensors can pass through membranes and into white blood cells called lymphocytes to detect metabolic changes, as well as early radiation damage or infection, in astronauts. Because of the amount of radiation, they are exposed to, astronauts on space missions are at a higher risk of acquiring cancer as a result of cell damage. Dendrimers, which are synthetic polymers that are created layer by layer into spheres with a diameter of less than 5 nanometers, are used to make the sensors. Because of their small size, these sensors are designed to be administered transdermally, or through the skin. There would be no need for injections or intravenous administration during space missions if this could be done and administered every few weeks. Blood samples would no longer be required to be obtained and evaluated if these sensors were used. This sensor would dramatically improve the living conditions for astronauts in space [21].

10.4.4   NANOSENSOR ELECTRONIC DEVICES USED IN THE MILITARY

A "deployable nanosensor"—a form of sensor that can be deployed—has not gotten a lot of attention. Sensors employed in the military or other forms of national security are the most common examples. One such sensor is the Sniffer STAR, a nano-enabled chemical sensor that may be integrated into a small unmanned aerial vehicle. This sensor is a compact, portable chemical detection system that combines a nanomaterial for sample collection and a concentration microelectromechanical (MEM)-based "chemical lab on-a-chip" detector [22].

10.4.5   HEALTH MONITORING ELECTRONIC DEVICES

Blood glucose monitors use a sophisticated chemical mechanism to determine the quantity of sugar in a sample of blood. The blood is combined with glucose oxidase in the test strip, which reacts with the glucose in the blood sample to produce gluconic acid. Ferrocyanide is formed when ferricyanide, another chemical in the test strip, combines with the gluconic acid. The ferrocyanide alters the current run through the blood sample

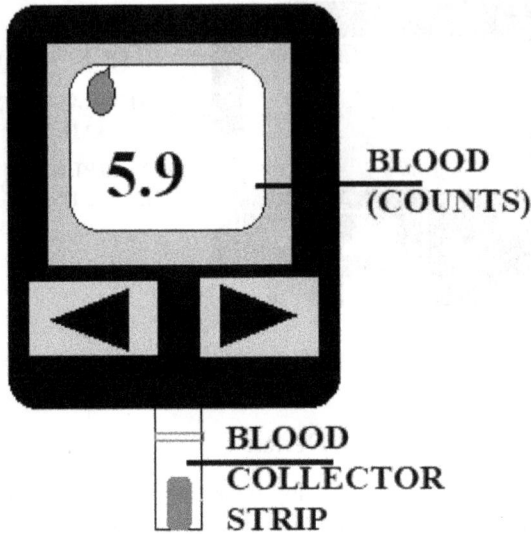

FIGURE 10.3   Blood glucose monitor device.

by the electrode within the test strip, allowing the concentration of blood glucose within the sample to be correctly detected within a reasonable margin of error [23].

10.4.6   INSECT KILLING DEVICES

Mosquitoes' antennas contain a battery of sensors, one of which is a chemical sensor. Carbon dioxide and lactic acid can be detected up to 36 meters away. When mammals and birds breathe, these gases are released. Mosquitoes also appear to be attracted to a chemical found in sweat [24].

10.4.7   TRANSPORTATION AND COMMUNICATION PURPOSE USABLE DEVICES

Nanosensors are utilized in transportation on land, at sea, in the air, and even in space. In the sphere of communications, nanosensors are expected to arise in wired and wireless technologies, as well as optical and RF technologies. They can be found in a variety of structures and institutions, including industries, offices, and even residences [25]. The study in the subject of wireless nanosensing is still ongoing, with the current focus being on communication mechanisms among nanosensor devices. Acoustic, nanomechanical, molecular, and electromagnetic communication are the four basic communication mechanisms used by nanomachines. Molecular and electromagnetic communication are two of the four communication strategies that fall under the category of wireless communication. The broadcast and receiving information contained in molecules is referred to as molecular communication, whereas electromagnetic communication refers to the transmission and reception of electromagnetic radiation from components made of various nanomaterials. For EM wave propagation at the nanoscale, graphene-based nano-antennas are utilized. Graphene is a one-atom-thick planner sheet made up of densely packed linked carbon atoms in a honeycomb crystal lattice. It is worth noting that transmitting at lower frequencies allows nanosensor devices to communicate over longer distances. However, because Nanosensor devices' energy efficiency is predicted to be low, Nanosensor devices will not communicate using megahertz frequencies, and higher energy waves can be used to control a large number of nanosensor devices deployed over a large area. As a result of these factors, it has been determined that nanosensor devices may be able to communicate in the terahertz band (0.1–10.0 THz) [26].

10.5 BENEFITS OF USING NANOSENSORS IN ELECTRONIC DEVICES

A variety of parameters has been set for detecting the efficiency and suitability of electronic devices. These parameters are based on nanosensors [27] whose suitability is determined by their sensitivity, selectivity, detection limit, response time, linearity, stability, and lifetime [28–31], as shown in Figure 10.4.

10.5.1 SENSITIVITY

Nanosensors attached to electronic devices increase the sensitivity of the device. The ratio between a physical parameter and the output signal generated by that physical parameter is commonly characterized as a sensor's sensitivity [32]. The general equation (Eq. 1).

$$\text{Sensitivity (S)} = \Delta Y / \Delta X \tag{1}$$

Where Y = output signal variation and X = input signal variation

It has been discovered that increasing the sensor's conductivity increases the sensor's sensitivity. Nanomaterial-based sensors have been found to have higher sensitivity than traditional sensors due to their increased surface area–to-volume ratio [33]. In

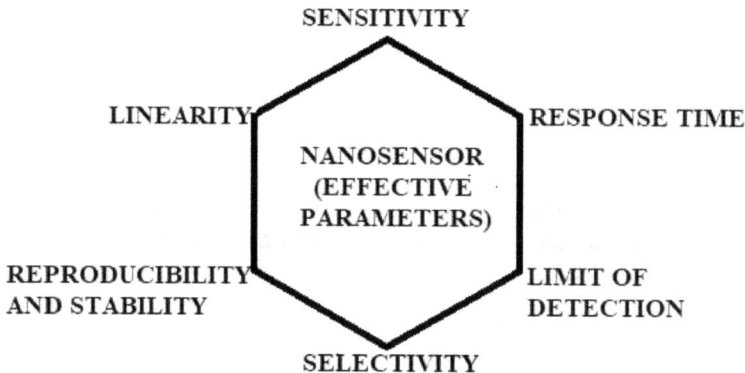

FIGURE 10.4   Effective parameters of nanosensors.

electrochemical and optical nanosensors, higher sensitivity is frequently attributed to improved conductivity arising from regulated dimensions and shapes leading to increased electrical and optical qualities [34]. This can considerably improve signal amplification and, as a result, the sensitivity of electrical gadgets [35].

10.5.2 LINEARITY IN ELECTRONIC DEVICES

The sensor's operating window, which distinguishes the actual measured curve from the model curve, is connected to the linearity. This is expressed as a range of permitted input and output estimates from the smallest to the largest [36]. The linear relationship described in Eq. 2 is a crucial fundamental equation that each sensor's linear calibration curve must satisfy.

$$Y = mX + c \tag{2}$$

Where Y is the output signal, X is the input signal, m is the sensor's slope/sensitivity, and c is the baseline offset constant parameter

In addition to the aforementioned linear connection, the calibration curve's correlation coefficient is important in determining the linearity of the developed sensor. Statistical approaches such as ANOVA or standard relative deviation, in particular, can be employed to analyze the linearity of the calibration curve(s), according to the FDA [37]. As a result of the reproducibility and consistency in the outcomes, electronic devices have been stated to have strong linearity [38].

10.5.3 LIMIT OF DETECTION (LOD)

The limit of detection (LOD), Eq. 3, which is defined as the smallest amount of analyte in a test sample that can be consistently differentiated from zero, is a critical parameter in sensors that combines sensitivity with theoretical instrument resolution [39,40].

$$LOD = 3 \times (SD / S) \tag{3}$$

Where SD = standard deviation of the blank solution and S = slope of the calibration curve

The creation of non-linear calibration curves has a significant impact on traditional sensors, resulting in an over/under assessment of the target analyte(s). The improved sensitivity of the target analyte employed in electronic device(s) is due to this linear calibration curve. The signal-to-noise ratio is also lowered when nanosensors are used in a device.

10.5.4 ENHANCES THE SELECTIVITY

In electronics, selectivity refers to a nanosensor's ability to differentiate the response of competing substances while generating the greatest output of the target analyte(s)

[41]. Low selectivity is a significant disadvantage for sensor deployment in environmental applications. The emergence of nanomaterial-based sensors, on the other hand, has successfully reformed these dynamics. To selectively target the analyte of interest from competing molecules within the complex sample matrix, nanomaterials with distinctive morphologies corresponding to the target analyte have been created [42]. Furthermore, the increase in binding sites caused by the increased surface-to-volume ratio of nanomaterials has been proven to improve the selectivity of nanosensors. As a result, nanosensor tuning is important for device selectivity and accuracy.

10.5.5 RESPONSE TIME

Nanomaterials have been proven to improve the selectivity of nanosensors with increasing the binding sites caused by the increased surface-to-volume ratio. As a result, nanosensor tuning is important for the device's selectivity and accuracy [43]. These input parameters, obviously, are dependent on the sensors, which do not take effect instantly; rather, they take a certain amount of time, known as the response time. The response times of nanosensors have been found to be faster than those of standard sensors, which is interesting [43]. This means that electronic device–enabled sensors react faster to changes in input parameters than traditional sensors.

10.5.6 GOOD REPRODUCIBILITY, REPEATABILITY, AND STABILITY

Three of the most significant criteria to examine when evaluating the performance of a nanosensor are its stability, reproducibility, and repeatability, which is expressed as inter-day and intra-day precision [44]. The sensor's capacity to resist lengthy operational hours in adverse conditions is also examined [45]. Traditional sensors appear to diminish in performance when used for a long time in hostile conditions; however, nanosensors often retain their compact structure even after operating for a long time without a substantial drop in performance [46–48].

Various types of nanosensors-based electronic devices are summarized in Table 10.1 with their applications.

10.6 ADVANCEMENT OF ELECTRONIC DEVICES AND CHALLENGES WITH NANOSENSORS

The goal of a sensor device is to detect events or changes in the environment and convey the data to its readout electronics or computer processors; hence, good electronics are a must for sensor devices. The sensing equipment used to measure chemical, physical, or biological factors were cumbersome in the beginning. They were also frequently inaccurate because the end user had to interpret and decode the sensor signal manually. Numerous recent developments in the sensing sector have been prompted by nanotechnological multidisciplinary advancements, offering many new solutions for highly constructed devices with great performance characteristics. Sensitivity, selectivity, resolution, accuracy, and precision limits of

TABLE 10.1
Nanosensors-Based Electronic Devices

Nanosensor	Material	Device	Application	References
Thermocouple sensors	Gold/Ni Thin Film	High temperature detection	To determine the temperature profile of solid structures under electron beam irradiation	[49]
Environmental monitoring	CNTs, graphene quantum dots, metal oxides, Lipid Membrane sensor	Pollution detector, Microfluidic device, Monitoring device	Detects heavy metal ions (Hg^{2+}, Pb^{2+}, As^{3+}, Ag^+), H_2O_2, organophosphate pesticides, toxic gases, industrial wastewater To detect pesticides and pathogens Rapid detection of insecticides, pesticides, hydrazines, naphthalene acetic acid, arochlor, toxins, polyaromatic hydrocarbons, etc.	[50] [51] [52]
Health monitoring nanosensor	Ni oxide	Chronic arthritic disease Osteoarthritis (OA)	Applied in the clinical diagnosis of OA severity and the evaluation of drug therapy	[53]
Animal health monitoring nanosensors	Quantum dot	Environment and physiology detector	Integrated livestock monitoring data will assist the agricultural industry and farmers to improve the animal production in the near future	[54]
Energy related sensor	Graphene	Energy-harvesting technology	Can be applied to triboelectrics nanogenerators (TENGs), photovoltaics, thermoelectrics, radio frequency (RF), and biofuel cells; becoming a growing trend in the formation of self-powered systems	[55]
Nanosensors in food industry and agriculture	Gold/cadmium selenide—zinc sulfide core—shell quantum dots	Contamination and food freshness detector device	Nanosensors can detect pathogenic bacteria, food-contaminating toxins, adulterant, vitamins, dyes, fertilizers, pesticides, taste and smell; food freshness can be monitored using time, temperature, and oxygen indicators; product authenticity and brand protection can be assessed using invisible nanobarcodes; overall, nanosensors with unique properties are improving food security	[56]

sensing devices are always being enhanced in electronic devices. The following are the most significant obstacles encountered in the creation of electronic devices based on nanosensors.

1. *Obtained chief parameters:* Obtaining a low limit of detection (LOD), suppressing non-specific adsorption of interfering species, and preserving the sensor's repeatability and stability in complicated actual matrices are all significant challenges for electronic devices [57].
2. *Economical:* Sensor devices face major technological obstacles in terms of lowering their cost, size, and energy consumption.
3. *Integration with sensors:* Another problem is the reliable integration of these innovative materials and structures into sensor devices, which is often overlooked and underestimated. Their integration must be scalable for commercial device manufacture, as otherwise, their utilization will be limited to laboratory studies with minimal socio-economic impact.
4. *To produce real-time evaluation data:* Another essential characteristic of sensors is that they produce data, which is commonly overlooked as part of the sensing field. These data must be processed and decoded, which is a major focus of sensor technology research. Using artificial intelligence, deep learning, or other ways to manage "big data," the detected data must be evaluated and changed in real time to offer final feedback to the end user. In this discipline, managing these increasingly massive data sets pouring from widely scattered and heterogeneous sources is becoming an increasingly difficult task. Due to all of the variables involved in this multidisciplinary field of sensing devices, it is extremely difficult to predict their real-time evolution, but significant advances have been made that have had a huge impact on society, improving product quality, food and environmental safety, disease diagnosis, medicine, health and wealth, process studies, and more. As new sensor technologies develop and extend the extent and scope of their impact on our lives, we can be certain that this tendency will continue [58].

10.7   CONCLUSION AND FUTURE PROSPECTS

Nanodevices, in the broadest sense, are the key enablers that will allow humanity to fully harness the technical potential of electronic, magnetic, mechanical, and biological systems. Although the best examples of nanodevices are currently connected with the information technology industry, their promise is considerably larger. Nanosensors, which are commonly employed in electronic devices, will have a huge impact on our abilities to increase energy conversion, regulate pollution, generate food, and improve human health and longevity in the future. Electronic devices based on nanosensor technology are quickly evolving and are the focus of global research efforts. Furthermore, achieving large-scale production is widely regarded as the most difficult task in the development of flexible electrical devices. Many of these breakthroughs in electronic devices will become a reality in the future.

REFERENCES

[1] Ben, G. Streetman, 18: Semiconductors and transistors. In Reference Data for Engineers: Radio, Electronics, Computer, and Communications; Edited by Wendy M. Middleton & Mac E. Van Valkenburg. Ninth Edition. Elsevier Publication: Boston, 2002, pp. 18-1–18-27.

[2] Jasinski, J. Chapter 6 Applications: Nanodevices, Nanoelectronics, and Nanosensors, pp. 77–96. DOI: 10.1007/978-94-015-9576-6_6.

[3] Arduini, F., Cinti, S., Scognamiglio, V., Moscone, D., & Palleschi, G. How cutting-edge technologies impact the design of electrochemical (bio) sensors for environmental analysis: A review. Anal. Chim. Acta. 2017;959, 15–42.

[4] Peng, H. Polymer materials for energy and electronic applications ‖ flexible electronic devices based on polymers. 2017, 325–354. doi:10.1016/B978-0-12-811091-1.00009–4.

[5] Thiruvengadam, M., Rajakumar, G., & Chung, I. M. Nanotechnology: Current uses and future applications in the food industry. 3 Biotech. 2018;8(1), 74.

[6] Nanotechnology Examples and Applications. www.nanowerk.com/nanotechnology-examples-and-applications.php.

[7] Arduini, F., Cinti, S., Scognamiglio, V., & Moscone, D. Nanomaterial-based sensors. In Handbook of Nanomaterials in Analytical Chemistry. Elsevier Publications: Newark, USA, 2020, pp. 329–359. https://doi.org/10.1016/B978-0-12-816699-4.00013-X.

[8] Eom, M., & Kim, B. I. The traffic signal control problem for intersections: A review. Eur. Transp. Res. Rev. 2020;12, 50. https://doi.org/10.1186/s12544-020-00440-8.

[9] Bayda, S., Adeel, M., Tuccinardi, T., Cordani, M., & Rizzolio, F. The history of nanoscience and nanotechnology: From chemical-physical applications to nanomedicine. Molecules (Basel, Switzerland). 2019;25(1), 112. https://doi.org/10.3390/molecules25010112.

[10] Kumar, H., Kuča, K., Bhatia, S. K., Saini, K., Kaushal, A., Verma, R., Bhalla, T. C., & Kumar, D. Applications of nanotechnology in sensor-based detection of food-borne pathogens. Sensors. 2020;20(7), 1966. https://doi.org/10.3390/s20071966.

[11] Jeevanandam, J., Barhoum, A., Chan, Y. S., Dufresne, A., & Danquah, M. K. Review on nanoparticles and nanostructured materials: History, sources, toxicity and regulations. Beilstein Journal of Nanotechnology. 2018;9, 1050–1074. https://doi.org/10.3762/bjnano.9.98

[12] Pooja, M., & Rana, C. P. Chapter 21: Modern applications of quantum dots: Environmentally hazardous metal ion sensing and medical imaging. In Handbook of Nanomaterials for Sensing Applications. Micro and Nano Technologies. Elsevier Publications: USA, 2021, pp. 465–503.

[13] Adam, T., & Dhahi, T. S. Nanosensors: Recent perspectives on attainments and future promise of downstream applications. Process Biochemistry. 2022 June;117, 153–173. https://doi.org/10.1016/j.procbio.2022.03.024.

[14] Mousavi, S. M., Hashemi, S. A., Zarei, M., Amani, A. M., & Babapoor, A. Nanosensors for chemical and biological and medical applications. Med Chem (Los Angeles). 2018;8, 205–217. https://doi.org/10.4172/2161-0444.1000515.

[15] Hilmani, A., Maizate, A., & Hassouni, L. Automated real-time intelligent traffic control system for smart cities using wireless sensor networks. Wireless Communications and Mobile Computing. 2020, Article ID 8841893, 28 pages. https://doi.org/10.1155/2020/8841893.

[16] Javaid, M., Haleem, A., Singh, R. P., Rab, S., & Suman, R. Exploring the potential of nanosensors: A brief overview. Sensors International. 2021;2, 100130. https://doi.org/10.1016/j.sintl.2021.100130.

[17] Deshpande, A. H., Weldode, J. M., & Pise, J. S. Applications of nanosensors in various fields: A review. International Journal of Management, Technology and Engineering. 2018;8. ISSN NO: 2249–7455.

[18] Bobacka, J. Electrochemical sensors for real-world applications. J Solid State Electrochem. 2020;24, 2039–2040. https://doi.org/10.1007/s10008-020-04700-4.

[19] Agrawal, S., & Prajapati, R. Nanosensors and their pharmaceutical applications: A review. International Journal of Pharmaceutical Sciences and Nanotechnology. 2012 January–March;4(4).

[20] Zhang, X., Guo, Q., & Cui, D. Recent advances in nanotechnology applied to biosensors. Sensors (Basel, Switzerland). 2009;9(2), 1033–1053. https://doi.org/10.3390/s90201033.

[21] Iosim, S., MacKay, M., Westover, C., & Mason, C. E. Translating current biomedical therapies for long duration, deep space missions. Precision Clinical Medicine. 2019;2(4), 259–269. https://doi.org/10.1093/pcmedi/pbz022.

[22] Tomar, S. Nanotechnology enabled sensor applications. CBW Magazine. 2014. www.idsa.in/cbwmagazine/NanotechnologyEnabledSensorApplications_stomar.

[23] Yoo, E. H., & Lee, S. Y. Glucose biosensors: An overview of use in clinical practice. Sensors (Basel, Switzerland). 2010;10(5), 4558–4576. https://doi.org/10.3390/s100504558.

[24] Nell, G. How Mosquitoes Sniff Out Human Sweat to Find Out. www.npr.org/sections/health-shots/2019/03/28/706838786/how-mosquitoes-sniff-out-human-sweat-to-find-us, acceded 28 March 2019.

[25] Nagel, D. J. Nanotechnology-Enabled Sensors: Possibilities, Realities, and Applications. www.fierceelectronics.com/person/david-j-nagel-phd.

[26] Rupani, V., Kargathara, S., & Sureja, J. A review on wireless nanosensor networks based on electromagnetic communication. International Journal of Computer Science and Information Technologies. 2015;6(2), 1019–1022.

[27] Graboski, A. M., Martinazzo, J., Ballen, S. C., Steffens, J., & Steffens, C. Nanosensors for water quality control. In Nanotechnology in the Beverage Industry. Elsevier: Amsterdam, The Netherlands, 2020, pp. 115–128.

[28] Fan, Y. Z., Tang, Q., Liu, S. G., Yang, Y. Z., Ju, Y. J., Xiao, N., Luo, H. Q., & Li, N. B. A smartphone-integrated dual-mode nanosensor based on novel green-fluorescent carbon quantum dots for rapid and highly selective detection of 2,4,6-trinitrophenol and pH. Appl. Surf. Sci. 2019;492, 550–557.

[29] Cheng, C., Zhang, R., Wang, J., Zhang, Y., Wen, C., Tan, Y., & Yang, M. An ultrasensitive and selective fluorescent nanosensor based on porphyrinic metal—organic framework nanoparticles for Cu2+ detection. Analyst. 2020;145, 797–804.

[30] Liu, J., Pan, L., Shang, C., Lu, B., Wu, R., Feng, Y., Chen, W., Zhang, R., Bu, J., & Xiong, Z. A highly sensitive and selective nanosensor for near-infrared potassium imaging. Sci. Adv. 2020;6, eaax9757.

[31] Shoaie, N., Daneshpour, M., Azimzadeh, M., Mahshid, S., Khoshfetrat, S. M., Jahanpeyma, F., Gholaminejad, A., Omidfar, K., & Foruzandeh, M. Electrochemical sensors and biosensors based on the use of polyaniline and its nanocomposites: A review on recent advances. Microchim. Acta. 2019;186, 465.

[32] Zhou, G., Wang, Y., & Cui, L. Biomedical sensor, device and measurement systems. Adv. Bioeng. 2015;177.

[33] Leal-Junior, A. G., Frizera, A., & Pontes, M. J. Sensitive zone parameters and curvature radius evaluation for polymer optical fiber curvature sensors. Opt. Laser Technol. 2018;100, 272–281.

[34] O'Riordan, A., & Barry, S. Electrochemical nanosensors: Advances and applications. Rep. Electrochem. 2016;6, 1.

[35] Tang, L., & Li, J. Plasmon-based colorimetric nanosensors for ultrasensitive molecular diagnostics. ACS Sens. 2017;2, 857–875.

[36] Moosavi, S. M., & Ghassabian, S. Linearity of calibration curves for analytical methods: A review of criteria for assessment of method reliability. In Calibration and Validation of Analytical Methods: A Sampling of Current Approaches. IntechOpen Ltd.: London, UK, 2018, pp. 109–127.

[37] Tiwari, J. N., Vij, V., Kemp, K. C., & Kim, K. S. Engineered carbon-nanomaterial-based electrochemical sensors for biomolecules. ACS Nano. 2016;10, 46–80.

[38] Bernal, E., & Guo, X. Limit of detection and limit of quantification determination in gas chromatography. Adv. Gas Chromatogr. 2014;3, 57–63.

[39] Hare, D. J., New, E. J., De Jonge, M. D., & McColl, G. Imaging metals in biology: Balancing sensitivity, selectivity and spatial resolution. Chem. Soc. Rev. 2015;44, 5941–5958.

[40] Schroeder, V., Savagatrup, S., He, M., Lin, S., Swager, T. M. Carbon nanotube chemical sensors. Chem. Rev. 2018;119, 599–663.

[41] De, A., Chen, S., & Carlen, E. T. Probe-free semiconducting silicon nanowire platforms for biosensing. In Semiconducting Silicon Nanowires for Biomedical Applications. Elsevier: Amsterdam, The Netherlands, 2014, pp. 229–265.

[42] Rowland, C. E., Brown III, C. W., Delehanty, J. B., & Medintz, I. L. Nanomaterial-based sensors for the detection of biological threat agents. Mater. Today. 2016;19, 464–477.

[43] Mnyipika, S. H., & Nomngongo, P. N. Square wave anodic stripping voltammetry for simultaneous determination of trace Hg(II) and Tl(I) in surface water samples using SnO2@MWCNTs modified glassy carbon electrode. Int. J. Electrochem. Sci. 2017;12, 4811–4827.

[44] Ghosh, A., Zhang, C., Shi, S. Q., & Zhang, H. High-temperature gas sensors for harsh environment applications: A review. Clean Soil Air Water. 2019;47, 1800491.

[45] Arduini, F., Cinti, S., Scognamiglio, V., Moscone, D. Nanomaterial-based sensors. In Handbook of Nanomaterials in Analytical Chemistry. Elsevier: Amsterdam, The Netherlands, 2020, pp. 329–359.

[46] Sharma, P., Pandey, V., & Sharma, M. M. M. A review on biosensors and nanosensors application in agroecosystems. Nanoscale Res Lett. 2021;16, 136. https://doi.org/10.1186/s11671-021-03593-0.

[47] Kumar, A. Nanosensors: Applications and challenges. International Journal of Science and Research (IJSR). 2019 July;8(7), 1472–1474. ISSN: 2319–7064

[48] Tshimangadzo, S., & Munonde, P. N. Nomngongo, nanocomposites for electrochemical sensors and their applications on the detection of trace metals in environmental water samples. Sensors. 2021;21, 131. https://dx.doi.org/10.3390/s21010131.

[49] Dachen Chu, D., Taner Bilirl, R., Pease, F. W., & Goodson, K. E. Thin film nano thermocouple sensors for applications in laser and electron beam irradiation. The 12th International Conference on Solid State Sensors, Actuators and Micrasystems, Boston, June 8–12, 2003.

[50] SuWenhe Wu, S., Gao, J., Lu, J., & Fan, C., Nanomaterials-based sensors for applications in environmental monitoring. Journal of Materials Chemistry, 2012;22, 18101–18110.

[51] Willner M. R., & Vikesland, P. J. Nanomaterial enabled sensors for environmental contaminants. Journal of Nanobiotechnoogyl. 2018;16, 95. https://doi.org/10.1186/s12951-018-0419-1

[52] Paraskevi, G., Dimitrios Nikolelis, N., Siontorou, C. G., & Karapetis, S. Lipid membrane nanosensors for environmental monitoring: The art, the opportunities, and the challenges. Sensors. 2018;18(1), 284. https://doi.org/10.3390/s18010284.

[53] Jin, P., Wiraja, C., Zhao, J., Zhang, J., Zheng, L., & Xu, C. Nitric oxide nanosensors for predicting the development of osteoarthritis in rat model. ACS Applied Material Interfaces. 2017;9, 25128–25137. doi:10.1021/acsami.7b06404.

[54] Younis, S., Zia, R., Tahir, N., Zunaira, S., Waheed, B., Khan, S., & Bajwa, S. Z. Chapter 24: Nanosensors for animal health monitoring in nanosensors for smart agriculture. Micro and Nano Technologies. 2022;509–529.

[55] Liu, Z., Li, H., Shi, B., Fan, Y., Wang, Z. L., & Li, Z. Wearable and implantable triboelectric nanogenerators. Advance Functional Material. 2019;3, 1808820. doi:10.1002/adfm.201808820.

[56] Kumar, V., Guleria, P., & Mehta, S. K. Nanosensors for food quality and safety assessment. Environmental Chemisrty Letters. 2017;15, 165–177. https://doi.org/10.1007/s10311-017-0616-4.

[57] Ferrag, C., & Kerman, K. Grand challenges in nanomaterial-based electrochemical sensors. Front. Sens. 2020. https://doi.org/10.3389/fsens.2020.583822.

[58] Comini, E. Achievements and challenges in sensor devices. Front. Sens. 2021;11. https://doi.org/10.3389/fsens.2020.607063.

11 DNA-Aptamer–Based Electrochemical Biosensors for the Detection of Thrombin
Fundamentals and Applications

*Mohammad Al Mamun, Yasmin Abdul Wahab,
M. A. Motalib Hossain, Abu Hashem,
and Mohd Rafie Johan*

CONTENTS

DOI: 10.1201/9781003263852-11

11.1 INTRODUCTION

A biosensor is an analytical tool that turns data into electrical signals by combining a biological sensing element (identification element), a transducer, and a substrate (conducting unit). In addition, as shown in Figure 11.1, there will be an electronic circuit that includes a signal amplifier, a microcontroller or processor, and a display unit. Biosensors have multiple advantages over lab-based apparatuses such as low cost, compact size, ease of use, and quick results.

The preferred biological recognition materials used in biosensors are typically in the form of enzyme, antibody, peptide, DNA, or aptamer [1, 2]. Through a process of interaction between the target analytes and sensing elements, results in corresponding signals (usually as current, potential, conductivity, impedance, etc.) are transduced with the support of a transducer to the detector. Aptamers are synthetic ss-DNA (single-stranded DNA, deoxyribonucleic acid) or ss-RNA (single-stranded RNA, ribonucleic acid), usually fewer than 100 base pairs that are chosen by the systematic evolution of ligands by exponential enrichment (SELEX). Many benefits and unique characteristics distinguish aptamers from other biological recognition elements, making them attractive recognition elements for biosensors [3–10]. They have many applications due to their selectivity, high affinity, ease of synthesis, long-term consistency at a wide scale of temperatures, pH and electrolyte concentrations, nontoxicity, less tendency to denaturation, prevalence in animal-based products, fast refolding capacity, extended shelf life, and inexpensiveness [11, 12]. More importantly, aptamers are simple to alter by introducing functional groups as needed, which is necessary for labeling or surface immobilization [13]. Incorporating various nanomaterials and composites [14–18] into DNA-aptamer–based biosensors

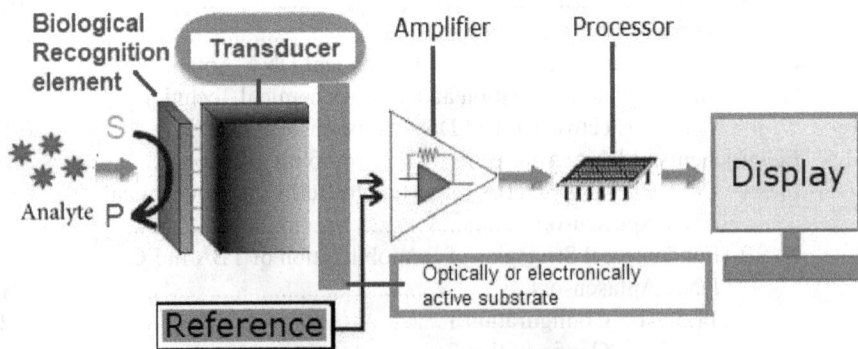

FIGURE 11.1   A simplified block diagram displaying the essential elements of a biosensors.

(referred to as aptasensors) increases transducing time, response time, signal-to-noise ratio, and detection limit, and modifies the architecture of the sensor interface at nanoscale [19].

A great deal of nanotechnological devices like nanomaterials-based DNA-aptasensors involve supramolecular concepts like host-guest chemistry and self-assembly [20]. The biosensors also involve different types of non-covalent interactions during fabrication and application in the practical fields: for example, carbon nanotube (CNT)/graphene-based DNA biosensors on gc (glassy carbon) substrate are fabricated based on following supramolecular concepts [1, 2, 21, 22]:

1. π–π interaction: When CNT/graphene particles (sp^2 carbon) are deposited on gc substrate (sp^2 hybridized).
2. π–π stacking: Pyrene molecules stacked on CNT/gc surface.
3. Self-assembly: When DNA immobilized on the CNT/gc surface.
4. Molecular recognition: When target analytes bind on aptamer surface through electrostatic interaction or host–guest interactions during DNA response through its conformational changes.
5. Selectivity: A host molecule determines whether one guest, or a family of guests, binds much more strongly than others. The ratio of equilibrium constants is used to measure selectivity. DNA biosensors also hold this concept since they are very selective to their target analytes.
6. Complementarity: Supramolecular interactions are what give DNA its structure. A hydrophobic driving force can help the two single DNA strands come together spontaneously in a rigorous self-assembly process.

Electrochemical DNA-aptasensors have shown great potential in identifying multiple analytes with high sensitivity, specificity, and cost-effectiveness [3–10] among various DNA biosensors which use various signal transduction methods such as electrochemistry [23–25], optics [26–28], and atomic force microscopy [29]. Proteins are a prominent class of biomarkers that are found in all living organisms [30]. They play important roles in a variety of biological processes, including immunological interaction, body tissue maintenance, molecular transit and storage, and hormone control [31].

Protein expression levels in serum, tissue, or saliva that are raised or depressed have shown tremendous promise as significant indicators of biological condition and have played an important role in clinical prognosis and diagnosis [32]. Thrombin, for example, is a blood coagulation serine protease that catalyzes the breakdown fibrinogen to fibrin monomer, which subsequently polymerizes to produce fibrin. It can influence endothelial cell activation, platelet aggregation, and other vital vascular biology responses by acting as a hormone. It is crucial for hemostasis, and coagulation abnormalities can lead to a variety of diseases [33], including liver disease, heart attack, stroke, deep vein thrombosis, pulmonary embolism, and leukemia, to name a few [34, 35]. COVID-19 patients have been found to have high thrombin levels at the time of diagnosis [36], which is linked to a syndrome known as COVID-19–associated coagulopathy (CoAC), which manifests as a variety of thromboembolic consequences in COVID-19 patients [37]. Thrombin is also thought to be a valuable biomarker in the diagnosis of pulmonary metastasis. As a result, developing a quick and precise approach for detecting thrombin is crucial in clinical practice. The cost-effective, sensitive, and quick detection of proteins using biosensors is a critical issue in basic research and clinical applications [38, 39]. Electrochemical methods have gotten a lot of interest in the creation of DNA biosensors because of their portability, low expense, easy instrumentation, superior sensitivity, and quick response [21, 40–43].

11.2 DIFFERENT TYPES OF BIOSENSORS

Biosensors are categorized into two principal types, as shown in Figure 11.2, depending on the biological element employed in the study and the signal transduction mechanism used [44]. Antibodies, DNA, and enzymes are the most frequent bio-recognition elements utilized in biosensors. Microorganisms (phage or bacteria), tissues, cell receptors, and other biological or bio-recognition elements are examples. The type of transduction mode in the sensor is the most used classification of biosensors. Optical biosensors, electrochemical biosensors, and mass-based biosensors are the three types of biosensors based on the method of transduction as discussed in the following subsections.

11.2.1 Optical Biosensors

Optical transductions are mainly based on light absorption or light emission by the recognition or sensing components of the biosensors. Optical fibers are vital components in optical biosensors [45]. Optical fibers enable recognition of sensing components based on light features such as scattering, absorption, fluorescence, and chemiluminescence. The non-electrical behavior of optical biosensors is one of their key advantages. By altering the wavelength of the light, they can evaluate many elements on a single layer, but the difficulties in the miniaturization of optical biosensors are the main obstacle for their commercialization.

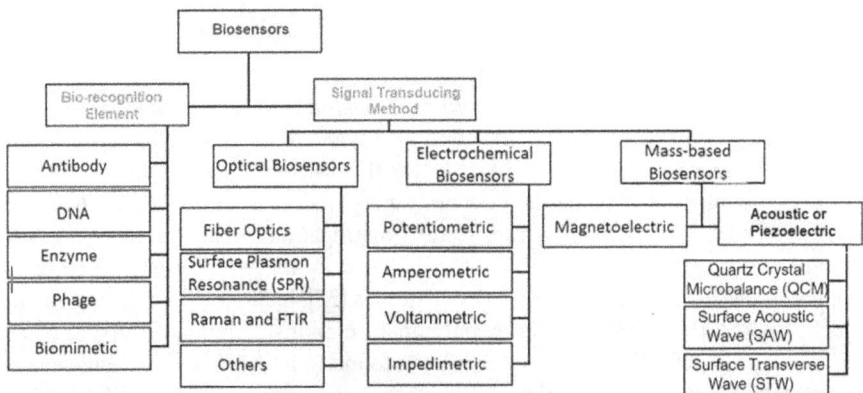

FIGURE 11.2 Classification of biosensors.

11.2.2 ELECTROCHEMICAL BIOSENSORS

Biological molecules are bound on a probing surface in electrochemical biosensors. A non-interfering membrane helps to keep the sensor molecules in place. The sensing molecules then respond to the substance to be identified in a way that generates an electrical signal proportionate to the quantity being evaluated. Potentiometric [46, 47], voltammetric, impedimetric [48, 49], amperometric [50, 51], and other types of transducers can be used in electrochemical biosensors to convert chemical information into a quantifiable electrical signal. This sort of biosensor has tremendous scope to build portable and miniaturized biosensor devices due to cost-effectiveness, user-friendliness, and versatility.

11.2.3 MASS-BASED BIOSENSORS

This type of biosensor can be classified as magneto-electric and piezoelectric biosensors. Magneto-electric biosensors monitor biological interactions as magneto-electric signals using paramagnetic or super-paramagnetic particles or crystals. Bioreceptors, which can be DNA (complementary to a sequence or aptamers), antibodies, or other molecules, are found on the surface of such particles or crystals. Some magnetic particle characteristics that can be measured as signals will be affected by the bioreceptor's binding. Piezoelectric mass-based biosensors can be subdivided as follows: quartz crystal microbalance (QCM) works on the concept of mass change, surface acoustic wave (SAW) biosensors work on the concept of sound vibrations or acoustics, and the surface transverse wave (STW) biosensors work on the basis of transverse waves [52], resulting from the interaction between the sensing elements and the targeted analytes. These types of biosensors are not as much popular as optical or electrochemical biosensors because of their limitations associating with cost, versatility, user-friendliness, and performance.

11.3 APPLICATIONS OF BIOSENSORS

Biosensors have grown increasingly essential in a variety of sectors (Figure 11.3), including medicine [53], clinical analysis [3], and general health monitoring [54], since their introduction in the early 1950s [55].

Biosensors have been used in a variety of industries, including industrial processing [56, 57], agriculture [10], food processing [9], pollution control [7], drug screening and forensics [5, 58]. The following subsections represent a short list of potential applications for biosensors.

11.3.1 MEDICINE, CLINICAL, AND DIAGNOSTIC APPLICATIONS

Medicine, clinical, and diagnostic applications are the major areas of attention for biosensors. In biochemical labs and clinics, electrochemical biosensors are commonly used to monitor and analyze glucose and lactic acid levels. In the realm of home health care, commercial biosensors are now becoming increasingly popular, notably for self-monitoring blood sugar with a glucometer [59]. The key advantage

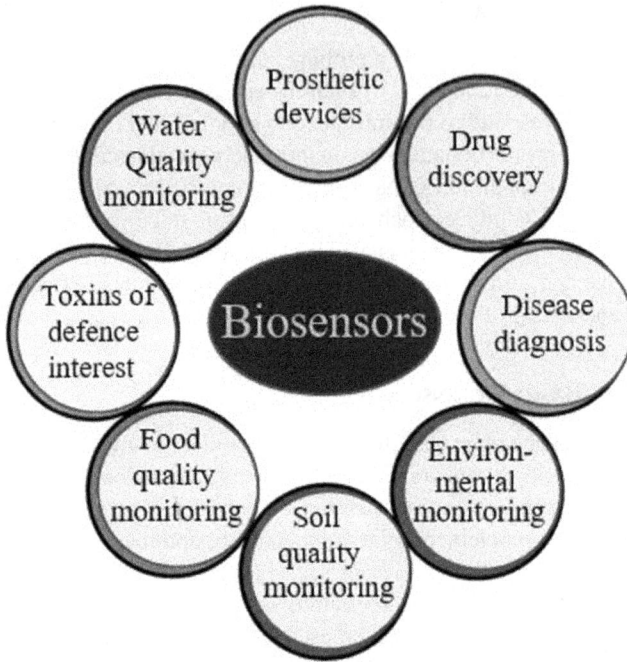

FIGURE 11.3   Different fields of application of biosensors.

of this technique is that blood samples cannot be tampered with, and the raw specimen is used for more precise results. Self-checking tools were previously single-use applications, which indicated that tests could only be conducted once, and the sensor would have to be discarded. However, advancements in this field urge the use of recyclable sensors to improve patient care.

11.3.2   ENVIRONMENTAL MONITORING

Environmental pollution analysis is one of the most important uses of biosensors. Water contamination monitoring, in particular, is an application at which biosensors excel. Numerous pollutants are constantly poisoning groundwater, resulting in deteriorating drinking water quality. In the fight against water pollution, biosensors with parts that can find nitrate and phosphate are becoming more common. The military can also use it to identify poisons and harmful biological specimens that could be used as bioweapons.

11.3.3   INDUSTRIAL APPLICATIONS

Fermentation is a widely used industrial technique employed in the production of dairy, wine, and relevant products. Large-scale microbes and cell cultures must be sustained for this purpose. Screening these elusive and expensive processes is critical to lowering manufacturing costs and ensuring safe fermentation. Biosensors are used to track and quantify the production of fermented foods.

11.3.4 Food Industry

Commercial biosensors capable of detecting alcohol, acids, carbohydrates, and other compounds are already on the market and are applied in the food sector to verify food quality by measuring alcohol, amino acids, carbohydrates, gases, and other compounds.

11.4 ELECTROCHEMICAL DNA-APTASENSORS

When electrochemical techniques are integrated with DNA-aptasensors to measure the signals caused by the contact between the DNA-aptamer recognition layer and the targeted analytes, they are termed "electrochemical DNA-aptasensors." The principal components of an electrochemical DNA-aptasensor are an electrode or substrate, a signal transduction layer, and a recognition layer. GCE (glassy carbon electrode), GPE (graphite pencil electrode), CPE (carbon paste electrode), PtE (platinum electrode), AuE (gold electrode), CuE (copper electrode), ITO (indium-tin oxide), ZnE (zinc electrode), SPCE (screen printed carbon electrode), BPGE (basal-plane pyrolytic graphite electrode), and BDDE (boron-doped diamond electrode) are used in different electrochemical DNA-aptasensors as substrate or electrode. Among them, the most common, highly available, and low-cost substrate is a carbon-based electrode. The second layer of an electrochemical DNA-aptasensor is built with either nanomaterials or nanocomposites, which generally transduce the signal to the detector. The third and final layer that interacts with the targeted analytes is the DNA-aptamer sequence, which is known as sensing elements or recognition elements of electrochemical DNA-aptasensors. Specific and selective target binding DNA-aptamer sequences (with different base pairs) are considered to be potent and very valuable recognition elements for a range of targets from metal ions [60, 61], tiny molecules, proteins, enzymes, antibiotics, cocaine, pesticides, and even to an entire organism, cells, and cancer biomarkers [10, 62–67]. They hold the benefits of a broad range of applications, relatively low cost, longer shelf life, stability at room temperature, diverse functionality, and easy operation as a comparison with traditional immunosensors (antibody-based biosensors). It can be specific to various components of the target. After the aptamers sequence is identified, this can be synthesized with high purity, relatively short time, and reproducibility. Chemical tags such as quenchers, fluorescence probes, redox probes, electrochemical indicators, enzymes, and nanoparticles can easily chemically modify aptamers. Nanocrystal tags, redox indicators, quantum dots, and enzymes are the most commonly utilized signal tags in DNA biosensor design. The mostly used redox probe/mediators are $[Ru(NH_3)_6]^{3+/2+}$, $Fe(CN)_6^{3-/4-}$ [60], DA (dopamine), or $[IrCl_6]^{2-/3-}$, HRP, ferrocene, etc., are used in solution and surface, whereas commonly used redox indicator/marker is methylene blue (MB), which changes color at a specific electrode potential and pH value [10, 62–67].

11.4.1 Nanomaterials Used in Electrochemical DNA-Aptasensors

Incorporating nanomaterials improves the analytical performance of DNA-aptasensors significantly, in line with developments in nanomaterial-based biosensors [22, 68]. The use of nanoparticles or nanocomposites as transducers in DNA-aptasensors significantly enhances the aptasensor's signal intensity or sensitivity. The development of nanoscience and nanotechnology has yielded different nanomaterials (as efficient electron transfer agents) such as carbon NMs [69] (fullerene, SWCNT [single walled carbon nanotube], c-SWCNT [carboxylated-single-walled carbon nanotube], GO [graphene oxide], rGO [reduced graphene oxide], MWCNT [multi-walled carbon nanotube], c-MWCNT [carboxylated-multi-walled carbon nanotube]), AuNPs (gold nanoparticles), AgNPs (silver nanoparticles), hybrid NPs (nanoparticles) such as graphene nanosheet doped with gold nanoparticles, polymer NMs, PtNPs (platinum NPs), palladium NPs, magnetic NPs, and metal oxide NPs such as ZnO, CeO, FeO, etc., are usually employed in the development of aptasensors for electrochemical recognition of various analytes. Those NPs, either in the form of spherical NPs or nanorod or nanotube or nanoporous materials, have improved electron transfer kinetics and a high adsorption capacity, which gives the DNA-aptasensors more stability and a higher surface-to-volume ratio for better immobilization of DNA-aptamer sequences. The features of metallic nanoparticles mix with those of polymers, carbon nanomaterials, and semiconductor oxides to enable highly sensitive recognition, high-capacity loading with biomolecules, and higher material durability in hybrid metallic nanoparticles. Materials like magnetic nanoparticles (Fe_3O_4), quantum dots (CdSe, CdS), and up-conversion nanoparticles (UPCNPs) have been utilized alongside metallic NPs in several bioanalytical concepts to increase the sensitivity of the test or perform a vital role in recognition.

11.4.2 Surface Characterization and Electrochemical Techniques Used in Electrochemical DNA-Aptasensors

Commonly used surface characterization techniques for electrochemical (EC) aptasensors are CV (cyclic voltammetry), EIS (electrochemical impedance spectroscopy), SEM (scanning electron microscopy), AFM (atomic force microscopy), and XRD (X-ray diffraction spectroscopy) (Figure 11.4) [70–74]. However, there are some techniques which are not frequently used, such as ESM (environmental scanning microscopy), Fourier-transform infrared spectroscopy (FTIR), transmission electron microscopy (TEM), scanning transmission electron microscopy (STEM), X-ray photoelectron spectroscopy (XPS), Raman spectroscopy (RS), surface plasmon resonance spectroscopy (SPRS), chronocolumetry (CC), chronoamperometry (CA), ultraviolet-visible spectroscopy (UV-vis), etc., for electrochemical aptasensor surfaces, but they are occasionally used for the characterization of surface materials after or before deposition on the surface [70, 75, 76].

All sorts of bioreceptors can be studied using electrochemical techniques [77]. Voltammetry—such as CV, DPV (differential pulse voltammetry), and SWV (square wave voltammetry)—is the most versatile electrochemical technique [21, 78]. It is used in electrochemical aptasensors frequently. Chemical and electrochemical

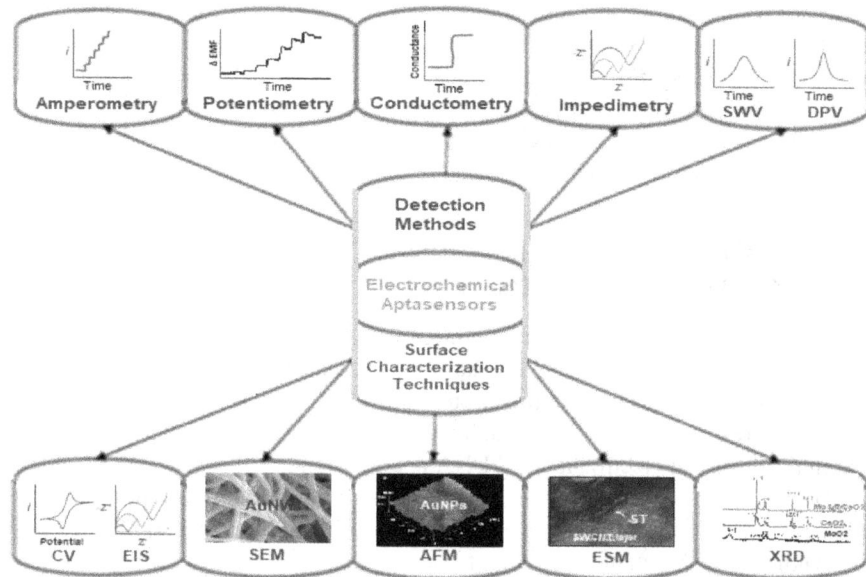

FIGURE 11.4 Surface characterization techniques and detection methods used in electrochemical aptasensor technology. SWV (square wave voltammetry), DPV (differential pulse voltammetry), CV (cyclic voltammetry), EIS (electrochemical impedance spectroscopy), SEM (scanning electron microscopy), AFM (atomic force microscopy), ESM (environmental scanning microscopy), XRD (X-ray diffraction), i: current, Z″: Imaginary impedance component, Z′: real impedance component, AuNW (gold nanowire), AuNPs (gold nanoparticles), ST (*Salmonella typhi*), SWCNT (single-walled carbon nanotubes).

reactions, as well as their coupling processes on the electrode surface, are frequently studied using the CV. SWV is a flexible and sensitive frequency-dependent electrochemical technique that is commonly utilized in quantitative and kinetic analyses of materials. Furthermore, the peak shape of SWV is straightforward for data processing. When compared to CV, DPV has a lower background current and a higher sensitivity. DPV is a detection technique used by almost all electrochemical DNA biosensors to determine target concentration.

Aside from them, EIS, amperometry, conductometry, and potentiometry are commonly used in electrochemical DNA biosensors. The electric field changes induced by the contact between bio-recognition elements and target alter impedance measurements in impedimetry. EIS is a frequency domain analysis method with a broad range of computed frequencies that provides more kinetic and electrode interface structural data than classic electrochemical approaches. At a constant applied voltage, amperometric measurements detect the current stimulated by electrochemical oxidation or reduction of electroactive species. Its advantages over typical optical sensors are speed, sensitivity (the capacity to detect even 10^{-9} M concentrations), and the ability to analyze turbid liquids. On the other hand, pH-sensing mechanisms require weakly buffered or non-buffered solutions, which is a drawback for this sort of sensor. Conductometric biosensors track variations in the electrical conductivity of a sample solution while a chemical or an enzymatic reaction with charged products progresses [79]. These measurements can be especially useful because they are quick and sensitive, albeit achieving appropriate specificity can be difficult [76]. Figure 11.4 lists several surface characterization and electrochemical methods generally used in electrochemical aptasensors.

11.5   ELECTROCHEMICAL DNA-APTASENSORS FOR THROMBIN DETECTION

Determining the amount of specific protein like thrombin in a sample is particularly fascinating and challenging. Because of the capacity to easily detect their low amounts in biological samples, this protein is a valuable diagnosis and monitoring tool for many disorders [80, 81]. Usually, thrombin is employed as a prototype protein in the design of electrochemical DNA-aptasensors.

The anti-thrombin DNA-aptamer sequence is the first example of an oligonucleotide developed in vitro to bind to a protein Figure 11.5a. Bock, et al. [82] described the synthetic DNA-aptamer coupling to human thrombin in 1992. In solution, the thrombin-binding DNA-aptamer (TBA) discovered has a well-defined folding

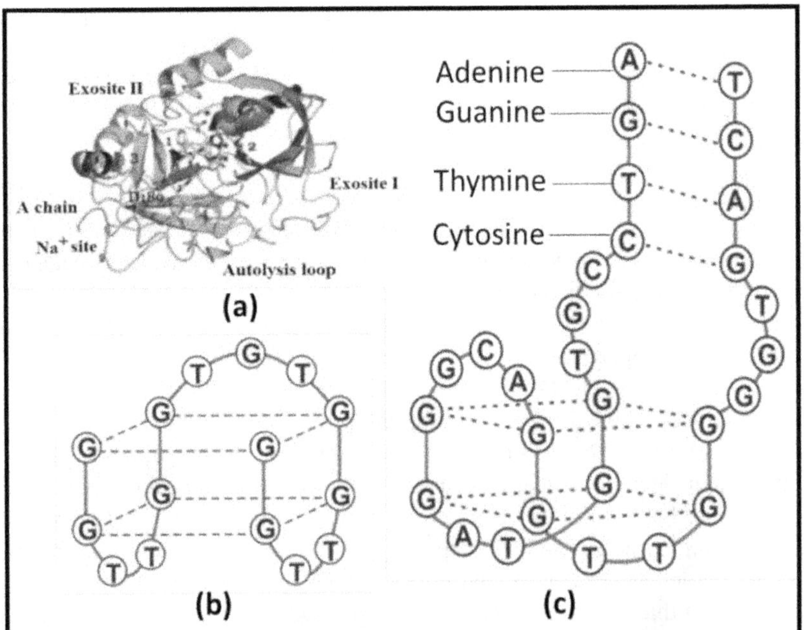

FIGURE 11.5   (a) Schematic 3-D model of thrombin with different binding sites; (b) G-quadruplex structure of thrombin-binding 14-mer DNA-aptamer; (c) G-quadruplex structure of thrombin-binding 29-mer DNA-aptamer.

structure [82]. As demonstrated in Figure 11.5b, this 15-mer DNA oligonucleotide (5′-GGT TGG TGT GGT TGG-3′) can make a strong intramolecular G-quadruple shape with an antiparallel alignment and a chair-like configuration [83]. This 15-mer thrombin aptamer binds to one of thrombin's binary anion binding positions (the fibrinogen-detection exosite) with a dissociation constant (K_d) of roughly 100 nm. Another 29-mer DNA oligonucleotide (5′-AGT CCG TGG TAG GGC AGG TTG GGG TGA CT-3′) is a thrombin-binding aptamer that attaches to the heparin-binding exosite of thrombin with a greater affinity of 0.5 nM K_d [84]. The G-quadruplex structure is also present in this 29-mer thrombin aptamer (Figure 11.5c). Two guanine quartets compose the center section of the structure.

11.5.1 CHARACTERISTICS OF TBA IMPORTANT TO BUILD ELECTROCHEMICAL DNA-APTASENSORS

Thrombin and the aptamers that bind to it are often used in electrochemical assays to show how the idea works because they have some unique properties.

1. Both DNA-aptamers are short oligos with 15- and 29-mer lengths. Synthesizing these short aptamers and adding reporter and/or capturing groups to them is relatively simple and affordable.
2. The two aptamers' K_d values differ by two orders of magnitude, which is beneficial and appropriate for designing electrochemical tests with varied binding affinities.
3. The two aptamers can bind to two different thrombin-binding sites without interfering with one another. As a result, establishing detection methods on an electrode utilizing a "sandwich" style is simple.
4. Both aptamers fold into a G-quadruplex structure, and binding to thrombin causes them to switch conformation. The G-quadruplex structure is also known as a DNAzyme, which can catalyze H_2O_2-mediated oxidation processes in the presence of hemin [85]. In a detection method, this reaction and the structural switch can be employed to generate electrochemical signals. As a result, the thrombin-binding aptamers can be used as affinity reagents as well as reporter probes.
5. Finally, thrombin is a protease enzyme that can cleave proteins and peptides with precision. Aptamers have different binding locations than enzyme catalytic sites Figure 11.5a. The enzymatic activity of thrombin could be sustained and used for the development of electrochemical sensing devices even without aptamer binding to thrombin.

11.5.2 FUNDAMENTAL STRATEGIES OF IMMOBILIZATION OF TBA IN EC DNA-APTASENSORS

Anti-thrombin DNA-aptamers have been applied as thrombin activity inhibitors and, to a lesser extent, as a bioreceptor for thrombin recognition [86]. Intense research has been carried out on electrochemical thrombin aptasensors, which are mainly based on three basic strategies or configurations (configurations 1, 2, and 3) [86], as illustrated in Figure 11.6. The recent progress of those three configurations has been made either by the incorporation of different NMs or their composites.

11.5.2.1 Configuration 1

In the most basic arrangement, a thiol terminal aptamer is bound on a gold electrode. The aptamer bound thrombin is identified by quantifying a thrombin catalyzed procedure that yields p-nitroaniline after incubation with thrombin. Optical detection of the p-nitroaniline product was possible, but electrochemical identification proved to be faster and more sensitive. Human α-thrombin is a highly selective serine protease that produces p-nitroaniline by hydrolyzing the thrombin chromogenic substrate, β-Ala-Gly-Arg-p-nitroaniline. The rate of synthesis of yellow-colored p-nitro aniline can be monitored using UV absorbance at 405 nm or electrochemical reduction of the nitro group. When an enzyme substrate is saturated, the rate of p-nitroaniline production is equivalent to the enzyme concentration. Biotin–aptamer is mounted on a streptavidin plate and treated with thrombin in the basic ELONA (enzyme-linked oligonucleotide assay) format. Incubation with the chromogenic substrate detects the bound thrombin. It has a sluggish response and requires at least 24 hours of incubation to distinguish between controls and samples. When a analogous method is utilized to identify the thrombin–aptamer complex calorimetrically in a homogeneous solution system rather than immobilized on a plate, a signal is produced in just around 30 minutes [87]. The fibrinogen exosite helps thrombin recognize substrates, cofactors, and inhibitors.

11.5.2.2 Configuration 2

The second arrangement uses a sandwich style to detect thrombin. Both of thrombin's electropositive exposits are able to be attaching the aptamer. The thiolated aptamer is mounted on a gold electrode once more before being treated with thrombin. The biotin-labeled aptamer is permitted to link to the other thrombin exosite in a second incubation stage. The biotin aptamer is treated with streptavidin-HRP before being used to create a peroxidase-labeled DNA-aptamer. Applying H_2O_2 and a diffusional osmium-based mediator, the HRP immobilized at the electrode surface is detected electrochemically. HRP activity is linked to the amount of thrombin by calibration.

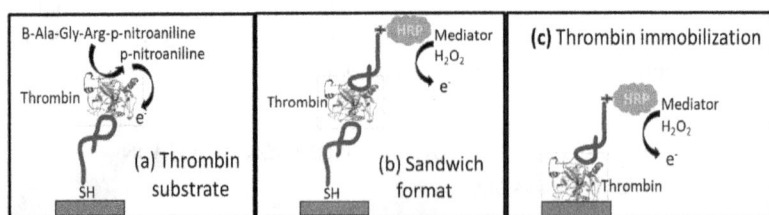

FIGURE 11.6 The three fundamental methodologies for designing an electrochemical aptasensor for thrombin recognition are depicted schematically: (a) thrombin as substrate; (b) sandwich format; (c) thrombin immobilization on the electrode surface, re-drawn [86].

Because of its many binding sites, thrombin can be used in a sandwich assay. Assay a second aptamer, this time HRP-labeled, incubated with already bound thrombin to complete the sandwich. HRP's enzymatic reaction makes it easy to detect electrochemically. HRP uses H_2O_2 as an oxidizing agent to catalyze the non-specific oxidation of a number of compounds. Previously, using an effective mediator (Med) and adding H_2O_2 to the electrocatalytic method shown in Figure 11.7 permitted electrochemical detection of peroxidase [88–91].

11.5.2.3 Configuration 3

In the third technique, thrombin is directly adsorbed on the electrode surface. The sensor is first incubated with biotin-labeled aptamer before being incubated with streptavidin-HRP. HRP is detected electrochemically once more, this time with H_2O_2 and a diffusional osmium-based mediator. Thrombin is exclusively recognized on the mercapto-treated surface and is not traceable on the unmodified surfaces, despite the fact that this arrangement hardly qualifies as a thrombin sensor.

The first technique used a thrombin-specific aptamer immobilized on an electrode surface among the three various combinations of electrochemical DNA-aptasensors. The p-nitroaniline reaction product formed by the thrombin catalyzed hydrolysis of β-Ala-Gly-Arg-p-nitroaniline is used to detect aptamer bound thrombin. The electrochemical method for detecting p-nitroaniline is 60 times quicker than the optical technique. The second method used two aptamers in a sandwich format to detect thrombin. The final method used thrombin anchored on a modified gold electrode and an HRP-labeled aptamer for recognition. The third technique attained the lower limit of detection, 3.5 nM, but the second setup generated more signal. The aptasensor could be utilized to diagnose lung metastasis in this setup, whereby the average clinical concentration of thrombin in patients with metastasis is 5.4 nM. These aptasensors produce findings in a matrix that is significantly less complicated than the clinical samples expected. While the electrodes are designed to detect thrombin, they may be affected by interference from other blood molecules, thereby increasing the system's detection limit. More investigations are required to diminish the limitation of non-specific adsorption to get more precise and accurate results. The effects of temperature, ionic strength, reusability, and portability of the aptasensor for thrombin detection can be examined in future. Furthermore, without labeling by enzymes either these aptasensors (the second and third strategies) are applicable to sense the thrombin or not, it may be investigated to avoid the expensive steps of enzyme labeling. After regeneration, the sensing ability of this aptasensor should be evaluated to get an idea about the shelf-life of the biosensor.

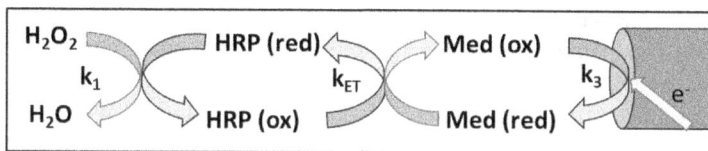

FIGURE 11.7   Electrocatalytic scheme for HRP label detection.

Although the enzyme-based recycling signal enhancement technique has received a lot of attention for boosting analytical sensitivity [92, 93], enzymes' relatively high cost and rigorous reaction conditions prevent them from being widely used. Furthermore, owing to its strong attraction between the aptamer target molecules and probe, rejuvenation of these aptasensors is difficult to achieve next to every test, preventing their recurrent use. Thus, developing a novel strategy to overcome the bottlenecks of existing electrochemical DNA-aptasensors—such as high cost or complicated preparation of electroactive DNA-aptamer probes, inadequate output signal resulting from signal-point label of signal molecules, and poor reusability after each investigation—remains a major challenge.

11.5.3 APPLICATIONS OF NANOMATERIALS IN ELECTROCHEMICAL DNA-APTASENSORS

In order to further enhance the performance of electrochemical DNA-aptasensors focusing on the sensitivity, reproducibility, LOD, antifouling ability, and robustness of the electrode surface, there are many researchers who introduce different NMs or their composites as the signal transducing elements in electrochemical DNA-aptasensors. Those NMs are attached either covalently or non-covalently with the DNA-aptamers sequences. In the non-covalent approach, negatively charged aptamers (those which hold negatively charged phosphate backbone) can be either Π-Π stacking or electrostatically assembled on the oppositely charged transduction layer in neutral buffer system with low osmolarity. For example, the 2D transducing materials such as graphene were functionalized with poly (sodium 4-styren sulfonate) and Fe_2O_3 (ferric oxide) that make the positively charged surface to adopt the negatively charged aptamers. However, the non-covalent approach regarding the Π-Π interactions [94, 95] were also used in different carbon NMs based EC DNA-aptasensors for thrombin detection but those biosensors seriously suffer from the non-specific interaction, non-homogeneous surface coverage and subsequent irreproducibility with low-stability of the apatasesnors surface [21, 96]. Those limitations have been overcome by the covalent approach. During the covalent functionalization process, the first step involves chemical or electrochemical functionalization of the bare electrode surface or the transduction layers with different functional groups of either oxygenated (e.g., carboxylic acid, -COOH, hydroxyl, -OH, aldehyde, -CHO) or amine (-NH$_2$) or thiol (-SH) groups. Those modified surfaces then covalently linked with either thiolated or -NH$_2$ terminated aptamers. Generally, amino-aptamers are covalently immobilized via the amino (-NH$_2$)- groups to the thiolic acids functionalized substrate followed by the terminal -COOH group is esterified for activation using 1-ethyl-3-[3-dimethylaminopropyl]-carbodiimide hydrochloride (EDC)/N-hydroxysulfosuccinimide (sulfo-NHS) solution [97, 98]. This approach can be applied to a range of substrates with Au (gold), Pt (platinum), glassy carbon (*gc*), carbon paste (*cp*), silicate, and polymer surfaces or different nanoparticles surfaces including single or multiwall carbon nanotubes, SWCNT or MWCNT, graphene, palladium NPs, and gold nanoparticles (AuNPs) or nanorods (AuNRs) [99, 100]. There are two terminals of aptamers—one is the 5′-end and the other the 3′-end—anyone can be modified with the suitable functional groups. However, depending on the particular aptamer a number of researchers suggest that the 3'-end is more appropriate for the biological target [101–103], because the exonucleases prefer this site as a primary target and its binding to the sensor surface would consecutively confer resistance to nucleases. Those covalent immobilization strategies are more advantageous to build a stable, sensitive, flexible, reproducible, and reusable recognition layer on the electrode surfaces [104–106].

11.6 CONCLUSIONS

The key elements of an electrochemical biosensor are conductive electrode or substrate, signal transduction component, and bio-recognition or sensing element that interacts with the targeted analyte to produce a signal. When the bio-recognition element is DNA-aptamer, this is termed as DNA-aptasensor. DNA-aptasensors with electrochemical signal transducing mode have been shown great potential in various analytes detection, offering a broad range of benefits. Electrochemical aptasensors combine the particular selectivity of aptamers with the remarkable sensitivity of electrochemical techniques and have consequently drawn a lot of interest for trace-level detection of selected analytes like thrombin. Nanomaterials and nanocomposites can improve the sensitivity, specificity, and antifouling capabilities of a thrombin aptasensor. In this chapter, we discuss different fundamental aspects with applications of biosensors and their recent progress, focusing on the most promising DNA-aptamer functionalized electrochemical biosensors for thrombin detection, considering its clinical significance.

11.7 ACKNOWLEDGMENT

This work was supported by the Universiti Malaya (grant number ST0172020).

REFERENCES

[1] A. Hashem *et al.*, "Nucleic acid-based electrochemical biosensors for rapid clinical diagnosis: Advances, challenges, and opportunities," *Critical Reviews in Clinical Laboratory Sciences*, pp. 1–22, 2021.

[2] M. Al Mamun, Y. A. Wahab, M. M. Hossain, A. Hashem, and M. R. Johan, "Electrochemical biosensors with aptamer recognition layer for the diagnosis of pathogenic bacteria: Barriers to commercialization and remediation," *TrAC Trends in Analytical Chemistry*, vol. 145, p. 116458, 2021.

[3] M. Jarczewska, Ł. Górski, and E. Malinowska, "Electrochemical aptamer-based biosensors as potential tools for clinical diagnostics," *Analytical Methods*, vol. 8, no. 19, pp. 3861–3877, 2016.

[4] M. Negahdary, A. Moradi, and H. Heli, "Application of electrochemical aptasensors in detection of cancer biomarkers," *Biomed. Res. Ther.*, vol. 6, pp. 3315–3324, 2019.

[5] E. Aydindogan, S. Balaban, S. Evran, H. Coskunol, and S. Timur, "A bottom-up approach for developing aptasensors for abused drugs: Biosensors in forensics," *Biosensors*, vol. 9, no. 4, p. 118, 2019.

[6] Y. Xiao, A. A. Rowe, and K. W. Plaxco, "Electrochemical detection of parts-per-billion lead via an electrode-bound DNAzyme assembly," *Journal of the American Chemical Society*, vol. 129, no. 2, pp. 262–263, 2007.

[7] G. March, T. D. Nguyen, and B. Piro, "Modified electrodes used for electrochemical detection of metal ions in environmental analysis," *Biosensors*, vol. 5, no. 2, pp. 241–275, 2015.

[8] L. Poltorak, E. J. Sudhölter, and M. de Puit, "Electrochemical cocaine (bio) sensing: From solid electrodes to soft junctions," *TrAC Trends in Analytical Chemistry*, vol. 114, pp. 48–55, 2019.

[9] Z. Li *et al.*, "Application of electrochemical aptasensors toward clinical diagnostics, food, and environmental monitoring," *Sensors*, vol. 19, no. 24, p. 5435, 2019.

[10] Y. Jiao *et al.*, "An ultrasensitive aptasensor for chlorpyrifos based on ordered mesoporous carbon/ferrocene hybrid multiwalled carbon nanotubes," *Rsc Advances*, vol. 6, no. 63, pp. 58541–58548, 2016.

[11] S. Tombelli, M. Minunni, and M. Mascini, "Analytical applications of aptamers," *Biosensors and Bioelectronics*, vol. 20, no. 12, pp. 2424–2434, 2005.

[12] Y. Xiao, R. Y. Lai, and K. W. Plaxco, "Preparation of electrode-immobilized, redox-modified oligonucleotides for electrochemical DNA and aptamer-based sensing," *Nature Protocols*, vol. 2, no. 11, p. 2875, 2007.

[13] S. Pang, T. P. Labuza, and L. He, "Development of a single aptamer-based surface enhanced Raman scattering method for rapid detection of multiple pesticides," *Analyst*, vol. 139, no. 8, pp. 1895–1901, 2014.

[14] T. Vo-Dinh, B. M. Cullum, and D. L. Stokes, "Nanosensors and biochips: Frontiers in biomolecular diagnostics," *Sensors and Actuators B: Chemical*, vol. 74, no. 1, pp. 2–11, 2001, https://doi.org/10.1016/S0925-4005(00)00705-X.

[15] T. Haruyama, "Micro-and nanobiotechnology for biosensing cellular responses," *Advanced Drug Delivery Reviews*, vol. 55, no. 3, pp. 393–401, 2003.

[16] K. K. Jain, "Nanodiagnostics: Application of nanotechnology in molecular diagnostics," *Expert Review of Molecular Diagnostics*, vol. 3, no. 2, pp. 153–161, 2003.

[17] C. Altay, R. H. Senay, E. Eksin, G. Congur, A. Erdem, and S. Akgol, "Development of amino functionalized carbon coated magnetic nanoparticles and their application to electrochemical detection of hybridization of nucleic acids," *Talanta*, vol. 164, pp. 175–182, 2017.

[18] A. Tabasi, A. Noorbakhsh, and E. Sharifi, "Reduced graphene oxide-chitosan-aptamer interface as new platform for ultrasensitive detection of human epidermal growth factor receptor 2," *Biosensors and Bioelectronics*, vol. 95, pp. 117–123, 2017.

[19] F. Dridi, M. Marrakchi, M. Gargouri, J. Saulnier, N. Jaffrezic-Renault, and F. Lagarde, "Nanomaterial-based electrochemical biosensors for food safety and quality assessment," in *Nanobiosensors*. Elsevier, 2017, pp. 167–204.

[20] K. Ariga and T. Kunitake, *Supramolecular chemistry-fundamentals and applications: Advanced textbook*. Springer Science & Business Media, 2006.

[21] M. Mamun and A. S. Ahammad, "Characterization of carboxylated-SWCNT based potentiometric DNA sensors by electrochemical technique and comparison with potentiometric performance," *Journal of Biosensors & Bioelectronics*, vol. 5, no. 3, p. 1, 2014.

[22] A. Hashem, M. M. Hossain, M. Al Mamun, K. Simarani, and M. R. Johan, "Nanomaterials based electrochemical nucleic acid biosensors for environmental monitoring: A review," *Applied Surface Science Advances*, vol. 4, p. 100064, 2021.

[23] K. Yang and C.-Y. Zhang, "Simple detection of nucleic acids with a single-walled carbon-nanotube-based electrochemical biosensor," *Biosensors and Bioelectronics*, vol. 28, no. 1, pp. 257–262, 2011.

[24] S. Shahrokhian, R. Salimian, and H. R. Kalhor, "A simple label-free electrochemical DNA biosensor based on carbon nanotube: DNA interaction," *RSC Advances*, vol. 6, no. 19, pp. 15592–15598, 2016.

[25] F. J. Hernandez and V. C. Ozalp, "Graphene and other nanomaterial-based electrochemical aptasensors," *Biosensors*, vol. 2, no. 1, pp. 1–14, 2012.

[26] W.-W. Zhao, J.-J. Xu, and H.-Y. Chen, "Photoelectrochemical DNA biosensors," *Chemical Reviews*, vol. 114, no. 15, pp. 7421–7441, 2014.

[27] D. Daems, W. Pfeifer, I. Rutten, B. Saccà, D. Spasic, and J. Lammertyn, "Three-dimensional DNA origami as programmable anchoring points for bioreceptors in fiber optic surface plasmon resonance biosensing," *ACS Applied Materials & Interfaces*, vol. 10, no. 28, pp. 23539–23547, 2018.

[28] V. Pavlov, Y. Xiao, B. Shlyahovsky, and I. Willner, "Aptamer-functionalized Au nanoparticles for the amplified optical detection of thrombin," *Journal of the American Chemical Society*, vol. 126, no. 38, pp. 11768–11769, 2004.

[29] B. Basnar, R. Elnathan, and I. Willner, "Following aptamer-thrombin binding by force measurements," *Analytical Chemistry*, vol. 78, no. 11, pp. 3638–3642, 2006.

[30] W. Song, H. Li, H. Liang, W. Qiang, and D. Xu, "Disposable electrochemical aptasensor array by using in situ DNA hybridization inducing silver nanoparticles aggregate for signal amplification," *Analytical Chemistry*, vol. 86, no. 5, pp. 2775–2783, 2014.

[31] R. Ahmad, H. Jang, B. S. Batule, and H. G. Park, "Barcode DNA-mediated signal amplifying strategy for ultrasensitive biomolecular detection on matrix-assisted laser desorption ionization time of flight (MALDI-TOF) mass spectrometry," *Analytical Chemistry*, vol. 89, no. 17, pp. 8966–8973, 2017.

[32] B. V. Chikkaveeraiah, A. A. Bhirde, N. Y. Morgan, H. S. Eden, and X. Chen, "Electrochemical immunosensors for detection of cancer protein biomarkers," *ACS Nano*, vol. 6, no. 8, pp. 6546–6561, 2012.

[33] H. Yousef, Y. Liu, and L. Zheng, "Nanomaterial-based label-free electrochemical aptasensors for the detection of thrombin," *Biosensors*, vol. 12, no. 4, p. 253, 2022.

[34] G. Shen, H. Zhang, C. Yang, Q. Yang, and Y. Tang, "Thrombin ultrasensitive detection based on chiral supramolecular assembly signal-amplified strategy induced by thrombin-binding aptamer," *Analytical Chemistry*, vol. 89, no. 1, pp. 548–551, 2017.

[35] C. Riccardi, E. Napolitano, C. Platella, D. Musumeci, and D. Montesarchio, "G-quadruplex-based aptamers targeting human thrombin: Discovery, chemical modifications and antithrombotic effects," *Pharmacology & Therapeutics*, vol. 217, p. 107649, 2021.

[36] E. Campello *et al.*, "Thrombin generation in patients with COVID-19 with and without thromboprophylaxis," *Clinical Chemistry and Laboratory Medicine (CCLM)*, vol. 59, no. 7, pp. 1323–1330, 2021.

[37] M. Ranucci *et al.*, "Covid-19-associated coagulopathy: Biomarkers of thrombin generation and fibrinolysis leading the outcome," *Journal of Clinical Medicine*, vol. 9, no. 11, p. 3487, 2020.

[38] N. L. Rosi and C. A. Mirkin, "Nanostructures in biodiagnostics," *Chemical Reviews*, vol. 105, no. 4, pp. 1547–1562, 2005.

[39] J. Wu, W. Pisula, and K. Müllen, "Graphenes as potential material for electronics," *Chemical Reviews*, vol. 107, no. 3, pp. 718–747, 2007.

[40] L. Li *et al.*, "Bacterial analysis using an electrochemical DNA biosensor with poly-adenine-mediated DNA self-assembly," *ACS Applied Materials & Interfaces*, vol. 10, no. 8, pp. 6895–6903, 2018.

[41] M. Lin, X. Yi, H. Wan, J. Zhang, F. Huang, and F. Xia, "Photoresponsive electrochemical DNA biosensors achieving various dynamic ranges by using only-one capture probe," *Analytical Chemistry*, vol. 92, no. 14, pp. 9963–9970, 2020.

[42] S. Han, W. Liu, M. Zheng, and R. Wang, "Label-free and ultrasensitive electrochemical DNA biosensor based on urchinlike carbon nanotube-gold nanoparticle nanoclusters," *Analytical Chemistry*, vol. 92, no. 7, pp. 4780–4787, 2020.

[43] Q. Hu, J. Kong, D. Han, L. Niu, and X. Zhang, "Electrochemical DNA biosensing via electrochemically controlled reversible addition: Fragmentation chain transfer polymerization," *ACS Sensors*, vol. 4, no. 1, pp. 235–241, 2019.

[44] Q. Wu, Y. Zhang, Q. Yang, N. Yuan, and W. Zhang, "Review of electrochemical DNA biosensors for detecting food borne pathogens," *Sensors*, vol. 19, no. 22, p. 4916, 2019.

[45] C.-T. Yang, L. Wu, P. Bai, and B. Thierry, "Investigation of plasmonic signal enhancement based on long range surface plasmon resonance with gold nanoparticle tags," *Journal of Materials Chemistry C*, vol. 4, no. 41, pp. 9897–9904, 2016.

[46] W. Tang *et al.*, "Label-free potentiometric aptasensing platform for the detection of Pb2+ based on guanine quadruplex structure," *Analytica Chimica Acta*, vol. 1078, pp. 53–59, 2019.

[47] N. F. Silva, C. M. Almeida, J. M. Magalhães, M. P. Gonçalves, C. Freire, and C. Delerue-Matos, "Development of a disposable paper-based potentiometric immunosensor for real-time detection of a foodborne pathogen," *Biosensors and Bioelectronics*, vol. 141, p. 111317, 2019.

[48] Q. Wang *et al.*, "Label-free aptamer biosensor for thrombin detection based on functionalized graphene nanocomposites," *Talanta*, vol. 141, pp. 247–252, 2015.

[49] C. Ocaña and M. del Valle, "Three different signal amplification strategies for the impedimetric sandwich detection of thrombin," *Analytica Chimica Acta*, vol. 912, pp. 117–124, 2016.

[50] Z. Su *et al.*, "Amperometric thrombin aptasensor using a glassy carbon electrode modified with polyaniline and multiwalled carbon nanotubes tethered with a thiolated aptamer," *Microchimica Acta*, vol. 184, no. 6, pp. 1677–1682, 2017.

[51] J. Zhao *et al.*, "Dual amplification strategy of highly sensitive thrombin amperometric aptasensor based on chitosan—Au nanocomposites," *Analyst*, vol. 137, no. 15, pp. 3488–3495, 2012.

[52] B. Rafique, M. Iqbal, T. Mehmood, and M. A. Shaheen, "Electrochemical DNA biosensors: A review," *Sensor Review*, vol. 39, no. 1, pp. 34–50, 2019.

[53] A. Mehlhorn, P. Rahimi, and Y. Joseph, "Aptamer-based biosensors for antibiotic detection: A review," *Biosensors*, vol. 8, no. 2, p. 54, 2018.

[54] M. Negahdary, A. Moradi, and H. Heli, "Application of electrochemical aptasensors in detection of cancer biomarkers," *Biomedical Research and Therapy*, vol. 6, no. 7, pp. 3315–3324, 2019.

[55] L. Trnkova, F. Jelen, J. Petrlova, V. Adam, D. Potesil, and R. Kizek, "Elimination voltammetry with linear scan as a new detection method for DNA sensors," *Sensors*, vol. 5, no. 6, pp. 448–464, 2005.

[56] S. Kiran and R. Misra, "Glucose biosensors: Progress, current focus and future outlook," *Materials Technology*, vol. 30, no. sup7, pp. B140–B149, 2015.

[57] S. Patel, R. Nanda, S. Sahoo, and E. Mohapatra, "Biosensors in health care: The milestones achieved in their development towards lab-on-chip-analysis," *Biochemistry Research International*, vol. 2016, 2016.

[58] L. Poltorak, E. J. Sudhölter, and M. de Puit, "Electrochemical cocaine (bio) sensing: From solid electrodes to soft junctions," *TrAC Trends in Analytical Chemistry*, vol. 114, pp. 48–55, 2019.

[59] A. Ahammad, A. Al Mamun, T. Akter, M. Mamun, S. Faraezi, and F. Monira, "Enzyme-free impedimetric glucose sensor based on gold nanoparticles/polyaniline composite film," *Journal of Solid State Electrochemistry*, vol. 20, no. 7, pp. 1933–1939, 2016.

[60] S. Amiri, A. Navaee, A. Salimi, and R. Ahmadi, "Zeptomolar detection of Hg2+ based on label-free electrochemical aptasensor: One step closer to the dream of single atom detection," *Electrochemistry Communications*, vol. 78, pp. 21–25, 2017.

[61] Y. Lin *et al.*, "Reagentless, electrochemical aptasensor for lead (II) detection," *Journal of New Materials for Electrochemical Systems*, vol. 20, no. 1, 2017.

[62] J. Song, S. Li, F. Gao, Q. Wang, and Z. Lin, "An in situ assembly strategy for the construction of a sensitive and reusable electrochemical aptasensor," *Chemical Communications*, vol. 55, no. 7, pp. 905–908, 2019.

[63] E. González-Fernández, N. de-los-Santos-Álvarez, M. J. Lobo-Castañón, A. J. Miranda-Ordieres, and P. Tuñón-Blanco, "Aptamer-based inhibition assay for the electrochemical detection of tobramycin using magnetic microparticles," *Electroanalysis*, vol. 23, no. 1, pp. 43–49, 2011.

[64] G. Shen, Y. Guo, X. Sun, and X. Wang, "Electrochemical aptasensor based on prussian blue-chitosan-glutaraldehyde for the sensitive determination of tetracycline," *Nano-Micro Letters*, vol. 6, no. 2, pp. 143–152, 2014.

[65] B. P. Crulhas, C. R. Basso, J. P. Parra, G. R. Castro, and V. A. Pedrosa, "Reduced graphene oxide decorated with AuNPs as a new aptamer-based biosensor for the detection of androgen receptor from prostate cells," *Journal of Sensors*, vol. 2019, 2019.

[66] K. Wang, M.-Q. He, F.-H. Zhai, R.-H. He, and Y.-L. Yu, "A novel electrochemical biosensor based on polyadenine modified aptamer for label-free and ultrasensitive detection of human breast cancer cells," *Talanta*, vol. 166, pp. 87–92, 2017.

[67] D. Jiang *et al.*, "Silver nanoparticles anchored on nitrogen-doped graphene as a novel electrochemical biosensing platform with enhanced sensitivity for aptamer-based pesticide assay," *Analyst*, vol. 140, no. 18, pp. 6404–6411, 2015.

[68] X. Hai, Y. Li, C. Zhu, W. Song, J. Cao, and S. Bi, "DNA-based label-free electrochemical biosensors: From principles to applications," *TrAC Trends in Analytical Chemistry*, vol. 133, p. 116098, 2020.

[69] S. Muniandy *et al.*, "Carbon nanomaterial-based electrochemical biosensors for foodborne bacterial detection," *Critical Reviews in Analytical Chemistry*, vol. 49, no. 6, pp. 510–533, 2019.

[70] J. Sabaté del Río, O. Y. Henry, P. Jolly, and D. E. Ingber, "An antifouling coating that enables affinity-based electrochemical biosensing in complex biological fluids," *Nature Nanotechnology*, vol. 14, no. 12, pp. 1143–1149, 2019.

[71] M. R. Hasan *et al.*, "Carbon nanotube-based aptasensor for sensitive electrochemical detection of whole-cell Salmonella," *Analytical Biochemistry*, vol. 554, pp. 34–43, 2018.

[72] M. Mamun, O. Ahmed, P. Bakshi, and M. Ehsan, "Synthesis and spectroscopic, magnetic and cyclic voltammetric characterization of some metal complexes of methionine:[(C5H10NO2S) 2MII]; MII= Mn (II), Co (II), Ni (II), Cu (II), Zn (II), Cd (II) and Hg (II)," *Journal of Saudi Chemical Society*, vol. 14, no. 1, pp. 23–31, 2010.

[73] M. Mamun, O. Ahmed, P. Bakshi, S. Yamauchi, and M. Ehsan, "Synthesis and characterization of some metal complexes of cystine:[Mn (C6H10N2O4S2)]; where MII= Mn (II), Co (II), Ni (II), Cu (II), Zn (II), Cd (II), Hg (II) and Pb (II)," *Russian Journal of Inorganic Chemistry*, vol. 56, no. 12, pp. 1972–1980, 2011.

[74] M. M. Hasan, M. E. Hossain, M. Mamun, and M. Ehsan, "Study of redox behavior of Cd (II) and interaction of Cd (II) with proline in the aqueous medium using cyclic voltammetry," *Journal of Saudi Chemical Society*, vol. 16, no. 2, pp. 145–151, 2012.

[75] R. Garcia and E. T. Herruzo, "The emergence of multifrequency force microscopy," *Nature Nanotechnology*, vol. 7, no. 4, pp. 217–226, 2012.

[76] A. L. Furst and M. B. Francis, "Impedance-based detection of bacteria," *Chemical Reviews*, vol. 119, no. 1, pp. 700–726, 2018.

[77] N. F. Silva, J. M. Magalhães, C. Freire, and C. Delerue-Matos, "Electrochemical biosensors for Salmonella: State of the art and challenges in food safety assessment," *Biosensors and Bioelectronics*, vol. 99, pp. 667–682, 2018.

[78] K. L. Rahman, M. Mamun, and M. Ehsan, "Preparation of metal Niacin complexes and characterization using spectroscopic and electrochemical techniques," *Russian Journal of Inorganic Chemistry*, vol. 56, no. 9, pp. 1436–1442, 2011.

[79] L. Su, W. Jia, C. Hou, and Y. Lei, "Review: Microbial biosensors," *Biosens. Bioelectron*, vol. 26, no. 5, pp. 1788–1799, 2011.

[80] A. Csordas *et al.*, "Detection of proteins in serum by micromagnetic aptamer PCR (MAP) technology," *Angewandte Chemie International Edition*, vol. 49, no. 2, pp. 355–358, 2010.

[81] S. M. Hanash, "Why have protein biomarkers not reached the clinic?," *Genome Medicine*, vol. 3, no. 10, pp. 1–2, 2011.

[82] L. C. Bock, L. C. Griffin, J. A. Latham, E. H. Vermaas, and J. J. Toole, "Selection of single-stranded DNA molecules that bind and inhibit human thrombin," *Nature*, vol. 355, no. 6360, pp. 564–566, 1992.

[83] R. F. Macaya, P. Schultze, F. W. Smith, J. A. Roe, and J. Feigon, "Thrombin-binding DNA aptamer forms a unimolecular quadruplex structure in solution," *Proceedings of the National Academy of Sciences*, vol. 90, no. 8, pp. 3745–3749, 1993.

[84] D. M. Tasset, M. F. Kubik, and W. Steiner, "Oligonucleotide inhibitors of human thrombin that bind distinct epitopes," *Journal of Molecular Biology*, vol. 272, no. 5, pp. 688–698, 1997.

[85] D.-M. Kong, J. Xu, and H.-X. Shen, "Positive effects of ATP on G-quadruplex-hemin DNAzyme-mediated reactions," *Analytical Chemistry*, vol. 82, no. 14, pp. 6148–6153, 2010.

[86] M. Mir, M. Vreeke, and I. Katakis, "Different strategies to develop an electrochemical thrombin aptasensor," *Electrochemistry Communications*, vol. 8, no. 3, pp. 505–511, 2006.

[87] K. Padmanabhan, K. Padmanabhan, J. Ferrara, J. E. Sadler, and A. Tulinsky, "The structure of alpha-thrombin inhibited by a 15-mer single-stranded DNA aptamer," *Journal of Biological Chemistry*, vol. 268, no. 24, pp. 17651–17654, 1993.

[88] T. J. Rydel *et al.*, "The structure of a complex of recombinant hirudin and human α-thrombin," *Science*, vol. 249, no. 4966, pp. 277–280, 1990.

[89] M. Vreeke, P. Rocca, and A. Heller, "Direct electrical detection of dissolved biotinylated horseradish peroxidase, biotin, and avidin," *Analytical Chemistry*, vol. 67, no. 2, pp. 303–306, 1995.

[90] M. Vreeke, R. Maidan, and A. Heller, "Hydrogen peroxide and. beta.-nicotinamide adenine dinucleotide sensing amperometric electrodes based on electrical connection of horseradish peroxidase redox centers to electrodes through a three-dimensional electron relaying polymer network," *Analytical Chemistry*, vol. 64, no. 24, pp. 3084–3090, 1992.

[91] I. Katakis and A. Heller, "L-. alpha.-glycerophosphate and L-lactate electrodes based on the electrochemical wiring of oxidases," *Analytical Chemistry*, vol. 64, no. 9, pp. 1008–1013, 1992.

[92] P. Miao, T. Zhang, J. Xu, and Y. Tang, "Electrochemical detection of miRNA combining T7 exonuclease-assisted cascade signal amplification and DNA-templated copper nanoparticles," *Analytical Chemistry*, vol. 90, no. 18, pp. 11154–11160, 2018.

[93] H. Cheng *et al.*, "Low background cascade signal amplification electrochemical sensing platform for tumor-related mRNA quantification by target-activated hybridization chain reaction and electroactive cargo release," *Analytical Chemistry*, vol. 90, no. 21, pp. 12544–12552, 2018.

[94] E. Heydari-Bafrooei, M. Amini, and M. H. Ardakani, "An electrochemical aptasensor based on TiO2/MWCNT and a novel synthesized Schiff base nanocomposite for the

ultrasensitive detection of thrombin," *Biosensors and Bioelectronics*, vol. 85, pp. 828–836, 2016.

[95] K. Park, "Impedance technique-based label-free electrochemical aptasensor for thrombin using single-walled carbon nanotubes-casted screen-printed carbon electrode," *Sensors*, vol. 22, no. 7, p. 2699, 2022.

[96] A. H. Loo, A. Bonanni, and M. Pumera, "Impedimetric thrombin aptasensor based on chemically modified graphenes," *Nanoscale*, vol. 4, no. 1, pp. 143–147, 2012.

[97] S. Balamurugan, A. Obubuafo, S. A. Soper, and D. A. Spivak, "Surface immobilization methods for aptamer diagnostic applications," *Analytical and Bioanalytical Chemistry*, vol. 390, no. 4, pp. 1009–1021, 2008.

[98] J. H. Kim, J.-H. Jin, J.-Y. Lee, E. J. Park, and N. K. Min, "Covalent attachment of biomacromolecules to plasma-patterned and functionalized carbon nanotube-based devices for electrochemical biosensing," *Bioconjugate Chemistry*, vol. 23, no. 10, pp. 2078–2086, 2012.

[99] J. Zhou, M. R. Battig, and Y. Wang, "Aptamer-based molecular recognition for biosensor development," *Analytical and Bioanalytical Chemistry*, vol. 398, no. 6, pp. 2471–2480, 2010.

[100] X. Liu, Y. Li, J. Zheng, J. Zhang, and Q. Sheng, "Carbon nanotube-enhanced electrochemical aptasensor for the detection of thrombin," *Talanta*, vol. 81, no. 4–5, pp. 1619–1624, 2010.

[101] A. Hayat and J. L. Marty, "Aptamer based electrochemical sensors for emerging environmental pollutants," *Frontiers in Chemistry*, vol. 2, p. 41, 2014.

[102] T. Hianik and J. Wang, "Electrochemical aptasensors—recent achievements and perspectives," *Electroanalysis: An International Journal Devoted to Fundamental and Practical Aspects of Electroanalysis*, vol. 21, no. 11, pp. 1223–1235, 2009.

[103] M. Velasco-Garcia and S. Missailidis, "New trends in aptamer-based electrochemical biosensors," *Gene Therapy and Molecular Biology*, vol. 13, no. 1, pp. 1–10, 2009.

[104] A.-E. Radi, "Electrochemical aptamer-based biosensors: recent advances and perspectives," *International Journal of Electrochemistry*, vol. 2011, 2011.

[105] C. Wang *et al.*, "The electromagnetic property of chemically reduced graphene oxide and its application as microwave absorbing material," *Applied Physics Letters*, vol. 98, no. 7, p. 072906, 2011.

[106] Y. Xu, G. Cheng, P. He, and Y. Fang, "A review: Electrochemical aptasensors with various detection strategies," *Electroanalysis: An International Journal Devoted to Fundamental and Practical Aspects of Electroanalysis*, vol. 21, no. 11, pp. 1251–1259, 2009.

12 Functional Nanomaterials for Biosensors and Bio-Applications

Dhanalakshmi Vadivel, and Daniele Dondi

CONTENTS

12.1 INTRODUCTION

Nanotechnology (Figure 12.1) is the major driving force of the biomaterials world for change in dimensions of thinking about human existence as evolved. Nanotechnology has different applications in diverse fields like energy, medicine, drugs, engineering, defense, fabrics, and cosmetics, but is not limited to these [1]. At the basic level explanation, nanoparticles (NPs) are sizes dependent from 1–100 nm. The definition is extended because carbon-based materials raise in the field of material even in a microscale, which also alters the idea of dimension. Nanoparticles can have different shapes, sizes, dimensions, and properties; in fact, they are not only in three dimensions (3D) but 2D, 1D, and 0D could also be possible. Their encapsulation, functionalization, and biocompatibility play a major role in the application of nanomaterials, including every nanomaterial having its properties. This chapter focuses on metal nanomaterials, carbon materials, metal oxides with carbon-based materials, DNA-modified nanoparticles, protein-modified nanoparticles, self-healing biomaterials, and nanomaterials in tissue engineering, showing different biosensors and future applications in the next level of research.

DOI: 10.1201/9781003263852-12

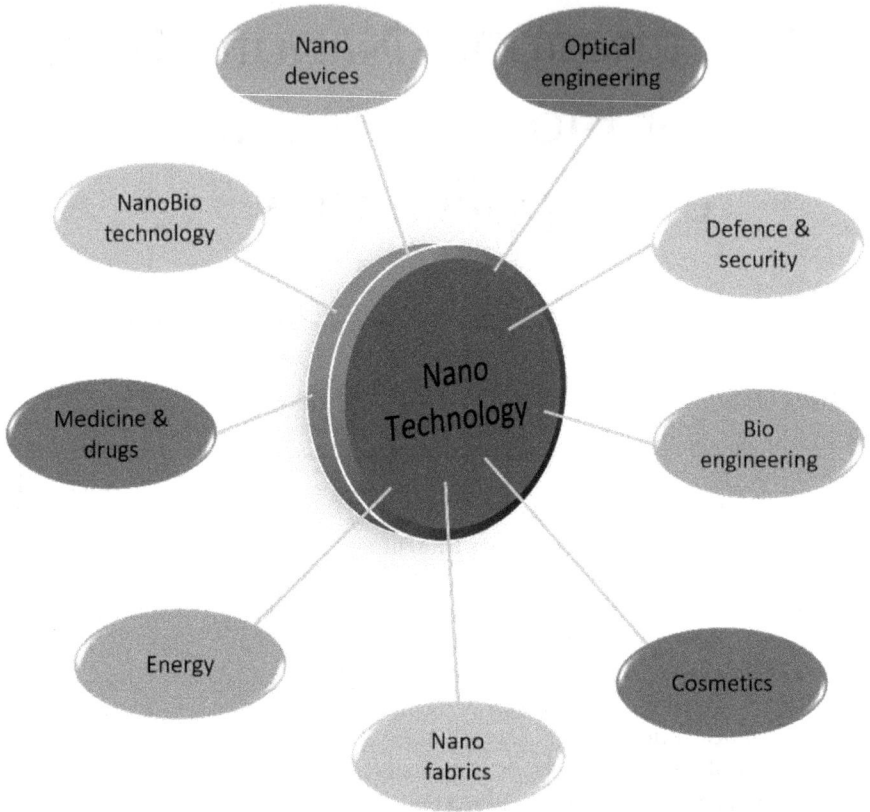

FIGURE 12.1   General nanotechnology applications.

The evolution of nanomaterials for almost two decades has been enormous due to different parameters responsible for numerous properties and applications of the material. A biosensor is derived from the coupling of a ligand-receptor binding reaction to a signal transducer. Much biosensor research has been devoted to the evaluation of the relative merits of various signal transduction methods including optical, radioactive, electrochemical, piezoelectric, magnetic, micromechanical, and mass spectrometric. NPs are currently being used for bioanalysis, including metallic (like gold and silver, Au and Ag) NPs [2], quantum dots (QDs) [3], magnetic NPs, lanthanide NPs [4,5], and silica NPs [6]. Different NPs have their unique properties and have been adapted to different applications in the bioanalysis field. it is necessary to fabricate stable, sensitive, and selective sensors. In nanoparticles, metal nanoparticles gain more attention in the earlier stage with their differential synthetic methods. The noble metal nanoparticles are much more interesting because of their stability and biocompatibility.

Surface-enhanced Raman spectroscopy is a promising technique for routine point-of-care diagnosis, as it allows selective detection of biomolecules via their vibration fingerprints and offers numerous advantages such as high sensitivity, rapid response, ease of use, and no or minimal sample preparation.

SERS detection of the analyte of interest is an easy and sensitive manner of detection. This could be possibly achieved in two different ways: these are the direct label-free technique and the indirect SERS label technique (Figure 12.2). In the direct label, analyte has the properties of SERS by itself when in contact with nanoparticles They are typically organic molecules and/or conjugated organic molecules like pesticides [7], dyes [8], and many explosives [9,10]. This is the most straightforward method of detection of an analyte in sensors. But it has the other disadvantage when it comes to the biosensor. Because biomolecules are normally less Raman active, almost no SERS properties are exhibited. Most of the biomolecules have indeed complex structures and the interaction with nanoparticles creates interference. For this reason, an indirect protocol is needed. In the indirect protocol, the sensing of biomolecules is achieved in a basic way that is relevant to the chapter of interest, biosensors [11]. Nanoparticles are coupled with antibodies, allowing the linkage with biomolecules of interest and organic dyes exhibiting the SERS property. Nanoparticles show SERS effect that is decreased when even a lower number of biomolecules link to the surface antibodies.

The detection limit to be 10^{-6} M, achieved for uric acid in aqueous solutions, was found by using colloidal nanoparticles as a SERS substrate. Another example reports the preparation of an integrated miniature three-electrode electrochemical surface-enhanced Raman spectroscopy chip (EC-SERS chip) which is suitable for rapid EC-SERS detection of biomolecules. In this experimental condition showing the result of 8.5×10^6, a strong enhancement factor with a very good EC-SERS regularity with a standard deviation of 1.41% is achieved for uric acid identification [12] by using SERS with a lower detection limit of approximately 0.2 μM for multiple characteristic peaks in the fingerprint region [13] by using Cucurbit[n]uril.

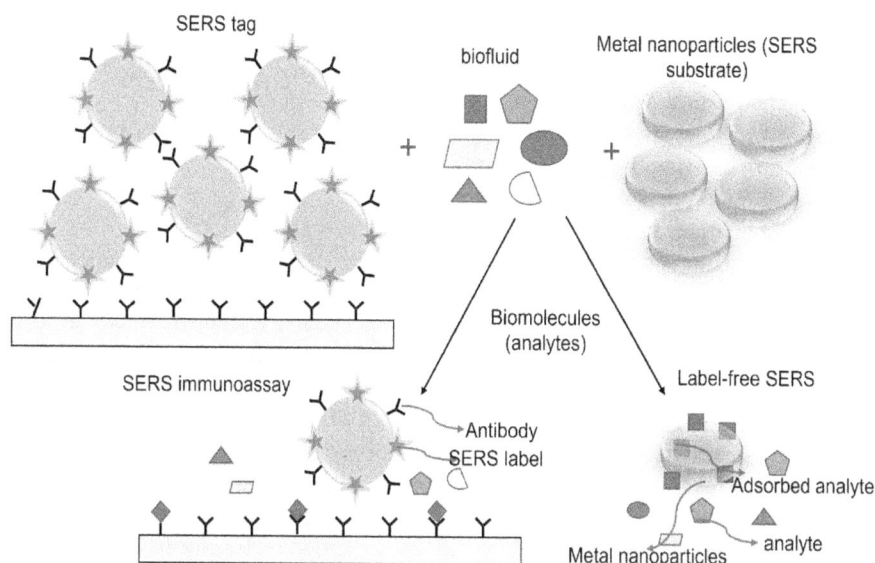

FIGURE 12.2 Metal nanoparticles in sensing for SERS materials and metal nanoparticles tagged with SERS and antibodies for selective binding.

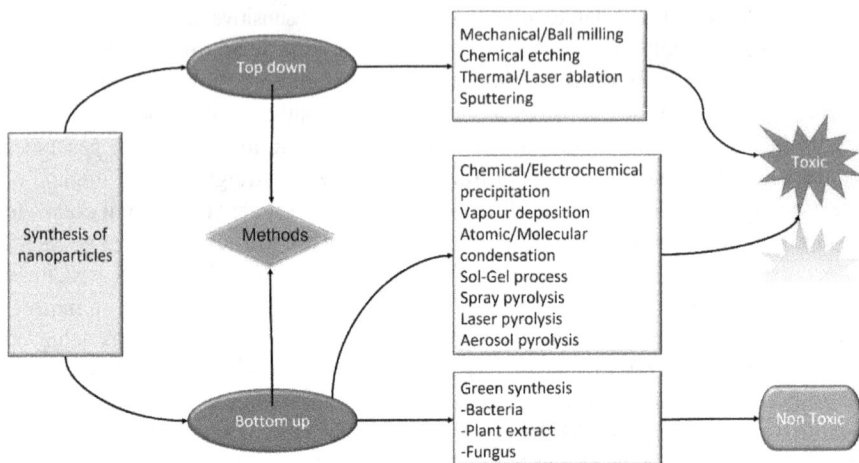

FIGURE 12.3   Preparation of nanomaterials.

12.2 METHODS OF NANOPARTICLES SYNTHESIS

The synthesis of nanoparticles was achieved in two major ways, top-down and bottom-up methods (Figure 12.3). Both approaches might use harmful chemicals, but in the bottom-up approach, there was a greenway synthesis which was also achieved. This method was highly preferred in this decade to avoid strong solvents and harsh conditions.

12.3 METAL NANOMATERIALS IN SENSING

A gold nanoparticle is the best example of noble metal nanoparticles which are treasured more in the development of nanomaterials and their applications in the biosensor and biomedical fields because of their higher stability, size, shape, and differential properties with their ease synthesis methods. Together with gold nanoparticles, gold nanorods (AuNRs) are also used for the development of new metallic surfaces to analyze and functionalize for biomedical applications such as biosensing and bioimaging tissue engineering. This wide selection is because of interesting optical properties and biomedical applications [14–17]. AuNRs can show the production of a robust surface plasmon optical phenomenon within the range of NIR. The AuNRs is an effective quencher with the immense absorption coefficients of 104-fold to 106-fold with respect to organic dyes within the specified limit of dyes concentration and molar extinction coefficient [18,19].

The magnitude of an optical phenomenon within the NIR region is often altered by tuning the features of size and ratio. Surface plasmon resonance (SPR) properties of metal films [20] are highly common in nanomaterials of gold. This surface plasmon property has the tendency of influence the dye molecules of interest. This way of approaching nanoscience for the analyte detection has vast applications and sensitivity because of small changes in the surface of nanoparticles can give a large observation in the analysis basically like in the UV-visible absorption and emission spectral analysis. The influence of analytes—for example, biomolecules like basic DNA, RNA, and nucleobases—have strong interference as the observation that make us understand the sensing.

SERS substrates–based noble metal nanoparticles (for example, Au NPs) have the tendency to increase the Raman signals of the analyte molecules up to ten orders of magnitude [21] by surface plasmon resonance through strong electromagnetic enhancement [22]. Au NPs of customized sizes can be achieved easily [23], and these are widely used in biomedical applications due to their unique properties

The highest seeking attention of the nanomaterial is because the molecular and plasmonic resonance hybridization between nanoparticles and dyes can help to obtain the distinct optical properties that completely depend on the relative spectral overlay of the dyes and gold nanoparticles or AuNRs. This overlay could be the donor-acceptor relationship, Forster resonance energy transfer (FRET) and energy transfer, etc., if it is a donor-acceptor principle, the emission of the donor as energy must have the overlay of acceptor absorption [24].

Au@SiO2 core-shell nanoparticles are used as a biosensor for uric acid and glucose with detection limits of 10^{-11}–10^{-4} M and 10^{-3}–10^{-12} M, respectively. Rhodamine 6G dye is used to evaluate the SERS-enhanced factor. The increased SERS signal was achieved by lowering the silica shell in Au@SiO2 (Figure 12.4). The optimum condition to achieve the highest SERS signal is a 1–2 nm silica shell thickness with 36 nm size gold nanoparticles [25].

Gold nanoparticles of different shapes counting AuNRs to have the inclination to quench the excited energy of fluorochromes such as dye molecules having a contact distance limit of 40 nm is also possible, since these candidates are under nanoparticle surface-energy transfer (NSET) conditional method. The quenching efficacy improved three-fold when particle size increased as of 5–70 nm. This kind of energy transfer method is highly for distinguishing hepatitis C virus (HCV) RNA sequence selectively (single-base mutations) in a sensitive way under homogeneity [26].

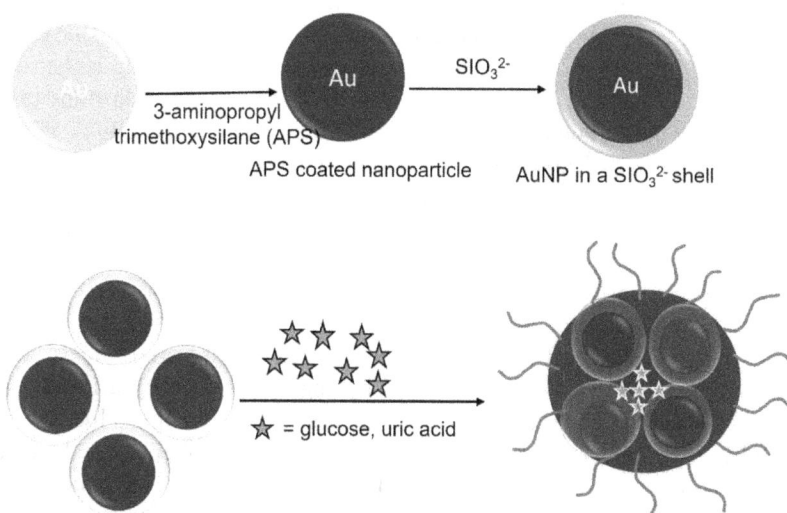

FIGURE 12.4 Gold nanoparticles in silica shell as a biosensor for glucose and uric acid.

AuNRs have enough stability to move in the blood, and this prolongs the period of availability that may promote AuNRs as drug carriers to deliver some water-soluble high impact as a chemotherapeutic agent against cancer, passive targeting of cancers owing to their enhanced permeability and retention (EPR) [27]. The AuNRs having the tendency to accumulate in the tumor cell sites can produce necessary heat by absorbing applied NIR (near IR) light having the selectivity to ablating tumors, so AuNRs by themselves act as a drug in the sense of using external light radiation called hyperthermic treatment [28–30].

12.4 CARBON-BASED NANOMATERIALS AS BIOSENSORS

Carbon is an important material which we cannot avoid when we speak about materials science. The scientific community is working on biosensors almost a decade not only to identify biomolecules in the field of medicine [31], but also the diverse field looks forward to the biosensing process on a small and industrial scale. In some bioprocessing [32] companies, industrial environmental monitoring [33,34] and food safety control [35] diverse field of agriculture [36]. Nowadays after the innovation of differential functional nanomaterials, carbon is a landmark that shows its potential as a biosensor in biological, biomedical, and clinical applications.

Tracking biological species in biosensors consists of very important aspects like the biorecognition component that can identify the analyte even if it is a mixture. A signal transducer is also an important part of the sensing component which can exhibit the output signals in optical, electrical, thermal, and electronic formats. Carbon allotropes (diamond, graphite) and synthetic carbon allotropes are important for their applications as sensors [37], tissue engineering [38], nanoelectronics [39], and other vital biomedical applications [40] because of their broad spectrum of chemical, biological, electronic, and mechanical properties. In the synthetic allotropes, carbon nanoparticles, carbon dot, graphene, and carbon nanotubes of the spectrum are included.

Carbon-based materials include also pyrolytic carbon [41], which is synthesized in a variety of procedures. In some cases, its produced by using the green method under pyrolytic conditions using an industrial waste product like Zeosil silica [42,43] and TiO_2 [44,45] as a substrate for the natural polymer lignin. Lignin is the second richest source of carbon. The produced nanosized pyrolytic carbon traps oxygen radicals [46] by the carbon radicals available. This could bring us to think about the application of these radicals in bio-applications. These materials not limited to bio-applications also served as a catalyst for photochemistry and green chemistry [44].

Carbon nanotubes (CNTs) are used as biosensors in the food industry, as diagnostic devices, and for enzymatic biosensing [47] because of their chemical stability, specific surface area, and electrical conductivity that is the important parameter for a biosensor. Nanoporous carbon materials are useful as biosensors having an increase in performance of biosensors due to higher surface area. Biosensor Au/CoNPs decorates nanoporous carbon material engaged to identify uric acid sensing [48]. Nanocarbon black is a nanosized material with a graphene-like structure chemically stable and is produced by petroleum products. This functional nanocarbon black has attention in the biosensor application due to its vital surface area and is used to detect the insecticide paraoxon. This constructed biosensor has a detection limit of 5μg/L^{-1}, which is useful to detect neurotoxic angiotensin-converting enzyme (ACE) inhibitor agents [49].

Fullerenes have an arrangement of closed carbon cages with pentagons and hexa-gons arrangement with sp^2 hybridized carbon atoms with different structures such as C70, C76, C82, C84, and C60. C60 is one of the 0D carbon nanoallotropes and has a simple and spherical structure with a diameter of about 0.71 nm which is highly studied [50]. The novel fabricated biosensor fullerene and molybdenum disulfide (MoS2) functionalized with polyamido amine was used to identify the Sul1 gene in *Salmonella typhimurium* in a sensitive way, specifically. This sensor is also useful to detect specific DNA sequences [51]. Diamond is an unavoidable allotrope of carbon having good biosensing application against glucose when combined, for example, with boron to produce boron-doped diamond (BDD). An amperometry biosensor is produced by combining it with platinum (Pt) nanoparticles and polyaniline [52].

Carbon dots have interesting optical, electrical, and catalytic [53] properties to construct an optical biosensor carbon dot exhibiting the properties of unique absorption characteristics [54]. The sensing ability of carbon dots with different nanocomposites is addressed in Table 12.1 with the detection limit of analytes.

Graphene oxide and reduced graphene oxide contains different functional group like epoxy, carbonyl, and hydroxyl groups which are utilized for biosensor applications

TABLE 12.1
Carbon Dot Sensing Ability as Biosensor [55]

Platform	Nanocomposite	Analyte	LOD	Linearity
Carbon	Fluorine and Nitrogen co-doped CDs	Catechol	0.014 µM	-
dots	Carbon dots-chitosan onto ITO glass substrate (CD-CH/ITO)	Vitamin 2 antigens	1.53 ng/mL	10–50 ng/mL
	Pd-Au@CDs nanocomposite modified glassy carbon electrode (Pd-Au@CDs/GCE)	DNA	1.82×10^{-17} mol/L	5.0×10^{-16}– 1.0×10^{-10} mol/L
	DNA-chitosan carbon dots deposited on the glassy carbon electrode (DNA/chiCD/GCE)	N-nitrosodimethylamine (NDMA) N-nitrosodiethanolamine (NDEA)	9.9×10^{-9} M 9.6×10^{-9} M	-
	Carbon dots (hyaluronic acid)/titanium dioxide/ copper (CD(HA)/TiO$_2$/ Cu^{2+})	Yrophosphates-alka line phosphatase (PPi-ALP)	2.31 cells/mL electrochemical 70.05 cells/mL optical	-
	Carbon dots dispersed in Polymethyl methacrylate (Cds-PMMA)	TNF-α	0.05 pg/mL	0.05–160 pg/ mL
	CDs/α-Fe$_2$O$_3$-Fe$_3$O$_4$	AFB1	0.5 pM	0.001–100.0 nM
	Carbon dots-black phosphorus nanohybrid (CDs-BP)	Ochratoxins A	0.03 fg/mL	0.1 fg/mL–10.0 ng/mL

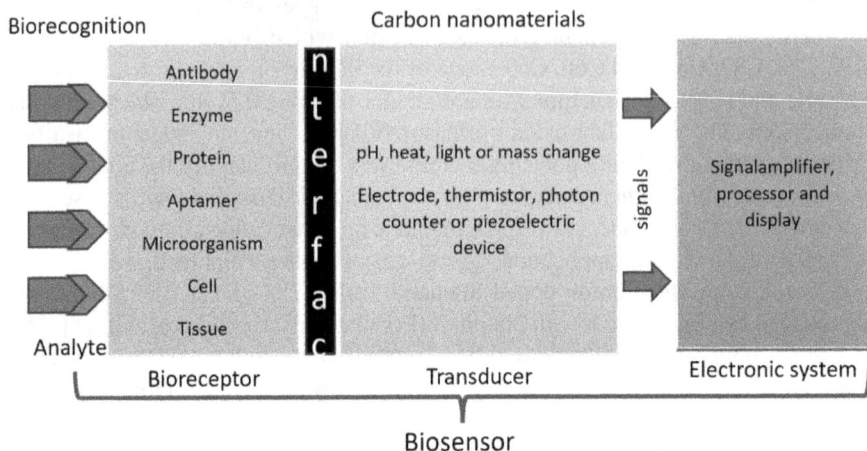

FIGURE 12.5 Construction of a biosensor for carbon materials.

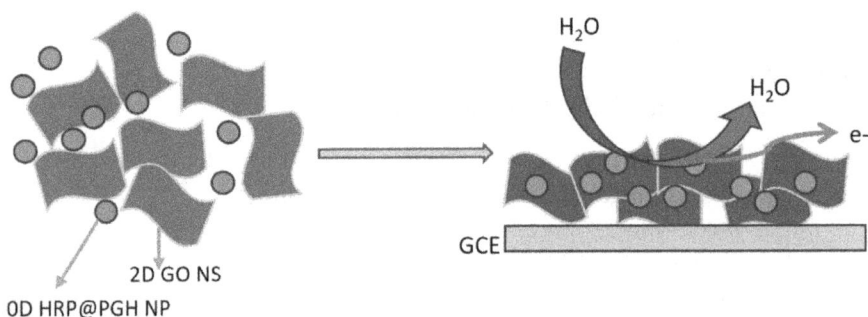

FIGURE 12.6 Reduced graphene oxides active with GCE.

(Figure 12.5) [56]. An enzymatic biosensor designed to detect H_2O_2 is fabricated by developing a biocomposite sensing film through the coating of 0D enzyme polymer nanoparticles with 0D on 2D graphene oxide nanosheets with 2D as conductive supports. Graphene oxidenano sheets are reduced into graphene nanosheets followed by photo-crosslinkage [57].

Copper nanocrystals over reduced graphene oxide sheets (Figure 12.7) are prepared by copper nanocrystals with reduced graphene oxide over and are used as a glucose sensor. The copper oxidation into copper oxide and its subsequent hydration imply formation of the electronic state that favors the glucose sensor, which may convert into gluconolactone. The obtained synthetic catalysis is considered as a novel candidate for the non-enzymatic glucose biosensor [58].

12.5 DNA-MODIFIED NANOPARTICLES

Nanoparticle shapes directed by biomolecules like DNA (Figure 12.8) have a larger influence on the properties of nanoparticles. DNA-modified nanomaterials can exhibit

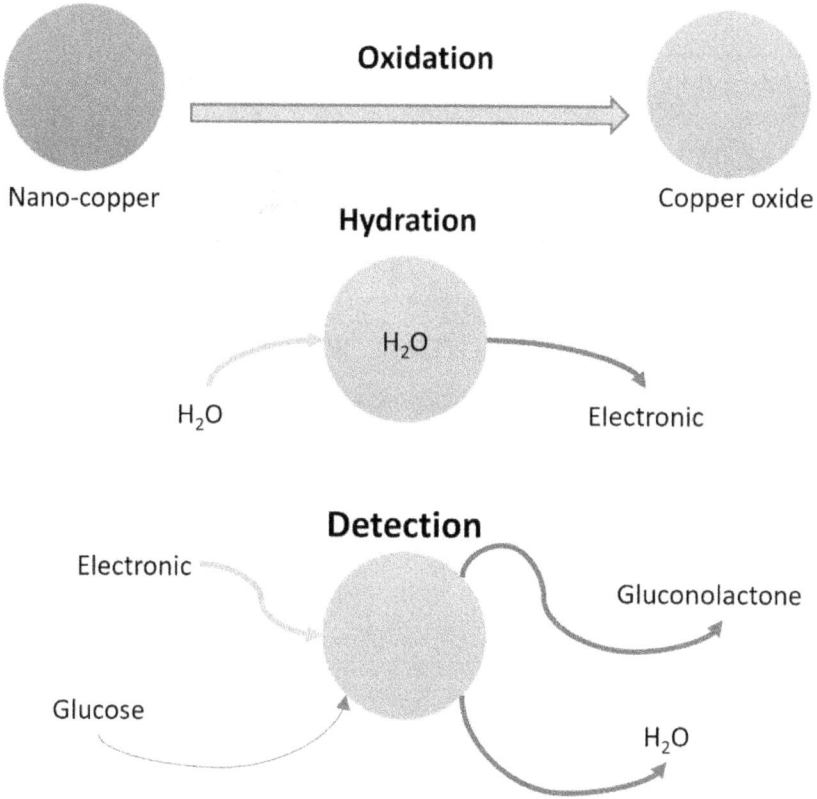

FIGURE 12.7 Detection of non-enzymatic glucose by copper oxide/graphene oxide materials.

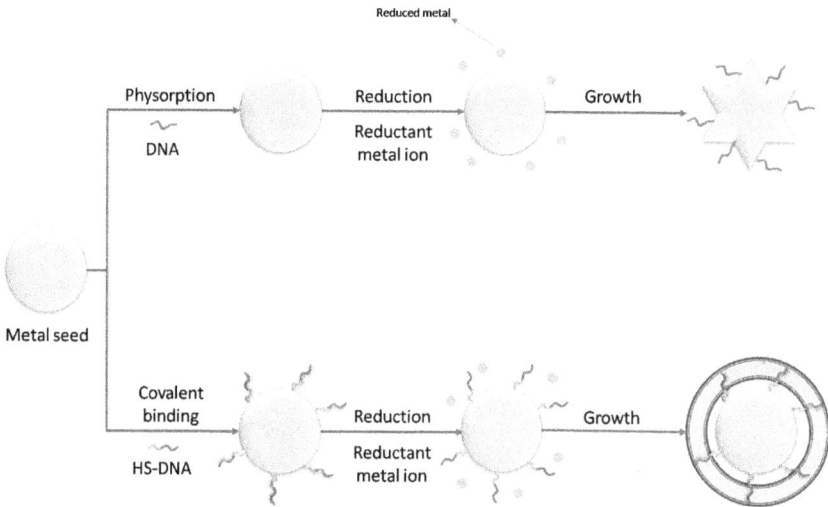

FIGURE 12.8 DNA-modified metal nanoparticles.

increased surface-enhanced Raman spectroscopy (SERS); this may cause the biomolecule to recognize analytes in very low amount.

DNA-aptamers specifically bind to different molecules such as protein and also small molecules. The DNA-functionalized nanoparticles have higher affinity to the surface proteins; also, this material has higher biocompatibility with respect to nanoparticles synthesized by other methods. DNA-functionalized gold nanodumbbells do not have significant cytotoxicity against breast tumor cells [59]. Thiolate DNA-modified spherical nanoparticles are used for single-cell imaging with no significant cell damage [60].

12.6 PROTEIN-FUNCTIONALIZED NANOPARTICLES

Drug delivery is one of the most important applications of NPs because most of the important drugs have issues with the solubility, mobility, or mode of administration. Nowadays, nanoparticles are used to transport insoluble drugs [61] in the bloodstream and reach the proper locations like cells and tissues. On the other hand, dendrimers [62], liposomes [63], and polymers [64] are also used as carriers for drug delivery. Nanoparticles encapsulated by biopolymer (Table 12.2) like protein-based nanoparticles are useful for medical applications because of their biodegradability and lower toxicity [65].

TABLE 12.2
The Protein Material Used for Nanoparticles Encapsulation [71]

Material	Advantage	Disadvantage
Silk protein fibroin	High stability Flexibility with high mechanical strength, suitable for various machining conditions Low immunogenicity Biodegradability Biocompatibility	Sericin may cause immunogenic reactions Slow degradation of silk II crystalline antiparallel β-sheet domain
Human serum albumin	High stability High solubility in physiological fluids Biodegradability Non-immunogenicity Non-toxicity Availability and readiness	Expensive cost
Gliadin	Biocompatibility Biodegradability Non-immunogenicity Non-toxicity High stability	Large particle size Rapid degradation speed
Gelatin	Biocompatibility Biodegradability Ease of bridge Safety	Low mechanical strength Rapid degradation speed

TABLE 12.2 *(Continued)*
The Protein Material Used for Nanoparticles Encapsulation [71]

Material	Advantage	Disadvantage
Legumin	Bioadhesive Wide surface area Small particle size Low immunogenicity High stability	Low yield
30Kc19	High stability Increased cell growth and viability Biodegradability Non-immunogenicity Non-toxic Enzyme-stabilizing property Cell-penetrating property	Low nanoparticle size and yield when using only 30Kc19α
Lipoprotein	Non-immunogenicity Biodegradability Biocompatibility Long circulation half-life Naturally targeting property	Difficult to separate native LDL
Ferritin	High stability pH stability Thermal stability Biodegradability	High cost

12.6.1 Silk Protein Fibroin

Fibroin is the major protein extracted from silk fibers which are produced by Bombyx mori silkworm, which is achieved by removing outer cover sericin by a thermo-chemical degumming process by sodium carbonate yielding 85% of protein [66]. Insoluble fibroin can be converted as accessible regenerative fibroin by calcium chloride or lithium bromide treatment [67]. Fibroin is composed of both heavy and light chains of polymers. The lighter chain consists of lighter amino acids of a smaller scale with a lower percentage of 15% Asp, 14% Ala, 11% Gly, 11% Ser, and traces of cysteine. In the case of heavier chains, crystalline β-sheets are formed through intermolecular hydrogen bonding of glycine and alanine moieties. This tends to make the hydrophobic property and mechanical properties with high tensile strength [68–70].

12.6.2 Human Serum Albumin

Human serum albumin (HSA) is a globular protein consisting of about 585 amino acids [72] having main units I, II, and III, with subunits A and B. Within the HSA, there are two main binding sites known as Sudlow's sites I and II, located in subunits IIA and IIIA, respectively [73]. HSA contains different amino acids that can act as active sites with their carbonyl, thiol, and amino groups able to form non-covalent interactions with different drugs and ligands [74,75]. In this way, HSA is a good candidate as a drug carrier. The featured behavior of HSA for ligands and drugs is its high stability and easy binding affinity [76]. At the concentration of up to 60 mg/ml of HSA, the presence of the cross-linking agent is considerably small. At this scale, the salt is added to the HSA solution and there is the occurrence of yellow color as bulk with higher particle size. So, the use of phosphate buffer and sodium chloride is not the right choice to make HSA nanoparticles; therefore, using the desolvation method is acknowledged for the synthesis of this kind of nanoparticles [77,78].

HSA nanoparticles are also used in the delivery of HEK 293T, whereby a plasmid was loaded onto HSA nanoparticles coupled with plasmid and cell-penetrating peptide [79]. These nanoparticles were prepared by using the desolvation method having the particle size up to 222.8 ± 42.4 nm. The loading efficacy of plasmid is found to be 78.3 ± 13.0%, which confirmed that surface modification of nanoparticles using cell-penetrating peptide did not affect the plasmid loading efficiency because it looks like competition but is not. The nanoparticle-mediated system shows almost no cytotoxicity when HSA nanoparticles are alone, but from lower concentration to higher concentration, the percentage of effect increased up to 50%. The established HSA nanoparticles are loaded with an anti-cancer drug specific for eye diseases called bevacizumab [80]. The resulting bevacizumab-loaded nanoparticles of HSA show an increase in the value of average particle size distribution to 300 nm even the modified nanoparticle synthesis starts with lower particle size with higher stability with a double phase released pattern with the initial rate of 400 μg/ml in the first five minutes followed by a sustainable release for more than 24 hours.

12.6.3 GLIADIN

Gliadin is a polymer constructed by single-chain polypeptides associated with intramolecular disulfide bonds which are prepared by the desolvation method having an average particle size distribution of 500 nm with a higher yield of 90% when compared to the initial protein. Gliadin protein is sensitive to heat and pH. The behaviors with salt are more similar to HSA nanoparticles, resulting in aggregation and instability [81]. These Gliadin polymers have specific properties of affinity in the upper gastrointestinal tract [82]. Each of the polypeptides has molecular weights up to 100 kDa [83]. The development of mucoadhesive gliadin nanoparticles loaded with amoxicillin is more active against *Helicobacter pylori* [84]. A number of studies concerning the interaction of the protein with polysaccharides show improved stability against environmental stress [85,86].

12.6.4 GELATIN

Gelatin is a water-soluble polymer having an average size of up to 220 kDa and it is solubilized in water at about 40°C [87]. Gelatin is obtained from thermal degradation, enzymatic degradation, acidic medium and hydrolyzing collagen in an alkaline medium [88]. Two different types of gelatins are available, named type A and B depending on the preparation technique [89–91]. Gelatin creates the supramolecular triple helical structure of amino acids like alanine, glycine, and proline [92]. Gelatin in general is recognized as safe because of its non-toxicity (food grade). The gelatin is used in drug administration and food, so the U.S. Food and Drug Administration (FDA) approved gelatin for commercial purposes in the market. Gelatin is used as an encapsulating agent for RNA, DNA, basic fibroblast growth factor, bone morphogenetic protein 2, and BSA. The nanoparticles encapsulated by gelatin have increased mechanical strength, lower decomposition rate, and improved solubility in an aqueous solution [93].

12.7 SELF-HEALING BIOMATERIALS VIA NANOMATERIALS

Self-healing is a quite common phenomenon in nature, like in soft and hard tissues. Even if it is still not possible to reach the level of quality present in living organisms, many important strategies have been studied and published. In the first reported materials having such capabilities, the inclusion of unreacted monomers in the polymer matrix was made. Monomers mix and react with the initiator present in the matrix when a breaking occurs. This approach has many drawbacks. First, the presence of monomers, mostly in the liquid state, reduces the mechanical strength of the specimen. Second, monomers and initiators are usually toxic and unstable materials, and their leakage from the polymerized object is unwanted [94]. One interesting strategy is the use of reversible chemical reactions forming dynamic covalent bonds such as using Diels-Alder reactions [95], disulfide bonds (mimicking proteins) [96], imines, boronates, and many others [97]. A second possibility is the exploitation of non-covalent interactions (hydrogen bonds, metal to ligand interactions, hydrophobic interactions, electrostatic interactions, host-guest interactions, etc.) [98,99]. The use of nanoparticles largely increases the mechanical stability of the material with a consequent increase in surface area, thus speeding up the reaction rates [100]. Furthermore, self-restoration of cracks happens more easily if nanoparticles are present, thanks to the presence of soft boundaries due to minimization of the interface area. Many materials, including concrete [101], have been studied and represent an interesting topic due to their importance. Many different strategies including nanoparticles of different nature (organic, inorganic) were used. Inorganic nanoparticles are widely used, both metal-derived and carbon-derived.

Healing materials can be divided into autonomic and non-autonomic systems. In the latter case (see Figure 12.9), the healing process is started by external stimuli or sources of energy (heat, light, mechanical and chemical). Gold nanoparticles are a good method for the conversion of light into heat, thus giving thermal energy needed for the repair (Figure 12.9).

The most interesting application is the self-healing in hydrogels, even if they are not among the materials with the best mechanical strength. The importance of hydrogels is, of course, the possibility to embed or interface them with cells and other biological materials (Figure 12.10). Moreover, their ability to change shape by using temperature, pH or other external stimuli is of great importance [102]. An amino-acid-based (11-(4-(pyrene-1-yl)butanamido)undecanoic acid) hydrogel containing carbon-based nanomaterials (like graphene, reduced graphene oxide [RGO] or single-walled nanotubes [SWCNT]) was reported by Roy and coworkers [103]. Self-healing double network hydrogels for tissue engineering are reported [104].

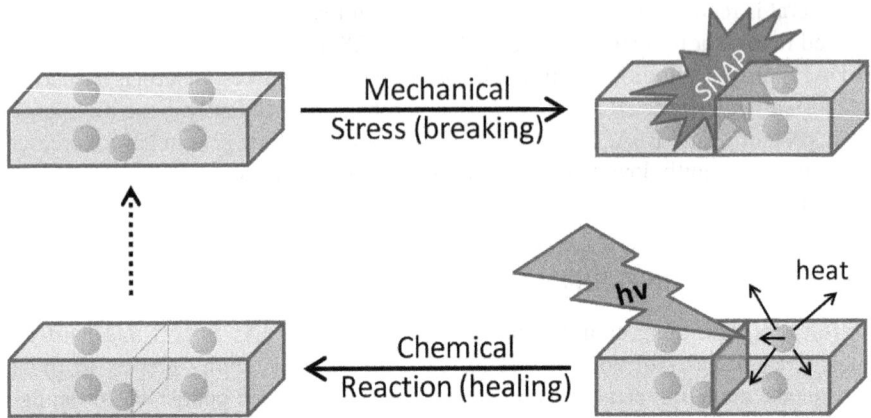

FIGURE 12.9 Example of a self-healing material triggered by light conversion into heat by using nanomaterials.

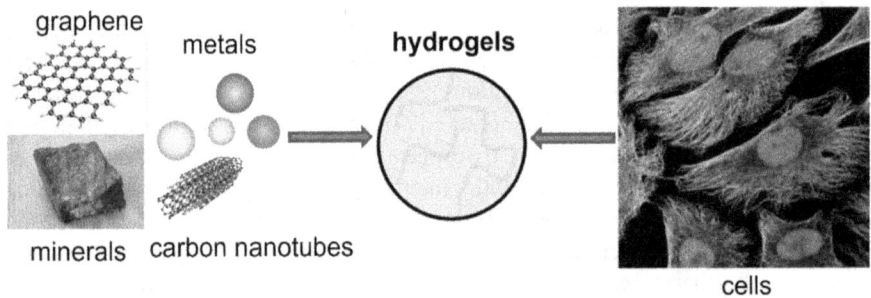

FIGURE 12.10 Metals, carbon nanotubes, minerals, and graphene in the formation of cells.

Janus nanocrystalline cellulose modified by polylysine and polydopamine showed interesting self-healing capacity when polymerized in acrylate hydrogel (Figure 12.10) [105]. The material showed high tensile strength (8.8 MPa), maintaining the biocompatibility.

12.8 NANOMATERIALS FOR TISSUE ENGINEERING

Hydrogel nanocomposites have many applications in biomedical engineering and tissue engineering [106,107] due to the presence of a high amount of water together with the possibility to include compatibilizers with cells. Nanomaterials are particularly interesting for cell growth due to their physical (size smaller than cells), tuneable chemical characteristics, and easy functionalization. Both 2D and 3D nanomaterials are suitable for cell compatibilizations in biomedical fields (Figure 12.11).

Among all the materials studied, a special case is the use of self-assembled monolayers (SAMs) [108] by which the bottom-up approach of self-assembly property, typical of functional nanomaterials, is exploited. Cells embedded in artificial

nanomaterials   cells

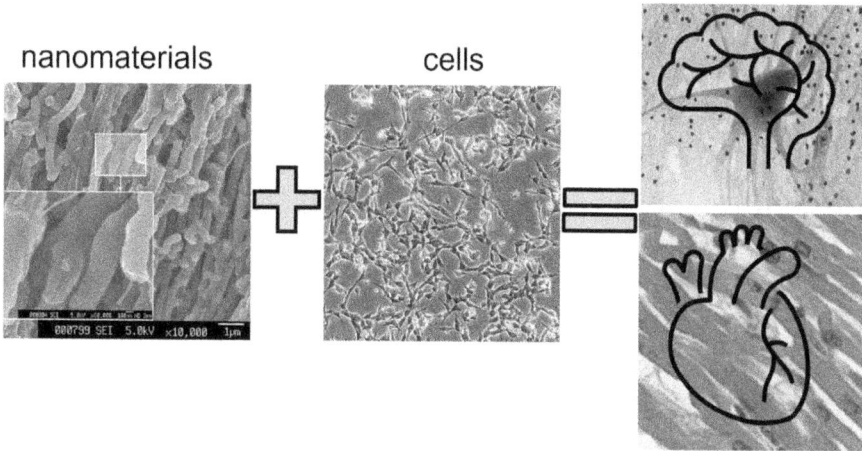

FIGURE 12.11   Neuron cell culture by the use of nanomaterials.

hydrogels can shed a light on the cell-sensing mechanisms, studying the response to mechanical stress [109]. It is in fact thought that some clinical diseases may result from altered cellular mechanotransduction [110]. Nanoparticles in general allow the inclusion and functionalization of growth factors, proteins, and specific cell-binding. Hydrogels, due to their intermediate mechanical resistance, can affect the proliferation, differentiation, migration, and morphologies of cells [111].

Nanomaterials can play an important role, too, just to cite the famous example of a neuron cell culture with a terrain containing CNTs [112] able to let scientists think about possible neuronal implants of connected neuronic networks. Another cutting-edge technology is the use of conductive nanomaterials for the generation of functional cardiac tissue [113]. Nanomaterials indeed allow scientists to interact with cells at their native size scales, allowing them to create biotic–abiotic interfaces for sensing. This opens to a next-generation medicine whereby it is possible to power implanted devices and replace damaged cells [114].

Polysaccharide nanocrystal-based (PN) nanomaterials derived from natural sources like cellulose, starch, and chitin represent an important class of materials for special applications [115]. Since these materials are derived from natural resources, they have less toxicity than other nanoparticles (metal, CNT, etc.), and they ideally have a mechanism for their degradation without giving toxic intermediates. For this reason, they are largely used in the biomedical field for uses like drug carriers, nanosponges, and bioimaging.

An interesting application is the preparation of mechanically adaptive nanomaterials starting from cellulose derivatives. These materials exhibit a stiffness depending on the number of hydrogen bonds formed in a percolative network of nanocellulose fibers [116]. When the number of hydrogen bonds is high, the material has an exceptional mechanical strength, but when a hydrogen bond disruptor is inserted (usually water), the material loses its strength. With opportune tuning, the materials can be switched on and off, thus modulating their stiffness according to the environment.

This can find applications like in the preparation of microelectrodes for brain voltage detection that can be absorbed by bodily fluids.

Finally, nanomaterials are useful for the preparation of a bioactive bone substitute having the proper mechanical properties [117]. This can be considered for the curing of bone fractures but also osteoporosis. Nanometer-sized ceramics, polymers, minerals (hydroxyapatite), metals, and composites might be used for these applications. This opens up an issue regarding the long-term toxicity of nanoparticles.

12.9 CONCLUSIONS

This chapter focuses on nanomaterials that are highly useful in the biomedical field due to their unique and broad spectrum of properties and applications. The traditional nanoparticles are metal nanoparticles, especially noble metal nanoparticles, which evolve in different ways like metal oxides, the mixture of metal oxides, magnetic nanoparticles, and carbon nanomaterials and micromaterials. Size controlling and functionalization is a choice with respect to the symmetry between the application we want and the nature of the nanomaterials. In the sense of biosensors, biocompatibility, and non-toxicity reasons, polymers are the focus of discussion. In upcoming research, nanomaterials with the presence of polymers—for example, hydrogel—can produce self-healing biomaterials. The combination of nanomaterials with tissue engineering results in microelectrodes for brain voltage detection and bioactive bone substitutes, which are the spectra of the future of nanoparticles in bio-applications proposed in this book chapter.

REFERENCES

[1] Eivazzadeh-Keihan, Reza, Ehsan Bahojb Noruzi, Elham Chidar, Mahdokht Jafari, Farahnaz Davoodi, Amir Kashtiaray, Mostafa Ghafori Gorab, et al. 2022. "Applications of Carbon-Based Conductive Nanomaterials in Biosensors." *Chemical Engineering Journal* 442 (August): 136183. https://doi.org/10.1016/j.cej.2022.136183.

[2] Daniel, Marie-Christine, and Didier Astruc. 2004. "Gold Nanoparticles: Assembly, Supramolecular Chemistry, Quantum-Size-Related Properties, and Applications toward Biology, Catalysis, and Nanotechnology." *Chemical Reviews* 104 (1): 293–346. https://doi.org/10.1021/cr030698+.

[3] Pinaud, Fabien, Xavier Michalet, Laurent A. Bentolila, James M. Tsay, Soren Doose, Jack J. Li, Gopal Iyer, and Shimon Weiss. 2006. "Advances in Fluorescence Imaging with Quantum Dot Bio-Probes." *Biomaterials* 27 (9): 1679–1687. https://doi.org/10.1016/j.biomaterials.2005.11.018.

[4] Steinkamp, Tanja, and Uwe Karst. 2004. "Detection Strategies for Bioassays Based on Luminescent Lanthanide Complexes and Signal Amplification." *Analytical and Bioanalytical Chemistry* 380 (1). https://doi.org/10.1007/s00216-004-2682-2.

[5] Aarnio, P. A., J. J. Ala-Heikkilä, T. T. Hakulinen, and M. P. Huhtinen. 2005. "High-Energy Proton Irradiation and Induced Radioactivity Analysis for Some Construction Materials for the CERN LHC." *Journal of Radioanalytical and Nuclear Chemistry* 264 (1): 51–60. https://doi.org/10.1007/s10967-005-0674-0.

[6] Qhobosheane, Monde, Swadeshmukul Santra, Peng Zhang, and Weihong Tan. 2001. "Biochemically Functionalized Silica Nanoparticles." *The Analyst* 126 (8): 1274–1278. https://doi.org/10.1039/b101489g.

[7] Zheng, Jinkai, and Lili He. 2014. "Surface-Enhanced Raman Spectroscopy for the Chemical Analysis of Food: SERS for Chemical Analysis of Food . . ." *Comprehensive Reviews in Food Science and Food Safety* 13 (3): 317–328. https://doi.org/10.1111/1541-4337.12062.

[8] Pilot, Roberto. 2018. "SERS Detection of Food Contaminants by Means of Portable Raman Instruments." *Journal of Raman Spectroscopy* 49 (6): 954–981. https://doi.org/10.1002/jrs.5400.

[9] Hakonen, Aron, Tomas Rindzevicius, Michael Stenbæk Schmidt, Per Ola Andersson, Lars Juhlin, Mikael Svedendahl, Anja Boisen, and Mikael Käll. 2016. "Detection of Nerve Gases Using Surface-Enhanced Raman Scattering Substrates with High Droplet Adhesion." *Nanoscale* 8 (3): 1305–1308. https://doi.org/10.1039/C5NR06524K

[10] Hakonen, Aron, Per Ola Andersson, Michael Stenbæk Schmidt, Tomas Rindzevicius, and Mikael Käll. 2015. "Explosive and Chemical Threat Detection by Surface-Enhanced Raman Scattering: A Review." *Analytica Chimica Acta* 893 (September): 1–13. https://doi.org/10.1016/j.aca.2015.04.010.

[11] Pilot, Roberto, Raffaella Signorini, Christian Durante, Laura Orian, Manjari Bhamidipati, and Laura Fabris. 2019. "A Review on Surface-Enhanced Raman Scattering." *Biosensors* 9 (2): 57. https://doi.org/10.3390/bios9020057.

[12] Huang, Chu-Yu, and Hung-Che Hsiao. 2020. "Integrated EC-SERS Chip with Uniform Nanostructured EC-SERS Active Working Electrode for Rapid Detection of Uric Acid." *Sensors* 20 (24): 7066. https://doi.org/10.3390/s20247066.

[13] Chio, Weng-I Katherine, Gemma Davison, Tabitha Jones, Jia Liu, Ivan P. Parkin, and Tung-Chun Lee. 2020. "Quantitative SERS Detection of Uric Acid via Formation of Precise Plasmonic Nanojunctions within Aggregates of Gold Nanoparticles and Cucurbit[n]Uril." *Journal of Visualized Experiments*, no. 164 (October): 61682. https://doi.org/10.3791/61682.

[14] Murphy, Catherine J., Anand M. Gole, John W. Stone, Patrick N. Sisco, Alaaldin M. Alkilany, Edie C. Goldsmith, and Sarah C. Baxter. 2008. "Gold Nanoparticles in Biology: Beyond Toxicity to Cellular Imaging." *Accounts of Chemical Research* 41 (12): 1721–1730. https://doi.org/10.1021/ar800035u.

[15] Huang, Xiaohua, Svetlana Neretina, and Mostafa A. El-Sayed. 2009. "Gold Nanorods: From Synthesis and Properties to Biological and Biomedical Applications." *Advanced Materials* 21 (48): 4880–4910. https://doi.org/10.1002/adma.200802789.

[16] Maltzahn, Geoffrey von, Ji-Ho Park, Amit Agrawal, Nanda Kishor Bandaru, Sarit K. Das, Michael J. Sailor, and Sangeeta N. Bhatia. 2009. "Computationally Guided Photothermal Tumor Therapy Using Long-Circulating Gold Nanorod Antennas." *Cancer Research* 69 (9): 3892–3900. https://doi.org/10.1158/0008-5472.CAN-08-4242.

[17] McLintock, Alison, Nathan Hunt, and Alastair W. Wark. 2011. "Controlled Side-by-Side Assembly of Gold Nanorods and Dye Molecules into Polymer-Wrapped SERRS-Active Clusters." *Chemical Communications* 47 (13): 3757. https://doi.org/10.1039/c0cc04353b.

[18] Dulkeith, E., M. Ringler, T. A. Klar, J. Feldmann, A. Muñoz Javier, and W. J. Parak. 2005. "Gold Nanoparticles Quench Fluorescence by Phase Induced Radiative Rate Suppression." *Nano Letters* 5 (4): 585–589. https://doi.org/10.1021/nl0480969.

[19] Jain, Prashant K., Kyeong Seok Lee, Ivan H. El-Sayed, and Mostafa A. El-Sayed. 2006. "Calculated Absorption and Scattering Properties of Gold Nanoparticles of Different Size, Shape, and Composition: Applications in Biological Imaging and Biomedicine." *The Journal of Physical Chemistry B* 110 (14): 7238–7248. https://doi.org/10.1021/jp0571700.

[20] Nikoobakht, Babak, and Mostafa A. El-Sayed. 2003. "Preparation and Growth Mechanism of Gold Nanorods (NRs) Using Seed-Mediated Growth Method." *Chemistry of Materials* 15 (10): 1957–1962. https://doi.org/10.1021/cm0207321.

[21] Yang, Jaeyoung, Mirko Palla, Filippo Giacomo Bosco, Tomas Rindzevicius, Tommy Sonne Alstrøm, Michael Stenbæk Schmidt, Anja Boisen, Jingyue Ju, and Qiao Lin. 2013. "Surface-Enhanced Raman Spectroscopy Based Quantitative Bioassay on Aptamer-Functionalized Nanopillars Using Large-Area Raman Mapping." *ACS Nano* 7 (6): 5350–5359. https://doi.org/10.1021/nn401199k.

[22] Liu, Yu, Haitao Zhang, Yijia Geng, Shuping Xu, Weiqing Xu, Jie Yu, Wenyuan Deng, Bo Yu, and Liping Wang. 2020. "Long-Range Surface Plasmon Resonance Configuration for Enhancing SERS with an Adjustable Refractive Index Sample Buffer to Maintain the Symmetry Condition." *ACS Omega* 5 (51): 32951–32958. https://doi.org/10.1021/acsomega.0c03923.

[23] Woźniak, Anna, Anna Malankowska, Grzegorz Nowaczyk, Bartosz F. Grześkowiak, Karol Tuśnio, Ryszard Słomski, Adriana Zaleska-Medynska, and Stefan Jurga. 2017. "Size and Shape-Dependent Cytotoxicity Profile of Gold Nanoparticles for Biomedical Applications." *Journal of Materials Science: Materials in Medicine* 28 (6): 92. https://doi.org/10.1007/s10856-017-5902-y.

[24] Prasad, Sonal, Andre Zeug, and Evgeni Ponimaskin. 2013. "Analysis of Receptor: Receptor Interaction by Combined Application of FRET and Microscopy." *Methods in Cell Biology* 117: 243–265. Elsevier. https://doi.org/10.1016/B978-0-12-408143-7.00014-1.

[25] Quyen, Tran Thi Bich, Wei-Nien Su, Kuan-Jung Chen, Chun-Jern Pan, John Rick, Chun-Chao Chang, and Bing-Joe Hwang. 2013. "Au@SiO$_2$ Core/Shell Nanoparticle Assemblage Used for Highly Sensitive SERS-Based Determination of Glucose and Uric Acid: Au@SiO$_2$ Core/Shell Nanoparticle Assemblage." *Journal of Raman Spectroscopy* 44 (12): 1671–1677. https://doi.org/10.1002/jrs.4400.

[26] Griffin, Jelani, Anant Kumar Singh, Dulal Senapati, Patsy Rhodes, Kanieshia Mitchell, Brianica Robinson, Eugene Yu, and Paresh Chandra Ray. 2009. "Size- and Distance-Dependent Nanoparticle Surface-Energy Transfer (NSET) Method for Selective Sensing of Hepatitis C Virus RNA." *Chemistry: A European Journal* 15 (2): 342–351. https://doi.org/10.1002/chem.200801812.

[27] Hashizume, Hiroya, Peter Baluk, Shunichi Morikawa, John W. McLean, Gavin Thurston, Sylvie Roberge, Rakesh K. Jain, and Donald M. McDonald. 2000. "Openings between Defective Endothelial Cells Explain Tumor Vessel Leakiness." *The American Journal of Pathology* 156 (4): 1363–1380. https://doi.org/10.1016/S0002-9440(10)65006-7.

[28] Rowe-Horwege R. W. 2006. "Systemic Hyperthermia." In: *Encyclopedia of Medical Devices and Instrumentation* (2nd Edition). J. G. Webster (Ed.). John Wiley and Sons, New York, USA, 42–62. ISBN: 978-0-471-26358-6

[29] Guerrini, Luca, Zuzana Jurasekova, Elena del Puerto, Liesbeth Hartsuiker, Concepcion Domingo, Jose Vicente Garcia-Ramos, Cees Otto, and Santiago Sanchez-Cortes. 2013. "Effect of Metal—Liquid Interface Composition on the Adsorption of a Cyanine Dye onto Gold Nanoparticles." *Langmuir* 29 (4): 1139–1147. https://doi.org/10.1021/la304617t.

[30] Rahman, Dewan S., Debdulal Sharma, and Sujit Kumar Ghosh. 2014. "Emission Behavior of Sudan Red 7B on Dogbone-Shaped Gold Nanorods: Aspect Ratio Dependence of the Metallic Nanostructures." *Chemical Physics* 429 (January): 27–32. https://doi.org/10.1016/j.chemphys.2013.11.020.

[31] Sireesha, Merum, Veluru Jagadeesh Babu, A. Sandeep Kranthi Kiran, and Seeram Ramakrishna. 2018. "A Review on Carbon Nanotubes in Biosensor Devices and Their Applications in Medicine." *Nanocomposites* 4 (2): 36–57. https://doi.org/10.1080/2055 0324.2018.1478765.

[32] Theuer, Lorenz, Judit Randek, Stefan Junne, Peter Neubauer, Carl-Fredrik Mandenius, and Valerio Beni. 2020. "Single-Use Printed Biosensor for L-Lactate and Its Application in Bioprocess Monitoring." *Processes* 8 (3): 321. https://doi.org/10.3390/pr8030321.

[33] Cennamo, Nunzio, Luigi Zeni, Paolo Tortora, Maria Elena Regonesi, Alessandro Giusti, Maria Staiano, Sabato D'Auria, and Antonio Varriale. 2018. "A High Sensitivity Biosensor to Detect the Presence of Perfluorinated Compounds in Environment." *Talanta* 178 (February): 955–961. https://doi.org/10.1016/j.talanta.2017.10.034.

[34] Meshram, B. D., A. K. Agrawal, Shaikh Adil, Suvartan Ranvir, and K. K. Sande. 2018. "Biosensor and Its Application in Food and Dairy Industry: A Review." *International Journal of Current Microbiology and Applied Sciences* 7 (2): 3305–3324. https://doi.org/10.20546/ijcmas.2018.702.397.

[35] Eivazzadeh-Keihan, Reza, Paria Pashazadeh, Maryam Hejazi, Miguel de la Guardia, and Ahad Mokhtarzadeh. 2017. "Recent Advances in Nanomaterial-Mediated Bio and Immune Sensors for Detection of Aflatoxin in Food Products." *TrAC Trends in Analytical Chemistry* 87 (February): 112–128. https://doi.org/10.1016/j.trac.2016.12.003.

[36] Kundu Monika, Krishnan Prameela, Kotnala Ravinder Kumar and Gajjala Sumana. 2019. "Recent Developments in Biosensors to Combat Agricultural Challenges and Their Future Prospects." *Trends in Food Science & Technology* 88 (June): 157–178. https://doi.org/10.1016/j.tifs.2019.03.024.

[37] Pashazadeh-Panahi, Paria, Mohammad Hasanzadeh, and Reza Eivazzadeh-Keihan. 2020. "A Novel Optical Probe Based on D-Penicillamine-Functionalized Graphene Quantum Dots: Preparation and Application as Signal Amplification Element to Minoring of Ions in Human Biofluid." *Journal of Molecular Recognition* 33 (5). https://doi.org/10.1002/jmr.2828.

[38] Eivazzadeh-Keihan, Reza, Ali Maleki, Miguel de la Guardia, Milad Salimi Bani, Karim Khanmohammadi Chenab, Paria Pashazadeh-Panahi, Behzad Baradaran, Ahad Mokhtarzadeh, and Michael R. Hamblin. 2019. "Carbon Based Nanomaterials for Tissue Engineering of Bone: Building New Bone on Small Black Scaffolds: A Review." *Journal of Advanced Research* 18 (July): 185–201. https://doi.org/10.1016/j.jare.2019.03.011.

[39] Wu, Ziping, Yonglong Wang, Xianbin Liu, Chao Lv, Yesheng Li, Di Wei, and Zhongfan Liu. 2019. "Carbon-Nanomaterial-Based Flexible Batteries for Wearable Electronics." *Advanced Materials* 31 (9): 1800716. https://doi.org/10.1002/adma.201800716.

[40] Eivazzadeh-Keihan, Reza, Fateme Radinekiyan, Hamid Madanchi, Hooman Aghamirza Moghim Aliabadi, and Ali Maleki. 2020. "Graphene Oxide/Alginate/Silk Fibroin Composite as a Novel Bionanostructure with Improved Blood Compatibility, Less Toxicity and Enhanced Mechanical Properties." *Carbohydrate Polymers* 248 (November): 116802. https://doi.org/10.1016/j.carbpol.2020.116802.

[41] Dondi, Daniele, Alberto Zeffiro, Andrea Speltini, Corrado Tomasi, Dhanalakshmi Vadivel, and Armando Buttafava. 2014. "The Role of Inorganic Sulfur Compounds in the Pyrolysis of Kraft Lignin." *Journal of Analytical and Applied Pyrolysis* 107 (May): 53–58. https://doi.org/10.1016/j.jaap.2014.02.002.

[42] Vadivel, D., and I. Malaichamy. 2018. "Pyrolytic Formation and Photoactivity of Reactive Oxygen Species in a SiO2/Carbon Nanocomposite from Kraft Lignin [Version 1; Peer Review: 2 Approved]." *F1000Research* 7: 1574. https://doi.org/10.12688/f1000research.16080.1

[43] Speltini, Andrea, Michela Sturini, Federica Maraschi, Elettra Mandelli, Dhanalakshmi Vadivel, Daniele Dondi, and Antonella Profumo. 2016. "Preparation of Silica-Supported Carbon by Kraft Lignin Pyrolysis, and Its Use in Solid-Phase Extraction of Fluoroquinolones from Environmental Waters." *Microchimica Acta* 183 (7): 2241–2249. https://doi.org/10.1007/s00604-016-1859-7.

[44] Dondi, D., and D. Vadivel. 2020. "Preparation of Catalysts from Renewable and Waste Materials." *Catalysts* 10: 662. https://doi.org/10.3390/catal10060662

[45] Vadivel, Dhanalakshmi, Diego Savio Branciforti, Andrea Speltini, Michela Sturini, Vittorio Bellani, Ilanchelian Malaichamy, and Daniele Dondi. 2020. "Pyrolytic Formation of TiO2/Carbon Nanocomposite from Kraft Lignin: Characterization and Photoactivities." *Catalysts* 10 (3): 270. https://doi.org/10.3390/catal10030270.

[46] Vadivel, Dhanalakshmi, Andrea Speltini, Alberto Zeffiro, Vittorio Bellani, Sergio Pezzini, Armando Buttafava, and Daniele Dondi. 2017. "Reactive Carbons from Kraft Lignin Pyrolysis: Stabilization of Peroxyl Radicals at Carbon/Silica Interface." *Journal of Analytical and Applied Pyrolysis* 128 (November): 346–352. https://doi.org/10.1016/j.jaap.2017.09.016.

[47] Lee, Myeongsoon, and Don Kim. 2020. "Exotic Carbon Nanotube Based Field Effect Transistor for the Selective Detection of Sucrose." *Materials Letters* 268 (June): 127571. https://doi.org/10.1016/j.matlet.2020.127571.

[48] Wang, Kaidong, Can Wu, Feng Wang, Minghao Liao, and Guoqiang Jiang. 2020. "Bimetallic Nanoparticles Decorated Hollow Nanoporous Carbon Framework as Nanozyme Biosensor for Highly Sensitive Electrochemical Sensing of Uric Acid." *Biosensors and Bioelectronics* 150 (February): 111869. https://doi.org/10.1016/j.bios.2019.111869.

[49] Arduini, Fabiana, Matteo Forchielli, Aziz Amine, Daniela Neagu, Ilaria Cacciotti, Francesca Nanni, Danila Moscone, and Giuseppe Palleschi. 2015. "Screen-Printed Biosensor Modified with Carbon Black Nanoparticles for the Determination of Paraoxon Based on the Inhibition of Butyrylcholinesterase." *Microchimica Acta* 182 (3–4): 643–51. https://doi.org/10.1007/s00604-014-1370-y.

[50] Georgakilas, Vasilios, Jason A. Perman, Jiri Tucek, and Radek Zboril. 2015. "Broad Family of Carbon Nanoallotropes: Classification, Chemistry, and Applications of Fullerenes, Carbon Dots, Nanotubes, Graphene, Nanodiamonds, and Combined Superstructures." *Chemical Reviews* 115 (11): 4744–4822. https://doi.org/10.1021/cr500304f.

[51] You, Huan, Zhaode Mu, Min Zhao, Jing Zhou, Yonghua Yuan, and Lijuan Bai. 2020. "Functional Fullerene-Molybdenum Disulfide Fabricated Electrochemical DNA Biosensor for Su11 Detection Using Enzyme-Assisted Target Recycling and a New Signal Marker for Cascade Amplification." *Sensors and Actuators B: Chemical* 305 (February): 127483. https://doi.org/10.1016/j.snb.2019.127483.

[52] Song, Min-Jung, Jong Hoon Kim, Seung Koo Lee, Jae-Hyun Lee, Dae Soon Lim, Sung Woo Hwang, and Dongmok Whang. 2010. "Pt-Polyaniline Nanocomposite on Boron-Doped Diamond Electrode for Amperometic Biosensor with Low Detection Limit." *Microchimica Acta* 171 (3–4): 249–255. https://doi.org/10.1007/s00604-010-0432-z.

[53] Hassanvand, Zahra, Fahimeh Jalali, Maryam Nazari, Fatemeh Parnianchi, and Carlo Santoro. 2021. "Carbon Nanodots in Electrochemical Sensors and Biosensors: A Review." *ChemElectroChem* 8 (1): 15–35. https://doi.org/10.1002/celc.202001229.

[54] Lin, Xiaofeng, Mogao Xiong, Jingwen Zhang, Chen He, Xiaoming Ma, Huifang Zhang, Ying Kuang, Min Yang, and Qitong Huang. 2021. "Carbon Dots Based on Natural Resources: Synthesis and Applications in Sensors." *Microchemical Journal* 160 (January): 105604. https://doi.org/10.1016/j.microc.2020.105604.

[55] Dreyer, Daniel R., Sungjin Park, Christopher W. Bielawski, and Rodney S. Ruoff. 2010. "The Chemistry of Graphene Oxide." *Chem. Soc. Rev.* 39 (1): 228–240. https://doi.org/10.1039/B917103G.

[56] Eivazzadeh-Keihan, Reza, Ehsan Bahojb Noruzi, Elham Chidar, Mahdokht Jafari, Farahnaz Davoodi, Amir Kashtiaray, Mostafa Ghafori Gorab, et al. 2022. "Applications of Carbon-Based Conductive Nanomaterials in Biosensors." *Chemical Engineering Journal* 442 (August): 136183. https://doi.org/10.1016/j.cej.2022.136183

[57] Xu, Sheng, Yayuan Liu, Wei Zhao, Qian Wu, Yanru Chen, Xuewen Huang, Zhijian Sun, Ye Zhu, and Xiaoya Liu. 2020. "Hierarchical 0D-2D Bio-Composite Film Based

on Enzyme-Loaded Polymeric Nanoparticles Decorating Graphene Nanosheets as a High-Performance Bio-Sensing Platform." *Biosensors and Bioelectronics* 156 (May): 112134. https://doi.org/10.1016/j.bios.2020.112134.

[58] Hsieh, Chien-Te, Wei-Hsun Lin, Yu-Fu Chen, Dong-Ying Tzou, Pei-Qi Chen, and Ruey-Shin Juang. 2017. "Microwave Synthesis of Copper Catalysts onto Reduced Graphene Oxide Sheets for Non-Enzymatic Glucose Oxidation." *Journal of the Taiwan Institute of Chemical Engineers* 71 (February): 77–83. https://doi.org/10.1016/j.jtice.2016.12.038.

[59] Zhang, Yifan, Tingjie Song, Tao Feng, Yilin Wan, Nicholas T. Blum, Chengbo Liu, Chunqi Zheng, et al. 2020. "Plasmonic Modulation of Gold Nanotheranostics for Targeted NIR-II Photothermal-Augmented Immunotherapy." *Nano Today* 35 (December): 100987. https://doi.org/10.1016/j.nantod.2020.100987.

[60] Kang, Jeon Woong, Peter T. C. So, Ramachandra R. Dasari, and Dong-Kwon Lim. 2015. "High Resolution Live Cell Raman Imaging Using Subcellular Organelle-Targeting SERS-Sensitive Gold Nanoparticles with Highly Narrow Intra-Nanogap." *Nano Letters* 15 (3): 1766–1772. https://doi.org/10.1021/nl504444w.

[61] Jabalera, Ylenia, Beatriz Garcia-Pinel, Raul Ortiz, Guillermo Iglesias, Laura Cabeza, José Prados, Concepcion Jimenez-Lopez, and ConsolaciónMelguizo. 2019. "Oxaliplatin: Biomimetic Magnetic Nanoparticle Assemblies for Colon Cancer-Targeted Chemotherapy: An In Vitro Study." *Pharmaceutics* 11 (8): 395. https://doi.org/10.3390/pharmaceutics 11080395.

[62] Jeong, Youngdo, Sung Tae Kim, Ying Jiang, Bradley Duncan, Chang Soo Kim, Krishnendu Saha, Yi-Cheun Yeh, et al. 2016. "Nanoparticle: Dendrimer Hybrid Nanocapsules for Therapeutic Delivery." *Nanomedicine* 11 (12): 1571–1578. https://doi.org/10.2217/nnm-2016-0034.

[63] Mao, Jie, Shujun Liu, Min Ai, Zhuo Wang, Duowei Wang, Xianjing Li, Kaiyong Hu, Xinghua Gao, and Yong Yang. 2017. "A Novel Melittin Nano-Liposome Exerted Excellent Anti-Hepatocellular Carcinoma Efficacy with Better Biological Safety." *Journal of Hematology & Oncology* 10 (1): 71. https://doi.org/10.1186/s13045-017-0442-y.

[64] Yang, Seon-Ah, Sungmoon Choi, Seon Mi Jeon, and Junhua Yu. 2018. "Silica Nanoparticle Stability in Biological Media Revisited." *Scientific Reports* 8 (1): 185. https://doi.org/10.1038/s41598-017-18502-8.

[65] Jacob, Joby, Józef T. Haponiuk, Sabu Thomas, and Sreeraj Gopi. 2018. "Biopolymer Based Nanomaterials in Drug Delivery Systems: A Review." *Materials Today Chemistry* 9 (September): 43–55. https://doi.org/10.1016/j.mtchem.2018.05.002.

[66] Numata, Keiji, and David L. Kaplan. 2010. "Silk-Based Delivery Systems of Bioactive Molecules." *Advanced Drug Delivery Reviews* 62 (15): 1497–1508. https://doi.org/10.1016/j.addr.2010.03.009.

[67] Melke, Johanna, Swati Midha, Sourabh Ghosh, Keita Ito, and Sandra Hofmann. 2016. "Silk Fibroin as Biomaterial for Bone Tissue Engineering." *Acta Biomaterialia* 31 (February): 1–16. https://doi.org/10.1016/j.actbio.2015.09.005.

[68] Wadbua, Paweena, Boonhiang Promdonkoy, Santi Maensiri, and Sineenat Siri. 2010. "Different Properties of Electrospun Fibrous Scaffolds of Separated Heavy-Chain and Light-Chain Fibroins of Bombyx Mori." *International Journal of Biological Macromolecules* 46 (5): 493–501. https://doi.org/10.1016/j.ijbiomac.2010.03.007.

[69] Ki, Chang Seok, Young Hwan Park, and Hyoung-Joon Jin. 2009. "Silk Protein as a Fascinating Biomedical Polymer: Structural Fundamentals and Applications." *Macromolecular Research* 17 (12): 935–942. https://doi.org/10.1007/BF03218639.

[70] Keten, Sinan, Zhiping Xu, Britni Ihle, and Markus J. Buehler. 2010. "Nanoconfinement Controls Stiffness, Strength and Mechanical Toughness of β-Sheet Crystals in Silk." *Nature Materials* 9 (4): 359–367. https://doi.org/10.1038/nmat2704.

[71] Langer, K., S. Balthasar, V. Vogel, N. Dinauer, H. von Briesen, and D. Schubert. 2003. "Optimization of the Preparation Process for Human Serum Albumin (HSA) Nanoparticles." *International Journal of Pharmaceutics* 257 (1–2): 169–180. https://doi.org/10.1016/S0378-5173(03)00134-0.

[72] Jahanban-Esfahlan, Ali, Siavoush Dastmalchi, and Soodabeh Davaran. 2016. "A Simple Improved Desolvation Method for the Rapid Preparation of Albumin Nanoparticles." *International Journal of Biological Macromolecules* 91 (October): 703–709. https://doi.org/10.1016/j.ijbiomac.2016.05.032.

[73] Hong, Seyoung, Dong Wook Choi, Hong Nam Kim, Chun Gwon Park, Wonhwa Lee, and Hee Ho Park. 2020. "Protein-Based Nanoparticles as Drug Delivery Systems." *Pharmaceutics* 12 (7): 604. https://doi.org/10.3390/pharmaceutics12070604

[74] Fasano, Mauro, Stephen Curry, Enzo Terreno, Monica Galliano, Gabriella Fanali, Pasquale Narciso, Stefania Notari, and Paolo Ascenzi. 2005. "The Extraordinary Ligand Binding Properties of Human Serum Albumin." *IUBMB Life (International Union of Biochemistry and Molecular Biology: Life)* 57 (12): 787–796. https://doi.org/10.1080/15216540500404093.

[75] Sudlow, G., D. J. Birkett, and D. N. Wade. 1975. "The Characterization of Two Specific Drug Binding Sites on Human Serum Albumin." *Molecular Pharmacology* 11 (6): 824–832.

[76] Bertucci, Carlo, and Enrico Domenici. 2002. "Reversible and Covalent Binding of Drugs to Human Serum Albumin: Methodological Approaches and Physiological Relevance." *Current Medicinal Chemistry* 9 (15): 1463–1481. https://doi.org/10.2174/0929867023369673.

[77] Caputo, Tania Mariastella, Angela Maria Cusano, Menotti Ruvo, Anna Aliberti, and Andrea Cusano. 2021. Current Pharmaceutical Biotechnology Publisher. Bentham Science Publishers, The Netherlands. https://doi.org/10.2174/1389201022666621082615 2311

[78] Elangovan, Akhilan, Dhananjay Suresh, Andrew O. Tarim, Anandhi Upendran, and Raghuraman Kannan. 2021 (2022). "Controlled Assembly of Gold and Albumin Nanoparticles to Formhybrid Multimeric Nanomaterials." *Polym Adv Technol* 33: 566–575. DOI: 10.1002/pat.5538566

[79] Mesken, J., A. Iltzsche, D. Mulac, and K. Langer. 2017. "Modifying Plasmid-Loaded HSA-Nanoparticles with Cell Penetrating Peptides: Cellular Uptake and Enhanced Gene Delivery." *International Journal of Pharmaceutics* 522 (1–2): 198–209. https://doi.org/10.1016/j.ijpharm.2017.03.006.

[80] Luis de Redín, Inés, Carolina Boiero, María Cristina Martínez-Ohárriz, MaiteAgüeros, Rocío Ramos, Iván Peñuelas, Daniel Allemandi, Juan M. Llabot, and Juan M. Irache. 2018. "Human Serum Albumin Nanoparticles for Ocular Delivery of Bevacizumab." *International Journal of Pharmaceutics* 541 (1–2): 214–223. https://doi.org/10.1016/j.ijpharm.2018.02.003.

[81] Joye, Iris J., Veronique A. Nelis, and D. Julian McClements. 2015. "Gliadin-Based Nanoparticles: Fabrication and Stability of Food-Grade Colloidal Delivery Systems." *Food Hydrocolloids* 44 (February): 86–93. https://doi.org/10.1016/j.foodhyd.2014.09.008.

[82] Ezpeleta, Isabel, Miguel A. Arangoa, Juan M. Irache, Serge Stainmesse, Christiane Chabenat, Yves Popineau, and Anne-Marie Orecchioni. 1999. "Preparation of Ulex Europaeus Lectin-Gliadin Nanoparticle Conjugates and Their Interaction with Gastrointestinal Mucus." *International Journal of Pharmaceutics* 191 (1): 25–32. https://doi.org/10.1016/S0378-5173(99)00232-X.

[83] Wu, Weihao, Xiangzhen Kong, Caimeng Zhang, Yufei Hua, and Yeming Chen. 2018. "Improving the Stability of Wheat Gliadin Nanoparticles: Effect of Gum Arabic

Addition." *Food Hydrocolloids* 80 (July): 78–87. https://doi.org/10.1016/j.foodhyd.2018. 01.042.

[84] Umamaheshwari, R. B., Suman Ramteke, and Narendra Kumar Jain. 2004. "Anti-Helicobacter Pylori Effect of Mucoadhesive Nanoparticles Bearing Amoxicillin in Experimental Gerbils Model." *AAPS PharmSciTech* 5 (2): e32. https://doi.org/10.1208/pt050232.

[85] Hu, Kun, and David Julian McClements. 2015. "Fabrication of Biopolymer Nanoparticles by Antisolvent Precipitation and Electrostatic Deposition: Zein-Alginate Core/Shell Nanoparticles." *Food Hydrocolloids* 44 (February): 101–108. https://doi.org/10.1016/j.foodhyd.2014.09.015.

[86] Liang, Hongshan, Qingrong Huang, Bin Zhou, Lei He, Liufeng Lin, Yaping An, Yan Li, Shilin Liu, Yijie Chen, and Bin Li. 2015. "Self-Assembled Zein-Sodium Carboxymethyl Cellulose Nanoparticles as an Effective Drug Carrier and Transporter." *Journal of Materials Chemistry B* 3 (16): 3242–3253. https://doi.org/10.1039/C4TB01920B.

[87] Holban, Alina Maria, and Alexandru Mihai Grumezescu. 2016. *Nanoarchitectonics for Smart Delivery and Drug Targeting.* Amsterdam, Paris [etc.: Elsevier/Academic Press]. http://international.scholarvox.com/book/88835246.

[88] Azarmi, Shirzad, Yuan Huang, Hua Chen, Steve McQuarrie, Douglas Abrams, Wilson Roa, Warren H. Finlay, Gerald G. Miller, and RaimarLöbenberg. 2006. "Optimization of a Two-Step Desolvation Method for Preparing Gelatin Nanoparticles and Cell Uptake Studies in 143B Osteosarcoma Cancer Cells." *Journal of Pharmacy & Pharmaceutical Sciences: A Publication of the Canadian Society for Pharmaceutical Sciences, Societe Canadienne Des Sciences Pharmaceutiques* 9 (1): 124–132.

[89] Madkhali, Osama, George Mekhail, and Shawn D. Wettig. 2019. "Modified Gelatin Nanoparticles for Gene Delivery." *International Journal of Pharmaceutics* 554 (January): 224–234. https://doi.org/10.1016/j.ijpharm.2018.11.001.

[90] Ninan, George, Joseph Jose, and Zynudheen Abubacker. 2011. "Preparation and Characterization of Gelatin Extracted from the Skins of Rohu (Labeo Rohita) and Common Carp (Cyprinus Carpio): Preparation and Characterization of Gelatin from the Skins of Carps." *Journal of Food Processing and Preservation* 35 (2): 143–161. https://doi.org/10.1111/j.1745-4549.2009.00467.x.

[91] Patel, Zarana S., Masaya Yamamoto, Hiroki Ueda, Yasuhiko Tabata, and Antonios G. Mikos. 2008. "Biodegradable Gelatin Microparticles as Delivery Systems for the Controlled Release of Bone Morphogenetic Protein-2." *Acta Biomaterialia* 4 (5): 1126–1138. https://doi.org/10.1016/j.actbio.2008.04.002.

[92] Elzoghby, Ahmed O. 2013. "Gelatin-Based Nanoparticles as Drug and Gene Delivery Systems: Reviewing Three Decades of Research." *Journal of Controlled Release* 172 (3): 1075–1091. https://doi.org/10.1016/j.jconrel.2013.09.019.

[93] Verma, Deepali, Neha Gulati, Shreya Kaul, Siddhartha Mukherjee, and Upendra Nagaich. 2018. "Protein Based Nanostructures for Drug Delivery." *Journal of Pharmaceutics* 2018 (May): 1–18. https://doi.org/10.1155/2018/9285854.

[94] Zhai, Lei, Ameya Narkar, and Kollbe Ahn. 2020. "Self-Healing Polymers with Nanomaterials and Nanostructures." *Nano Today* 30 (February): 100826. https://doi.org/10.1016/j.nantod.2019.100826.

[95] Heo, Yunseon, Mohammad H. Malakooti, and Henry A. Sodano. 2016. "Self-Healing Polymers and Composites for Extreme Environments." *Journal of Materials Chemistry A* 4 (44): 17403–17411. https://doi.org/10.1039/C6TA06213J.

[96] Canadell, Judit, Han Goossens, and Bert Klumperman. 2011. "Self-Healing Materials Based on Disulfide Links." *Macromolecules* 44 (8): 2536–2541. https://doi.org/10.1021/ma2001492.

[97] Cho, Seungwan, Sung Yeon Hwang, Dongyeop X. Oh, and Jeyoung Park. 2021. "Recent Progress in Self-Healing Polymers and Hydrogels Based on Reversible Dynamic B—O Bonds: Boronic/Boronate Esters, Borax, and Benzoxaborole." *Journal of Materials Chemistry A* 9 (26): 14630–14655. https://doi.org/10.1039/D1TA02308J.

[98] Blaiszik, B. J., S. L. B. Kramer, S. C. Olugebefola, J. S. Moore, N. R. Sottos, and S. R. White. 2010. "Self-Healing Polymers and Composites." *Annual Review of Materials Research* 40 (1): 179–211. https://doi.org/10.1146/annurev-matsci-070909-104532.

[99] Kanu, Nand Jee, Eva Gupta, Umesh Kumar Vates, and Gyanendra Kumar Singh. 2019. "Self-Healing Composites: A State-of-the-Art Review." *Composites Part A: Applied Science and Manufacturing* 121 (June): 474–486. https://doi.org/10.1016/j.compositesa.2019.04.012.

[100] Thakur, Vijay Kumar, and Michael R. Kessler. 2015. "Self-Healing Polymer Nanocomposite Materials: A Review." *Polymer* 69 (July): 369–383. https://doi.org/10.1016/j.polymer.2015.04.086.

[101] Huseien, Ghasan Fahim, Kwok Wei Shah, and Abdul Rahman Mohd Sam. 2019. "Sustainability of Nanomaterials Based Self-Healing Concrete: An All-Inclusive Insight." *Journal of Building Engineering* 23 (May): 155–171. https://doi.org/10.1016/j.jobe.2019.01.032.

[102] Salmaso, Stefano, Paolo Caliceti, Vincenzo Amendola, Moreno Meneghetti, Johannes Pall Magnusson, George Pasparakis, and Cameron Alexander. 2009. "Cell Up-Take Control of Gold Nanoparticles Functionalized with a Thermoresponsive Polymer." *Journal of Materials Chemistry* 19 (11): 1608. https://doi.org/10.1039/b816603j.

[103] Roy, Subhasish, Abhishek Baral, and Arindam Banerjee. 2013. "An Amino-Acid-Based Self-Healing Hydrogel: Modulation of the Self-Healing Properties by Incorporating Carbon-Based Nanomaterials." *Chemistry: A European Journal* 19 (44): 14950–14957. https://doi.org/10.1002/chem.201301655.

[104] Talebian, Sepehr, Mehdi Mehrali, Nayere Taebnia, Cristian Pablo Pennisi, Firoz Babu Kadumudi, Javad Foroughi, Masoud Hasany, et al. 2019. "Self-Healing Hydrogels: The Next Paradigm Shift in Tissue Engineering?" *Advanced Science* 6 (16): 1801664. https://doi.org/10.1002/advs.201801664.

[105] Cao, Linlin, Da Tian, Bencai Lin, Wenxiang Wang, Liangjiu Bai, Hou Chen, Lixia Yang, Huawei Yang, and Donglei Wei. 2021. "Fabrication of Self-Healing Nanocomposite Hydrogels with the Cellulose Nanocrystals-Based Janus Hybrid Nanomaterials." *International Journal of Biological Macromolecules* 184 (August): 259–270. https://doi.org/10.1016/j.ijbiomac.2021.06.053.

[106] Tutar, Rumeysa, Andisheh Motealleh, Ali Khademhosseini, and Nermin Seda Kehr. 2019. "Functional Nanomaterials on 2D Surfaces and in 3D Nanocomposite Hydrogels for Biomedical Applications." *Advanced Functional Materials* 29 (46): 1904344. https://doi.org/10.1002/adfm.201904344.

[107] Gaharwar, Akhilesh K., Nicholas A. Peppas, and Ali Khademhosseini. 2014. "Nanocomposite Hydrogels for Biomedical Applications: Nanocomposite Hydrogels." *Biotechnology and Bioengineering* 111 (3): 441–453. https://doi.org/10.1002/bit.25160.

[108] Lombardo, Domenico, Pietro Calandra, Luigi Pasqua, and Salvatore Magazù. 2020. "Self-Assembly of Organic Nanomaterials and Biomaterials: The Bottom-Up Approach for Functional Nanostructures Formation and Advanced Applications." *Materials* 13 (5): 1048. https://doi.org/10.3390/ma13051048.

[109] Girard, Philippe P., Elisabetta A. Cavalcanti-Adam, Ralf Kemkemer, and Joachim P. Spatz. 2007. "Cellular Chemomechanics at Interfaces: Sensing, Integration and Response." *Soft Matter* 3 (3): 307. https://doi.org/10.1039/b614008d.

[110] Ingber, Donald. 2003. "Mechanobiology and Diseases of Mechanotransduction." *Annals of Medicine* 35 (8): 564–577. https://doi.org/10.1080/07853890310016333.

[111] Manish K. Jaiswal, Janet R. Xavier, James K. Carrow, Prachi Desai, Daniel Alge, and Akhilesh K. Gaharwar. 2016. "Mechanically Stiff Nanocomposite Hydrogels at Ultralow Nanoparticle Content." *ACS Nano* 10(1): 246–256. https://doi.org/10.1021/acsnano.5b03918

[112] Fabbro, Alessandra, Susanna Bosi, Laura Ballerini, and Maurizio Prato. 2012. "Carbon Nanotubes: Artificial Nanomaterials to Engineer Single Neurons and Neuronal Networks." *ACS Chemical Neuroscience* 3 (8): 611–618. https://doi.org/10.1021/cn300048q.

[113] Zhou, Jin, Jun Chen, Hongyu Sun, Xiaozhong Qiu, Yongchao Mou, Zhiqiang Liu, Yuwei Zhao, et al. 2015. "Engineering the Heart: Evaluation of Conductive Nanomaterials for Improving Implant Integration and Cardiac Function." *Scientific Reports* 4 (1): 3733. https://doi.org/10.1038/srep03733.

[114] Ho, Dean, Andrew O. Fung, and Carlo D. Montemagno. 2006. "Engineering Novel Diagnostic Modalities and Implantable Cytomimetic Nanomaterials for Next-Generation Medicine." *Biology of Blood and Marrow Transplantation* 12 (1): 92–99. https://doi.org/10.1016/j.bbmt.2005.09.013.

[115] Lin, Ning, Jin Huang, and Alain Dufresne. 2012. "Preparation, Properties and Applications of Polysaccharide Nanocrystals in Advanced Functional Nanomaterials: A Review." *Nanoscale* 4 (11): 3274. https://doi.org/10.1039/c2nr30260h.

[116] Shanmuganathan, Kadhiravan, Jeffrey R. Capadona, Stuart J. Rowan, and Christoph Weder. 2010. "Biomimetic Mechanically Adaptive Nanocomposites." *Progress in Polymer Science* 35 (1–2): 212–222. https://doi.org/10.1016/j.progpolymsci.2009.10.005.

[117] Balasundaram, Ganesan, and Thomas J. Webster. 2006. "Nanotechnology and Biomaterials for Orthopedic Medical Applications." *Nanomedicine* 1 (2): 169–176. https://doi.org/10.2217/17435889.1.2.169.

13 Functional Nanomaterials for Health and Environmental Issues

Estelle Leonard and Erwann Guenin

CONTENTS

13.1 INTRODUCTION TO ONE HEALTH

"One Health" is a concept developed in the first decade of the 2000s by the World Health Organization (WHO) as an approach for programs, policies, legislation and research in which multiple sectors have to communicate and work together to achieve better public health outcomes (Figure 13.1). (Marais et al., 2012; World Health Organization, 2017) The main areas in which One Health is particularly relevant are

DOI: 10.1201/9781003263852-13

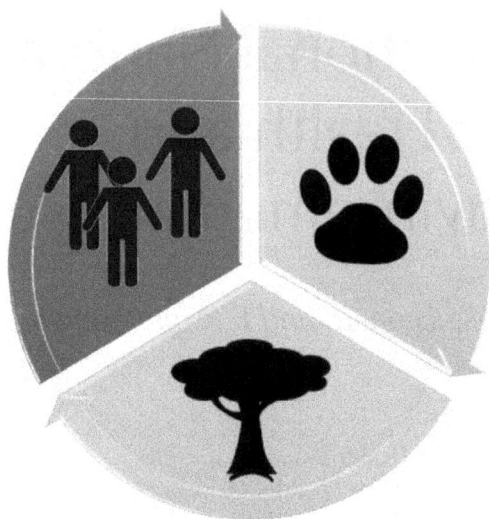

FIGURE 13.1 One Health illustration.

about food safety, zoonosis control, antibiotic resistance (Robinson et al., 2016) and more globally the interactions between animal and human health and how a collaborative and interdisciplinary approach could help to solve emerging global problems (Gibbs, 2014).

Historically, some warnings led scientists to envision the setup of this concept. One example was the huge escape of bats from fire in Malaysian forests, leading to Nipah disease. This virus was first recognized in 1999 during an outbreak among pig farmers in Malaysia. Transmission is thought to have occurred via unprotected exposure to secretions from the pigs, or unprotected contact with the tissue of a sick animal.

Another warning was the chlordecone splitting in the environment. Indeed, the environmental, sanitary and social consequences of the use of chlordecone in the banana plantations of Guadeloupe and Martinique have already been the subject of numerous studies, but the environmental, aquatic, vegetal, animal and human impact is still a huge problem today.

If nanoparticles can have health issues (Kreyling et al., 2006), and thus could not be compatible with the One Health concept, they are a large part of green chemistry, which is fundamentally linked to the respect of the environment even if their monitoring must be largely assessed (Albrecht et al., 2006). They also have numerous applications as sensors, especially for biochemistry (El-Ansary and Faddah, 2010), detection of water pollutants (Prosposito et al., 2020) and even for health monitoring (Nag and Mukhopadhyay, 2019).

Making nanoparticles and One Health compatible is therefore quite possible, and this chapter will be divided into two parts to highlight this consistency. The first will be dedicated to nanoparticles for health and medicine, including those for drug delivery, imaging, biosensing and theranostics. The other will treat nanomaterials for the environment, and particularly for dyes depollution, such as azo adsorption and degradation methods.

13.2 NANOMATERIALS SENSORS FOR HEALTH

13.2.1 Introduction

Since Richard Feynman's talk at Cal Tech in 1959 entitled "There's Plenty of Room at the Bottom" (Feynman, 1960) and the 1974 invention of the term "nanotechnology" by Pr. Norio Tanigushi to describe the production of materials at the nanoscale (Taniguchi, 1974), the development of nanomaterials has known an incredible growth in many different domains. Though not reaching the first expectations envisioned by the founders of the Foresight Institute, nanomaterials have lately notably modified several aspects of our daily life (Perkin and Gubala, 2017) such as for example in electronics. In medicine also, the applications of nanomaterials have permitted great advances and a new domain has even been created designed by the specific term of nanomedicine (Modi et al., 2021). This new market has rapidly flourished and is now anticipated to reach USD 250–500 million in the next five years, depending on the analysis with a compound annual growth rate (CAGR) of more than 10% (Farjadian et al., 2018; Gadekar et al., 2021). Nevertheless, following several scandals such as the asbestos crisis, the development of nanotechnology has been lately challenged by health issues questioning of society (Salamanca-Buentello and Daar, 2021), but the recent COVID-19 management has put the spotlight back on nanomedicine through the use of lipid nanovaccines, nanobased diagnostic systems or nanocontaining protective equipment (Abd Ellah et al., 2020; Mujawar et al., 2020; Weiss et al., 2020). In this section, we thus are going to study why nanomaterials are stirring so much interest in the biomedical domain. First, it is important to comprehend what are the roots of the utilization of nanotechnology in this domain—that is to say, what is the nanoscale bringing—but also, how to prepare these particular materials, and then to study their utilization in two distinct biomedical domains: imaging and biosensing.

13.2.2 Introduction to Nanomaterials in Medicine

13.2.2.1 Properties at the Nanoscale

First of all, when talking about nanomaterials, it is important to recall the definition of the nanoworld. Though for scientist the term "nanosized" refers easily to size at the nanometer scale (10^{-9} m) for the general population, such a small scale remains difficult to comprehend, and the following legal definition given by the European Union does not help to the understanding.

> A natural, incidental or manufactured material containing particles, in an unbound state or as an aggregate or as an agglomerate and where, for 50% or more of the particles in the number size distribution, one or more external dimensions is in the size range 1 nm–100 nm.
>
> (EU Commission Recommendation of 18 October 2011
> on the definition of nanomaterial)

That is why comparisons are often given, such as the "1/50,000 of the width of a human hair." But even simplifications like this one cannot fully explain the nanoscale. To better understand, one has just to remember early courses in chemistry when we learned that the scale to measure atomic level is the Angstrom and that typical atomic bound are a few angstroms (1.54 Å for a C–C bound). Thus, when compare to the chemical world, the nanoscale is relatively close to the size of molecules (1 nm is 10 Å) and the best way to visualize this proximity is to draw molecules at the surface of a 5 nm nanoparticle as was done by R. A. Sperling and W. J. Parak (Sperling and Parak, 2011), as shown in Figure 13.2. Hence, it must be kept in mind that nanomaterials are close to the atomic world.

The second comparison to be drawn to better understand the nanoscale and its implications for medicine is that nanomaterials are close to biological molecules in size. Once again, it is more explicit when comparing representation of current biological molecules and a nanoparticle (Sperling and Parak, 2011), as shown in Figure 13.3. This time, the objects are of similar size and it is clear to understand that such a nano-object will have a particular interaction with the molecular biology. In this

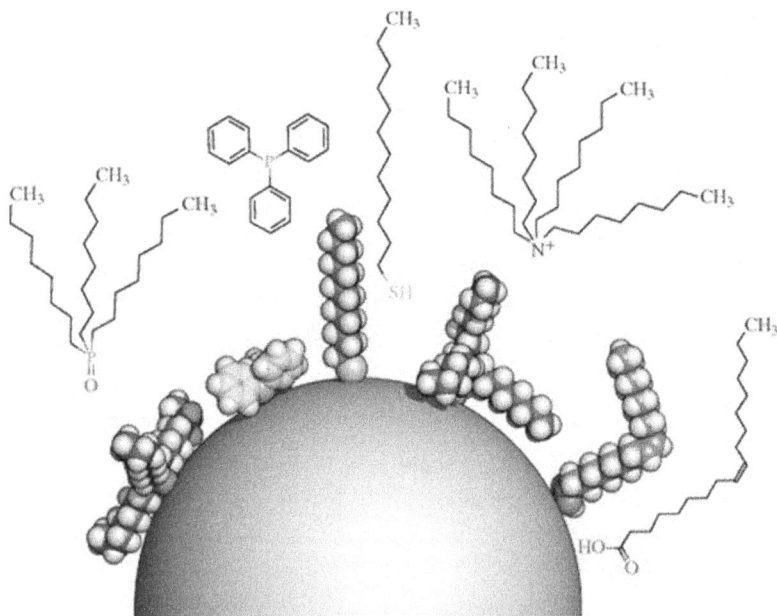

FIGURE 13.2 A nanoparticle of 5 nm core diameter with different hydrophobic ligand molecules both drawn to scale. The particle is idealized as a smooth sphere; the schematic molecule structures are not drawn to scale. Left to right: trioctylphosphine oxide (TOPO), triphenylphosphine (TPP), dodecanethiol (DDT), tetraoctylammonium bromide (TOAB) and oleic acid (OA). The spatial conformation of the molecules is only shown schematically as derived from their chemical structure and space-filling models.

Source: (Sperling and Parak, 2011; reused with permission. Copyright [2010] The Royal Society Publishing

case relations appear more as a matter of surface interaction and could explain the interest of nanomaterials in health science.

Important surface properties of nanomaterials are linked to a simple fact: as an object get smaller, it has more surface area—that is to say that the more the size is reduced, the more the number of atoms at the surface is getting close to the total number of atoms in the bulk material. Moreover, the relation between size and surface area per unit of volume does not decrease linearly. For example, when considering a sphere of R radius, the surface-to-volume ratio varies in 1/R and it clearly evolves differently when reaching value below 100 nm and more and more atoms are present on the surface compared to the total number of atoms (Pineiro et al., 2020) (Figure 13.4). This property of nanomaterial smaller than 100 nm is of great interest when considering surface interaction. First, it will allow for greatly diminishing the matter needed for a same application as surface and interfaces play a large role in material properties: catalysis, sensing, etc. For example, a cube of 1 cm dimension

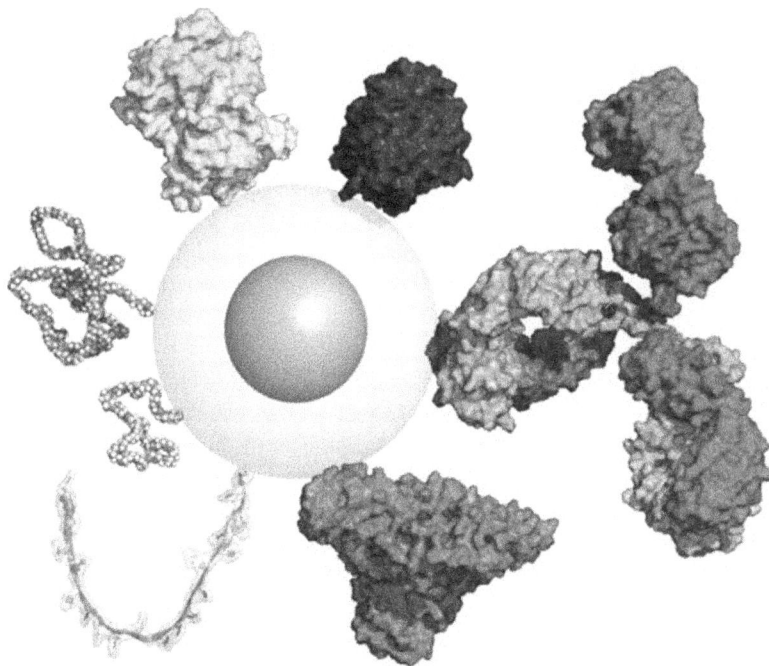

FIGURE 13.3 Relative size of nanoparticles and biomolecules, drawn to scale. Schematic representation of a nanoparticle with 5 nm core diameter, 10 nm shell diameter, with polyethylene glycol (PEG) molecules of 2000 and 5000 g mol^{-1} (on the left, light gray), streptavidin (green), transferrin (blue), antibody (IgG, purple), albumin (red), single-stranded DNA (20mer, cartoon and space filling). Proteins are crystal structures taken from the Protein Data Bank (www.rcsb.org) and displayed as surfaces; PEG and DNA have been modeled from their chemical structure and space filling.

Source: (Sperling and Parak, 2011; reused with permission. Copyright [2010] The Royal Society Publishing)

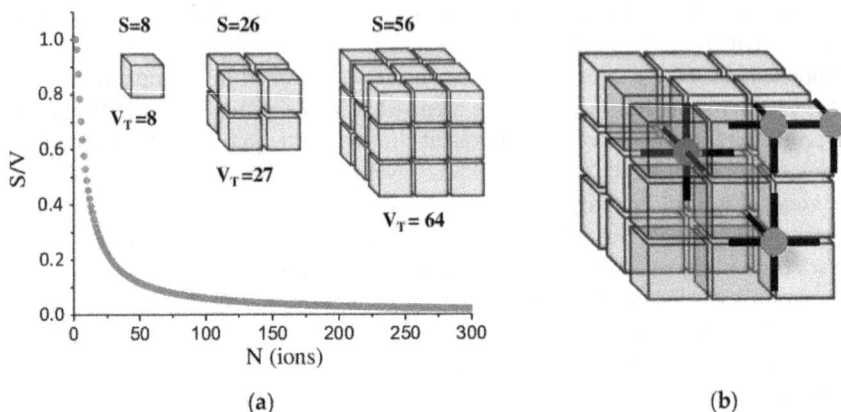

(a)   (b)

FIGURE 13.4   (a) Ratio of surface-to-volume ions in a cubic lattice. For very small nanoparticles (NPs), surface ions can amount up to large percentages of the totality and (b) corner, edge, surface and inner ions with different coordination numbers.

Source: (Reused from Pineiro et al., 2020; https://doi.org/10.3390/magnetochemistry6010004, open access article distributed under the Creative Commons Attribution License)

will have a total surface area of 6 cm², whereas the same cube cut into small cubes of 10 nm will end up with a total surface area of 600,000 cm².

But increasing the surface is not the only benefit from going to nanosize because when reducing the size, the nanomaterials are getting closer to object with a definite number of atoms (as example a gold nanocrystal of 2 nm radius contains around 2,300 atoms). Hence, nanomaterials are so small that they have properties mimicking the atomic level, and they therefore behave as a single object compared to bulk material in which appears as domains. In fact, at the nanoscale, a quantum effect (Kumar and Kumbhat, 2016) due to the confinement of the movement of electrons becomes predominant and affects the physical properties of nanomaterials such as their optical, electrical and magnetic behaviors (Sengupta and Sarkar, 2015). Moreover, these properties became highly dependent on their size and can be also dependent on the shape of the nanomaterial. The usual examples given of this phenomenon are the plasmonic properties of gold nanoparticles (Sarfraz and Khan, 2021), the semiconductor effect of quantum dots (QDots) (Arquer et al., 2021) or the superparamagnetic property of iron oxide nanoparticles (Devi et al., 2019). These new properties added with the enhanced surface volume ratio for nanomaterials are at the heart of the increased utilization of nanomaterials in health, allowing them to have improved efficiency in domains such as imaging and biosensing (Pirzada and Altintas, 2019).

13.2.2.2   Synthesis of Nanomaterials

There are two main strategies for production of nanomaterials: top-down and bottom-up approaches (Baig et al., 2021; Su and Chang, 2017) (Figure 13.5). In the top-down strategy, nanomaterials are produced from parent bulk material that is broken down until the correct size is reached. In the bottom-up approach, nanomaterials are built atom by atom (or molecule by molecule). Top-down methodologies such as ball milling, flame synthesis, laser ablation methods, plasma technology and lithography are generally used in industry. For example, nanolithography is widely utilized in electronic or optic industries for the fabrication of high resolution nanopatterning on a broad range of materials. However, with the exception of the most recent methodologies, which allow for low-cost mass production on a large scale, they do not allow for perfect control of size homogeneity.

On the other hand, bottom-up strategy starting from simple building blocks theoretically allows for perfect control over the production of nanomaterials with well-defined size, shape and highly homogeneous size dispersion. This is the case in the liquid syntheses (also known as wet synthesis) widely used for the preparation of inorganic nanoparticles (Cushing et al., 2004). This methodology is also used in the industry; nevertheless, it often suffers from poor scalability. Several techniques can be employed in liquid synthesis depending on the precursors, the reactants and the nature of the nanomaterial produced. The most common liquid syntheses abundantly described in the literature are chemical precipitation synthesis (LaGrow et al., 2019), solvothermal synthesis (Li et al., 2015) and sol-gel synthesis (Bokov et al., 2021). Though the mechanisms of the nano-object formation are well comprehended, such as the nucleation and growth mechanism described by Lamer (Vreeland et al., 2015), the main difficulties in these syntheses is to be able to effectively control the growth of the nanocrystals and afford a stable and homogeneous suspension in solution. This stability is governed most of the time by the surface coating of the nanomaterial.

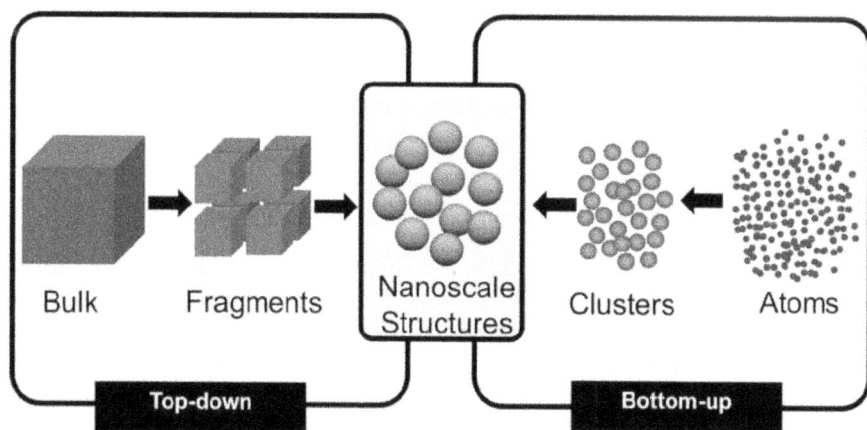

FIGURE 13.5 "Top-down" and "bottom-up" synthesis of nanofabrication.

Source: (Reused with permission of Rawat, 2015; Content from this work may be used under the terms of the Creative Commons Attribution 3.0 license. Any further distribution of this work must maintain attribution to the author[s] and the title of the work, journal citation and DOI)

13.2.2.3 Coating of Nanomaterials

Nanomaterial coating has three principal functions: the stabilization of the nano-crystals produced during the synthesis, the stabilization of the nano-objects in the medium in which it will be dispersed for its applications, and to allow further functionalization of the nanomaterials. First, as explained, coating of the nanoparticle allows for their stability in their solution medium during wet synthesis (Polte, 2015). Sometimes nanoparticles are produced with bare surfaces, but most of the time, a coating agent is added during the synthesis to control the stability of the nanocrystals and avoid coalescence or aggregation. Sometimes the coating agent is also a reactant, as for example in the Turkevich synthesis of gold nanoparticles (Wuithschick et al., 2015) whereby the citrates act both as reductant and stabilizing agent.

The stability of nanomaterials in solution is governed by the law that generally governs the stability of a suspension of colloidal particles, that is to say the DLVO theory (Hotze et al., 2010). This theory built up and explains that colloidal particles in suspension are unstable and tend to agglomerate because of attractive van der Waals forces existing at short interparticle distances. Other attractive forces can be considered such as magnetic forces, such as in the case of iron or cobalt oxide nanoparticles. To ensure stability, therefore, particles should be repulsed from each other by other forces. Most of the time, the repulsion between particles is driven by electrostatic interaction. This force can be due to the intrinsic surface potential of bare particles but is often enhanced by the adjunction of charged coating agents such as in the case of the Turkevich synthesis previously cited whereby ascorbate at the surface of synthetized gold nanoparticles creates strong negative surface potential. Another form of stabilization afforded by coating of the nanoparticles is steric repulsions (compression and interpenetration phenomenon) based on the formation of a layer of macromolecules onto the surface of nanomaterials (Lozsan et al., 2005).

Hence, coating agents are necessary for stabilization during synthesis of the nanomaterials but are also indispensable to stabilize them during their applications (Guerrini et al., 2018). If destabilized during their utilization, nanoparticles will agglomerate and therefore lose their highly favorable surface volume ratio. Moreover, as their properties (optical, magnetic and electrical) are closely related to their size, such agglomeration would be detrimental. Note that in some cases—notably with plasmonic particles—this particularity could be utilized as a detection method.

Such stabilization during their utilization is a great challenge in nanomedicine, whereby nanomaterials are used in complex biological media (Guerrini et al., 2018). Tailoring of the surface will be highly necessary in the presence of macromolecules such as proteins and high concentrations of salts that can be found, for example, in serum (Schubert and Chanana, 2019). Finally, another important aspect of the coating is the functionalization of the surface of the nanomaterial, as it can allow for their specific targeting or their affinity for a specific molecule.

Nanosurface functionalization is definitely an important aspect in the preparation of nanomaterials for health applications (Nam et al., 2013). Though there are as many various coating agents as exist various natures of nanomaterials, coating methodologies can be classified in three different types: small molecules, polymers and silica shell.

Small molecules are generally the first choice with polymers for the coating of nanomaterials, as they can be added easily directly in the synthesis medium, sometimes playing a double role coating and reducing agent. A simple way to understand the affinity of a coating agent for a surface is to go back to hard and soft acid and base theory. In the great diversity of small molecules, some functionality will be clearly devoted to one type of surfaces. For example, thiols or phosphines that are usual coating agents for pure metallic surfaces (gold, silver, platinum, etc.) or semiconductors such as QDots and carboxylate or phosphonate are usually utilized for metal oxides. Among polymers, a great diversity also exists, but with an important utilization of polyvinylpyrrolidone (Koczkur et al., 2015) praised for its amphiphilic nature. Great importance is also given to carbohydrates for their biocompatibility, stability and biodegradability (Kang et al., 2015). For example, carbomethyl dextran was utilized for several nanoparticle formulations used in the diagnostic domain such as Feridex®. But while often described as bringing increased stability compare to small molecules, polymers can also render difficult the approach of molecules nearby the surface and therefore impair some sensing capability of nanomaterials.

The last great methodology for nanoparticle coating is silica coating, which is probably the most universal method of the three as it can be employ on nearly all surfaces (pure metals, oxide, QDots, etc.) (Liu and Han, 2010). Silica coating is derived from the Stöber sol-gel synthesis of silica mesoporous particles. Here the silanization is taking place at the surface of a given nanomaterials by reacting silicon alkoxides. The most used systems are obtained through condensation of silicon alkoxides onto the bare surfaces of oxide nanoparticles or already coated particles bearing carboxylate or hydroxide functions. The control of the shell thickness is afforded by the operating conditions and functional groups can also be added to the surface by using functionalized building blocks such as the widely used 3-aminopropyltriethoxysilane (APTES).

Whatever the solution used to coat the nanomaterials, this coating is generally only the first step of the functionalization process, as tailoring the surface of nanomaterials has become a playground for surface chemists who have developed numerous efficient and elegant strategies to allow for the adjunction of new capabilities to these materials (Erathodiyil and Ying, 2011; Sapsford et al., 2013).

13.2.3 APPLICATIONS IN MEDICINE

13.2.3.1 Imaging

Medical imaging is an important part of the arsenal to fight against numerous pathologies. It is crucial in the early diagnosis of diseases, in the monitoring of the evolution of the pathology and also often in the following of treatments. Since the first development of X-ray radiography in the early 19th century, several imaging techniques have been developed and are used routinely in clinics. Among different imaging modalities, the most common are probably magnetic resonance imaging (MRI) computed tomography (CT), positron emission tomography (PET), single photon emission computed tomography (SPECT) and fluorescence imaging. Each of these techniques have their specificity in terms of spatial resolution and sensitivity, and can even in some cases be combined to couple several features. To get more accurate imaging information, contrast agent can be utilized allowing for example to enhance differentiation between healthy and pathogenic tissue. These contrast agents are usually small molecules or organometallic complexes—but ones that often present high toxicity and a short blood circulation lifetime and poor biodistribution.

Owing to their new properties brought at the nanoscale, some nanomaterials have been developed since the end of the 20th century as new contrast agents for medical imaging (Han et al., 2019). In addition, nanoparticulate contrast agents present several assets that can improve their use as imaging agents. First, their small size and the ability to tightly control it by varying synthesis parameters can contribute to a better biodistribution. This size, along with enhanced blood circulation, also allows for their passive targeting by enhanced permeability and retention (EPR) effect (Alasvand et al., 2017). Moreover, the ability to perfectly tune the surface of the nanomaterials permits the addition of new functionalities to the contrast agents. Active targeting is therefore possible by addition of targeting molecules to increase the spatial localization in a pathologic tissue or to target a specific type of cell sur-expressing a receptor. Finally, modification of the chemical nature of the surface varying surface charge and presence of polymers such as polyethylene glycol (PEG) can influence biodistribution, retention, clearance or modify blood circulation half-life, rendering the contrast agent stealth.

Magnetic resonance imaging (MRI) is probably one of the best known imaging techniques providing high spatial resolution without the use of ionizing radiation and is widely used in the detection of various cancers, for example. Two different modalities exist for MRI, and contrast agents are categorized in T1 or T2 contrast agents relatively to the relaxation time of water proton considered. Paramagnetic gadolinium (Gd^{3+}) chelates belongs to T1 contrast agents (having a bright contrast) and are the gold standard used in modern MRI. But a certain lack of sensitivity and toxicity concerns has driven the development of other alternatives. Among these alternatives, iron oxide nanoparticle, having a superparamagnetic behavior and therefore behaving as T2 contrast agents (having a black contrast) have attracted great attention at the end of the 20th century (Stephen et al., 2011). Several such contrast agents have reached the market being approved by the U.S. Food and Drug Administration and have been used during many years (Feridex®, Resovist®, Combidex®, Lumiren®) but not reaching the signal intensity obtained by T1 contrast agents (Fabian et al., 2014). Alternatively, manganese oxide nanoparticles are also studied as a T1 contrast agent, as well as gadolinium oxide nanoparticles that can have a dual T1/T2 modality. More recently, advances in controlling the size, stability, composition and coating thickness have allowed iron oxide nanoparticles to get T1 contrast in MRI (Estelrich et al., 2015).

Computed tomography (CT) is another medical imaging techniques praised for its speed, low cost and high image resolution. Contrast agents used in CT need a high X-ray attenuation that is reached, for example, by some metallic nanoparticle such as gold, tantalum oxide or zirconium oxide (Han et al., 2019). Gold nanoparticles have attracted great interest for their ease of production and functionalization allowing targeting, as well as for their unchallenged biocompatibility. They have not yet reached the clinical market, but gold nanoparticle radiosensitizers are already commercialized (AuroVist™) for in vivo research. Other approaches are the utilization of nanoparticles as vectors to support iodine since iodine-based contrast agents have been used for a long time in clinical applications.

The use of nanoparticles as support in imaging techniques is also well developed in nuclear imaging such as *positron emission tomography (PET)* and *single photon emission computed tomography (SPECT)*. These two techniques possess high sensitivity but less resolution than MRI and are widely used for molecular imaging and early diagnostics. The use of nanomaterials able to be targeted specifically has therefore generated much interest in this domain in recent years. Though radionuclides, essential to these methodologies (for example, ^{67}Ga and ^{99m}Tc for SPECT and ^{68}Ga and ^{64}Cu for PET), can be introduced directly during the synthesis of the nanomaterials, due to their limited half-life, preparation of radiotracers is preferred with a post-synthetical introduction of the radionuclides (Goel et al., 2017).

Concerning *fluorescence* imaging, though its use is limited by the depth of analysis compared to other imaging methodologies, it is certainly the most common method used in the biosciences due to its low cost and the fact that it does not use ionizing radiation. Fluorescence imaging is based on the emission of fluorescent radiation in the near infra-red (NIR) spectrum whereby the tissue does not completely stop it. A great number of different nanomaterials have been used in fluorescence imaging (Wolfbeis, 2015) based on two different approaches. The first one is the simple use of nanomaterials as support for fluorophores, as for example fluorescently labeled silica or polymeric or dendrimeric nanoparticles. The second one is the utilization of nanomaterials intrinsically fluorescent such as QDots (Michalet et al., 2005), carbon dots or fluorescent upconversion nanoparticles, metallic nanoparticles co-doped by a trivalent lanthanide ion such as Eu(III), Yb(III), Er(III) Michalet et al., 2005)

Finally, it is important to note that owing to their versatility, nanomaterials can be utilized for different imaging techniques—and thus, it is possible to develop

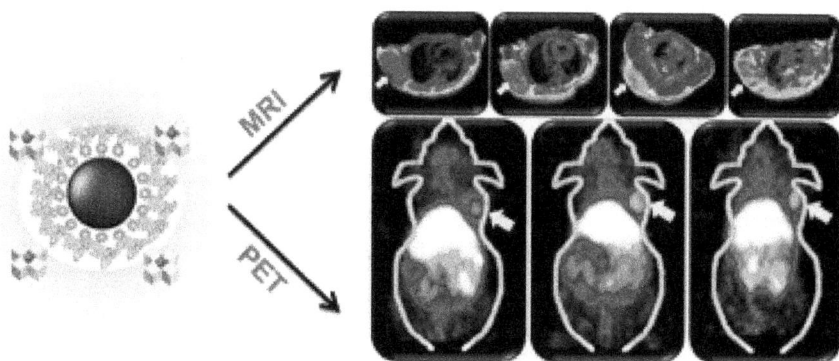

FIGURE 13.6 Left: Schematic illustration of 64Cu labeled MnO nanoparticles with a human serum albumin coating; right upper panel: MR images on U87MG xenograft animal model acquired at 0, 1, 4 and 24 hours after 64Cu-HSA-MONPs injection; right lower panel: PET images taken at 1, 4 and 24 hours after 64Cu-HSA-MONP injection.

Source: (Adapted from Huang et al. [2010]. Copyright [2010] Royal Society of Chemistry)

multimodal imaging agents with PET/CT, PET/MRI, PET/US and CT/MRI modalities (Figure 13.6). In addition, the imaging capacity can be coupled to a therapeutic application. Hence, several nanomaterials have therapeutic capabilities. It can be due to their ability to act as a drug delivery system or due to their intrinsic physical properties such as photothermal therapy (PTT) of gold nanoparticles or magnetic hyperthermia of iron oxide nanoparticles. Therefore, this dual applicability of nanomaterials has given rise to a new domain of research stirring increasing interest known as theranostic (Lee et al., 2012; Li et al., 2016).

13.2.3.2 Biosensing

Biosensors are sensing analytical apparatuses based on a biological recognition mechanism that is transduced into quantifiable signals. The biological mechanism is generally highly specific and sensitive and uses macromolecules intimately matching with each other, such as antibody-antigen or enzyme substrate recognition. Biosensing is usually developed to detect physiological modifications in biological fluids or apparition of pathogenic molecules and therefore is used in early detection or follow-up of various pathogenic disorders. Unfortunately, some diagnostic systems can present several drawbacks such as high costs and long signal treatment, decreasing their interests. Therefore, nanotechnology has also been intensively applicated to this area owing to the high surface ratio of nanomaterials, the new physical properties gained at the nanoscale and the possibility to couple biomolecules to the surface of nanomaterials (Lakshmipriya et al., 2021). In addition to these important assets, the great robustness of nanomaterials and the ability to shape up composite materials including nanomaterials open up the development of point-of-care personal medicine and other improvements such as for example wearable devices (Peng et al., 2020).

Among nanomaterials used in biosensing devices, gold nanoparticles are far from the most utilized (Aldewachi et al., 2018). As described earlier at the nanoscale level, nanomaterials acquire new properties that are not present in their bulk counterparts. Gold nanoparticles at the difference of macroscopic gold present a plasmonic effect that is the consequence of their small size. This effect is due to the collective oscillations of particles' free electrons at the surface which produce a localized surface plasmon resonance (LSPR). This surface plasmon resonance (SPR) enhances the electrical field near the surface, which can greatly improve detection properties in some applications such as surface-enhanced Raman spectroscopy that will be discussed later in this chapter. Another effect is the optical extinction that displays maximum absorption at the plasmon resonance frequency and is closely related to the size, form or agglomeration state of the particles (Szunerits et al., 2014). Both of these properties that are consequences of the nanoscale are exploited in biosensors using gold nanomaterials.

The easiest way to benefit from plasmonic properties of gold nanoparticles in biosensing is to use color changes that occur when they aggregate: *colorimetric sensing*. The principle is to tune the surface of gold nanoparticles with molecules or biomolecules able to interact with an analyte. When mixed together, the analyte induces the aggregation of the nanoparticles by cross-linking them and modifying the color of the colloidal solution (Figure 13.7A). A second methodology can also be employed by reversing the phenomenon, aggregating the nanoparticles prior to the test and then removing the crosslinker by interaction with the analyte (Figure 13.7B). The detection is therefore simple (can be done naked-eye) without the use of costly equipment and can be extremely sensitive and selective depending on the (bio)molecules utilized for the detection. This methodology can be applied to a wide range of analytes, such as deoxyribonucleic acid (DNA) for example. DNA detection using single-strand

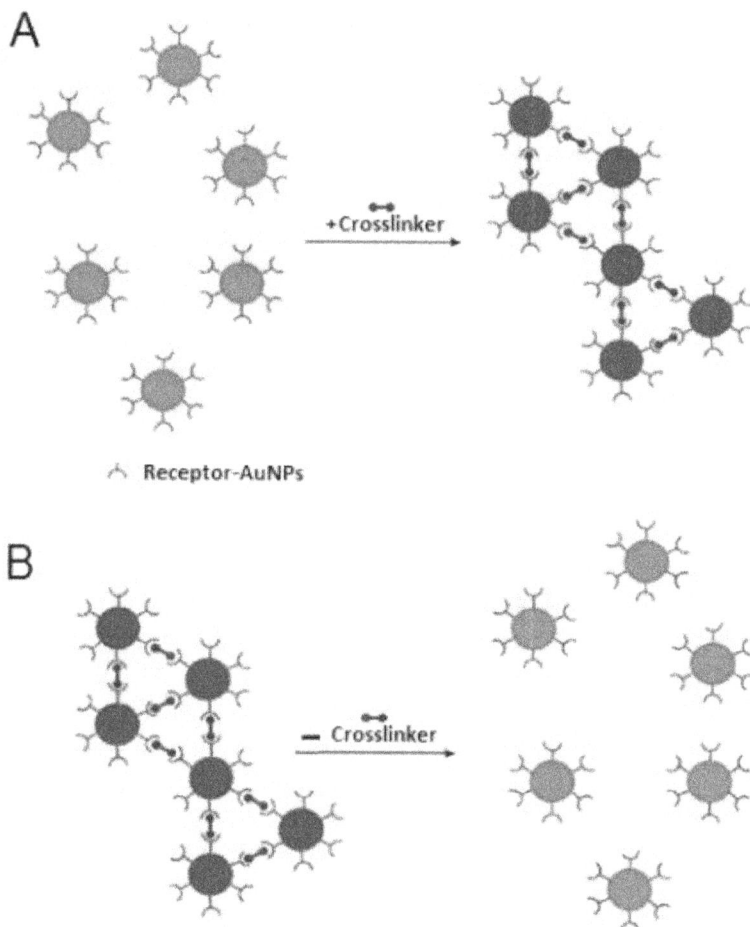

FIGURE 13.7 Schematic illustration of targeted substrates detection by binding to AuNPs using crosslinkers: A) addition of the analyte crosslinker results in the aggregation of the AuNPs with the color of the colloidal solution changing from red to blue; B) removal of the crosslinker induced by the presence of an analyte with resultant dispersion of aggregated AuNPs and accompanying color change from blue to red.

Source: (Aldewachi et al., 2018; reused with permission. Copyright [2018] Royal Society of Chemistry)

DNA–conjugated Au NPs was first introduced by Mirkin et al. (1996) and was widely applied for colorimetric sensing to discriminate different DNA sequences.

LSPR properties of gold nanomaterials have further been applied to biosensing in a large array of materials from the well-known and large product range of BIACORE (Jason-Moller et al., 2006) to more recent research exploiting different design strategies such as dynamic light scattering and hyper-Rayleigh scattering-based sensing, two-photon photoluminescence (TPPL)-based sensing, chiroplasmonic activity-based sensing and *surface-enhanced Raman scattering (SERS)*. This latest methodology appears probably as the most exiting one, owing to its highly sensitive detection characteristic with the theoretical possibility of single molecule detection. SERS is a technique that uses metallic colloids (or rough surfaces) such as gold nanoparticles at the surface of which the detection is executed. Due to plasmonic properties and charge transfer between particles and the analyte, an intensification of the Raman signal by up to 10^6 can be reached allowing unprecedented sensitivity. SERS detection has been of special interest for detection of small molecules, proteins, DNA (Kahraman et al., 2017) and even viruses (Ambartsumyan et al., 2020).

These properties have also largely been applied in lateral flow immunoassay strips testing that was deployed for several clinical detection processes (Byzova et al., 2020; Chen et al., 2020). The best known example is probably the pregnancy test detecting human chorionic gonadotrophin (HCG). The pink colored line appearing to assess the positivity is often due to gold nanoparticles. Lateral flow immunoassay using gold nanoparticles are also utilized for human immunodeficiency virus (HIV) detection and more recently in several COVID-19 tests. These tests are generally employing a detection strategy derived from the most used clinical technique to evaluate the presence of a given biomarker in a sample: enzyme-linked immunosorbent assay (ELISA). ELISA strips are generally coated with antibodies for the selected antigen and the analyte is incubated with other antibodies then bound to horseradish peroxidase (HRP) that will oxidize the color-changing substrates, thus enabling the revelation (Figure 13.8a). Several forms of ELISA have been improved by nanomaterial-based architectures to enhance detection (Figure 13.8b). Nanoparticles decorated with antibodies are utilized and allow for increasing binding sites, therefore enhancing the selectivity of detection. Visualization can be realized using catalytic properties of the nanoparticles by reaction with a color-changing substrate, but nanoparticles can also be directly used as the detected object owing to their plasmonic properties such as gold nanoparticles, as seen previously, or even due to their magnetic properties in the case of iron oxide–based immunoassays (Gao et al., 2020). In this later case, dedicated detectors have been developed possessing high sensitivity (Orlov et al., 2013).

Though gold nanoparticles clearly dominate the domain of nanobiosensing, other types of nanomaterials have also been utilized, including magnetic nanoparticles as described previously, and also QDots for their fluorescent properties. In fact, QDots whose development in nanomedicine have been hampered by their toxicity can be efficiently utilized in in-vitro fluorescence assays (Chern et al., 2019). Hence, they have many advantages compared to classical fluorescent molecules: they show high intensity of fluorescence, high stability and absorption and emission wavelength can be finely tuned by control of their size. Moreover, they have the ability to functionalized. QDots can be used as simple fluorescent tags in ELISA-linked immunoassays

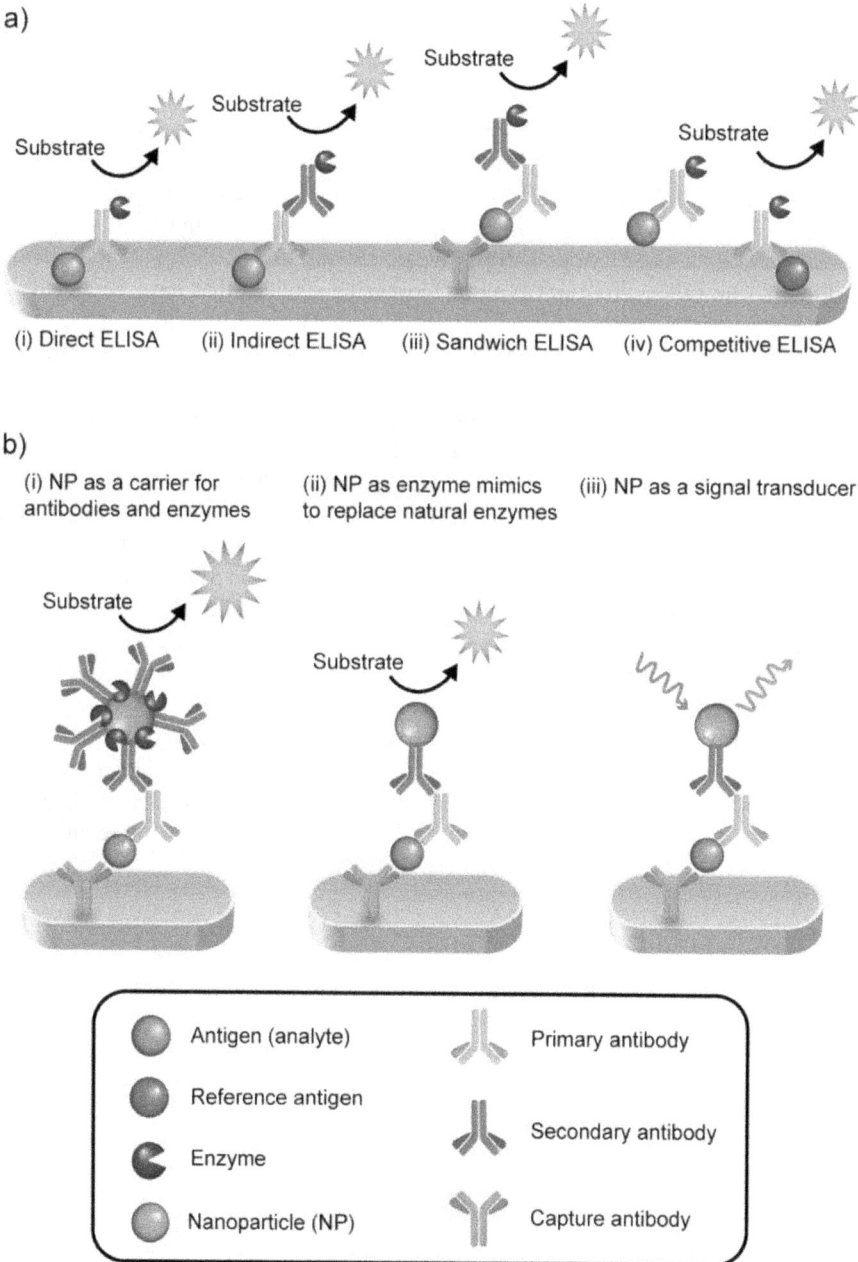

FIGURE 13.8   Showing: a) overview of direct, indirect, sandwich and competitive ELISA; b) schematic illustration of the roles of nanoparticles (NPs) in improving the performance of traditional ELISA by acting as: (i) carriers to load enzymes and antibodies for signal amplification, (ii) enzyme mimics to replace natural enzyme labels, and (iii) signal transducers to provide fluorescence or luminescence signals as an alternative output.

Source: (Gao et al., 2020; reprinted with permission. Copyright [2020] American Chemical Society)

but can further be combined with another chromophore in more complex detection systems such as sensors using Förster resonance energy transfer (FRET). FRET is based on energy transfers between two chromophores and is highly sensitive to changes in the distance between the two parts. FRET systems can be based on the apparition of a fluorescent signal when both chromophores are close or on contrary on photobleaching between the two. QDot-based FRET sensors have been implemented for detection of several biomolecules (Figure 13.9) and biological processes such as oligonucleotide hybridization or antibody sensing, enzyme activity or competition (Medintz and Mattoussi, 2009; Wegner et al., 2013).

Finally, the other important type of biosensing devices that have benefit from the development of nanotechnology are the electrochemical biosensors (Cho et al., 2020). These detectors are generally based on the coating of electrodes with biocatalysts, most commonly enzymes. Metallic nanoparticles—due to their high conductivity, possible catalytic activity and of course their high surface-to-volume ratio—have been extensively applied for the development of electrochemical biosensors. But once again gold nanomaterials have a place of choice among other materials because of improved electron transfer observed from the enzyme to the electrode in their presence (Cabaj and Sołoducho, 2014). Carbon nanotube (CNT) is another type of nanomaterial that has attracted also a lot of attention as an electrochemical biosensor. The high mechanical resistance and the electrical conductivity of CNT are some important assets, in particular as other types of nanomaterials display a wide range of functionalization methodologies to bind biomolecules onto CNT walls. Electrochemical biosensors based on nanomaterials have been efficiently utilized for the development of glucose or peroxide detection but also in more complex assays, for example in the detection of imunoglobuline, pathologic antigens (for diphteria, cancer) or hormones such as progesterone or HCG (Cho et al., 2020; Pingarron et al., 2008). and with miniaturization of devices. they are more and more studied for point-of-care testing in modern medicine.

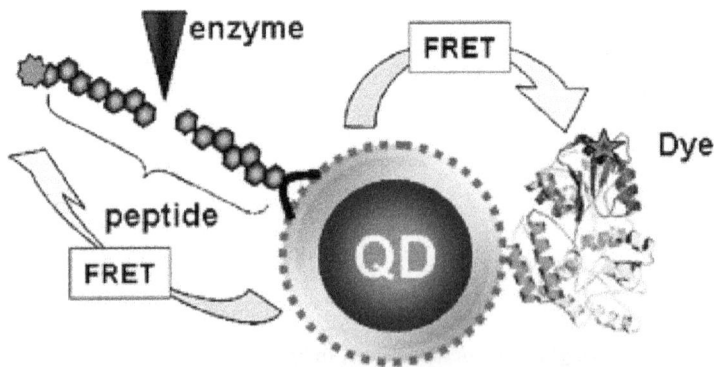

FIGURE 13.9   Schematic of FRET detection with QDots

Source: (Medintz and Mattoussi, 2009; reused with permission. Copyright [2009] Royal Society of Chemistry)

13.2.4 Conclusion

From this rapid overview, it is clear that nanotechnology has greatly affected health science and more precisely applications in imaging and biosensing. Due to their small size and enhanced surface-to-volume ratio, nanomaterials are able to efficiently enhance detection devices. Moreover, the new properties gained at the nanoscale allow for the emergence of new materials able to act, for example, as imaging contrast agents. Last but not least, the development of nanotechnology has been followed by the development of methodologies for the fine tuning of coating and functionalization of nanomaterials. This improvement of surface chemistry renders possible the stabilization of these nanomaterials in complex media for control of the bio-availability as well as the precise attachment of biomolecules for targeting or biosensing purposes. Development of more and more sophisticated nano-objects is therefore possible, paving the way to multimodality and theranostic approaches.

Nevertheless, though unprecedent improvements have been made in the production of nanomaterials, in some cases scaling up to an industrial level is still a bottleneck. In this area, the development of continuous flow synthesis—as well as the implementation of microfluidic systems for the detection devices—appears as part of the solution. Finally, nanomaterials toxicity evaluation remains a key point for their further development and determination of the balance between benefits and risks. An improvement of the dialogue between scientists and the public is also more and more necessary to allow for the acceptance of these technologies by as many people as possible.

13.3 NANOMATERIALS SENSORS ENVIRONMENT REMEDIATION

13.3.1 Introduction

Nanomaterials are widely used for air, soil and water depollution. We can cite, for example, the use of titanium dioxide for self-cleaning building surfaces (Calia et al., 2017), for depollution of chromium-contaminated surfaces (Gossard and Lepeytre, 2017) or even for air-depollution (Fanou et al., 2021) influenced by gold particles (Luna et al., 2019a).

However, we can focus this chapter on an environmental health problem known as water dyes pollution. Indeed, sensing, adsorbing and degrading dyes-polluted aquatic environment by dye-containing effluents is fundamental. Dyes affect aquatic fauna and flora, knowing that 12–14% of 7.10^5 tons of produced dyes are discarded to wastewaters annually (Auta and Hameed, 2011). This discharge of dyed wastewater into aquatic systems also reduces the light penetration in the water, and thus lowers the photosynthetic activity of aquatic plants (Said et al., 2013; Subasioglu and Bilkay, 2009).

In the textiles, leather and cosmetics domain, azobenzene dyes are major source of wastewater pollution (Drumond Chequer et al., 2013), and their removal from aquatic environments is a crucial point for human and environment health improvement.

Among the usual treatments for wastewater dyes (Banat et al., 1996), coagulation and adsorption are the most commonly used methods, but the amount of sludge thus created is a serious problem itself, so solutions based on nanomaterials can be found to prevent the dispersion of dyes in rivers and thus in the environment—but also, as dyes are usually seen by ultraviolet (UV)/visible spectrophotometry or colorimetry, the degradation or adsorption of these dyes by nanomaterials acts as sensors while analyzing the decrease of solution coloring, or the gas production in case of complete degradation. Thus, by sensing kinds of dyes from polluted water (sometimes really selectively), scientists can also—thanks to nanomaterials—degrade them and follow their degradation. An example can be given with the use of TiO$_2$/Na-HZSM-5 nano-composite photocatalyst for the decomposition of an organic azo dye (Guo et al., 2009).

13.3.2 Azo Dyes Environmental Remediation by Nanomaterials

Azo dyes are characterized by the presence of one or more azo bonds (—N=N—), in association with one or more aromatic structures. They are mainly resistant against light and oxidants (Kirk et al., 1993). Their main use is in textile coloring, but in a more discrete domain, their capability of isomerization is of great interest as polymers, additives or nanomotors. However, they or their metabolites can exhibit some mutagenicity, even though their structure can imply more or less toxicity with maximum nontoxic dose from 25–5,000 µg (Chung et al., 1981).

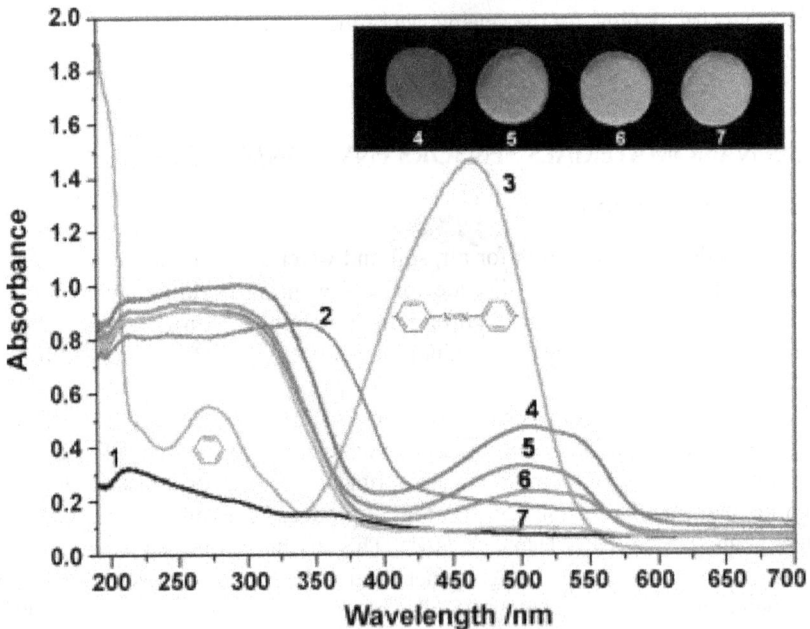

FIGURE 13.10 DRUV–vis spectra of nano-HZSM-5 (1), self-prepared TiO$_2$ (2), 20 mg/l methyl orange (MO) solution (3), and MO-adsorbed TiO$_2$/χNa-HZSM-5 (from 4–7, χ = 0, 0.6, 1.1, and 2.6 wt.%, respectively). Insert: photographs of TiO$_2$/χNa-HZSM-5 nanocomposites adsorbing MO.

Source: (Guo et al., 2009; reused with permission. Copyright [2009] Elsevier)

13.3.2.1 Methyl Orange

Among them, methyl orange (MO, Figure 13.11) is a color indicator used in chemistry to mark the presence of an acidic medium (pink/red) or a basic medium (yellow/orange). The modification of the conjugated system by the protonation of the double bond causes a change in the absorption maximum, so it is used for acid-base determinations. However, this commercial molecule can be used as an azobenzene model for monitoring chemical degradation. Indeed, examples can be found in the literature when titanium dioxide (TiO_2) is used as MO degradation sensor when coated on activated carbon (Li et al., 2006). Especially while illuminated by solar light, TiO_2 can lead to the formation of superoxide or hydroxyl radicals (Folli et al., 2012; Li et al., 2012), which involves dye degradation (Al-Qaradawi and Salman, 2002). The decrease of maximum absorption under UV/visible spectrophotometry allows following its degradation precisely. Also, an electrochemical response can be a good sensor for MO detection. That is how prepared trigon-like CeO_2 structures exhibit excellent sensing performance toward MO by cyclic voltammetry performances. Highly concentrated 100.0 mg/l dye on different electrodes at a scan rate of 50 mV/s were tested, and it was found that the CeO_2 trigons can greatly improve the electron transfer ability (Zhang et al., 2010).

TiO_2 exists in different forms, micrometric and nanometric. It is used in powder form for its ultraviolet absorption properties, but also as a food coloring. For example, the register of annual declaration of nanomaterials managed by France's Agency for Food, Environmental and Occupational Health and Safety (ANSES, 2021) reports more than 10,000 tons of TiO_2 are produced and imported into that country each year. But in addition to size, other physico-chemical properties specific to TiO_2 can also vary and influence its uses and toxicity: (i) the shape: spherical, elongated or fibrous; (ii) the coating; and (iii) the crystallinity: rutile, anatase or brookite. So using less titanium dioxide and enhancing its performance is of great interest.

For this purpose, it was demonstrated that incorporating Cu(I) on photocatalytic MO degradation was six-fold enhanced over surfactant-templated mesoporous TiO_2 (Trofimovaite et al., 2018). Indeed, isolated Cu(I) species suppress charge recombination and calcination increases mesopore size and nanocrystalline order, and these facts allow the activity enhancement.

The other opportunity to lower the potential toxicity for degrading MO is to combine or to replace TiO_2 with zinc oxide (ZnO), as the substitution of TiO_2 with ZnO used for photodegradation is ascribed to the photodegradation mechanism of ZnO

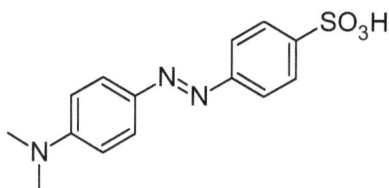

FIGURE 13.11   Methyl orange structure.

being similar to that of TiO_2. When combined, the shape and size of the composite nanoparticles can be manipulated from spherical TiO_2/ZnO composite nanoparticles to cubic composite nanoparticles, hexagonal nanorods and nanobelts. Then, the photocatalytic degradation of MO shows significant variation in rate with the best obtained when TiO_2 and ZnO are combined in nanocomposites (Liao et al., 2008). Moreover, zinc oxide is one of the two filters recognized as safe and effective by the U.S. Food and Drug Administration, even if it is also known as biocide, and thus its safety for the environment is not guaranteed. This is why the WHO recommended upper limit for Zn ions is ~5 ppm (Krenkel, 2012). In this context, nano-ZnO catalyst are of great interest as MO sensors and degradation (Chen et al., 2011; Shen et al., 2018), but those with residual Zn^{2+} not far from the recommended WHO threshold limits were able to decrease MO in solution from 20 ppm to less than 5 ppm, under 19 mW/cm² solar simulated light, leading to complete mineralization of methyl orange of the degraded MO (Zyoud et al., 2015). ZnO nanomushrooms were also found to degrade dyes such as MO. Under optimum conditions, over 92% photodegradation of MO was achieved in 210 minutes with MO concentration of 1.5×10^{-5} M and a zinc oxide nanomushroom amount of 0.2 g/100 ml. The photocatalytic decomposition of MO was examined by measuring the absorbance at regular time intervals 465 nm (Kumar et al., 2013).

In the case of dyes degradation, the products of degradation have to be analyzed precisely to avoid the environment release of more toxic products than the dye itself. Thanks to iron-containing catalysts (i.e., Fe_3O_4, $Fe–Fe_3O_4$, and Fe^0) the MO degradation under microwave irradiation was performed and the gas/light molecules of degradation were precisely analyzed (Siddique et al., 2022). Indeed, some gas were analyzed by the hotspots produced by $Fe–Fe_3O_4$ nanoparticles leading to a high yield of hydrogen production (around 90 mol%), carbon dioxide (less than 10 mol%) and traces of carbon monoxide plus methane. This led the authors to provide a hypothetic pathway of MO degradation inside the catalyst bed, under microwave.

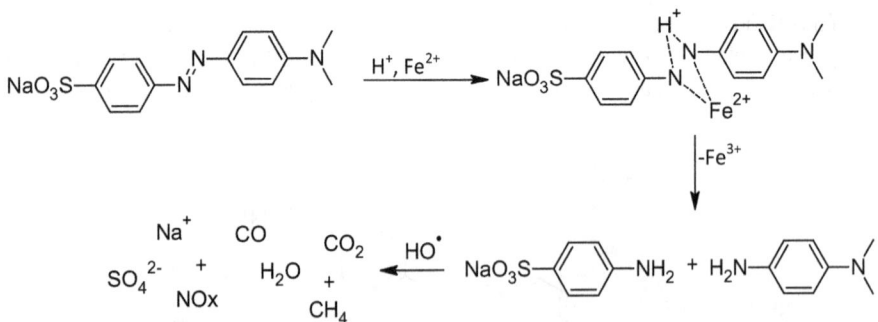

FIGURE 13.12 Provided degradation pathway hypothesis of MO by the catalyst under microwave irradiation.

Source: (Siddique et al., 2022)

MoS_2/TiO_2 composite can also have a strong catalytic effect on MO, and in this context, as MO contains a sulfur element, the reaction product SO_4^{2-} could be detected by adding $BaCl_2$ during the catalytic tests (Hu et al., 2010). This way, the reaction mechanism was proposed during the degradation of MO (Figure 13.13).

These complex degradation products were analyzed partially but precisely during photocatalytical degradation of MO in 2002 (Baiocchi et al., 2002). From this publication, it can be deduced that while a hydroxyl radical is generated, the first hydroxylation occurs at the ring containing the amino group. Then, the second hydroxyl radical is introduced in the ring containing the sulfonate group. This hypothesis was mainly proposed thanks to MS/MS analytical techniques.

Energy technologies can also enhance the results in MO degradation. Examples can be given whereby ultrasound acts on the enhancement of Ag/TiO2 catalyst, as cavitation increases the mass transfer rate in aqueous phase and the release rate of reactive radicals (Thompson and Doraiswamy, 1999), and as ultrasounds permit the constant refreshing of the photocatalyst surface (Wang et al., 2008).

Another way of sensing and depolluting wastewater from MO is to selectively adsorb it. With this technique, MO is not degraded, and it can be desorbed to be

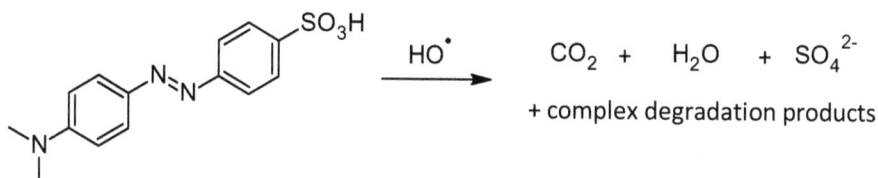

FIGURE 13.13   Proposed reaction mechanism.

Source: (Hu et al., 2010)

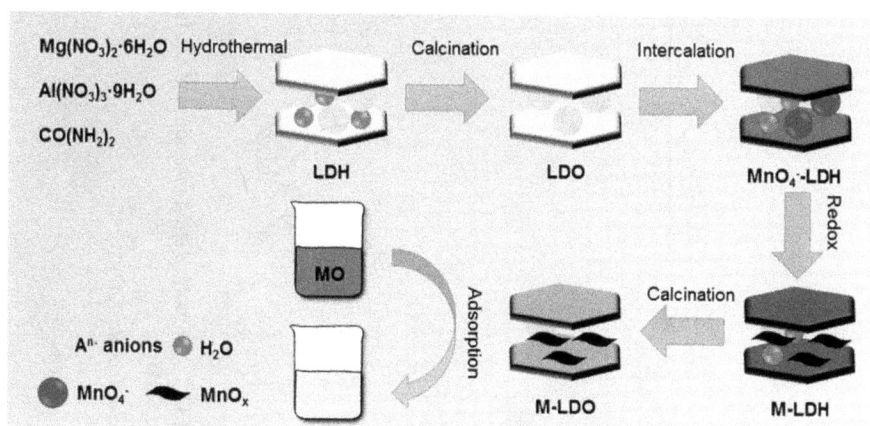

FIGURE 13.14   Schematic of the synthesis process of the as-prepared samples.

Source: (Peng et al., 2016; reused with permission. Copyright [2016] Royal Society of Chemistry)

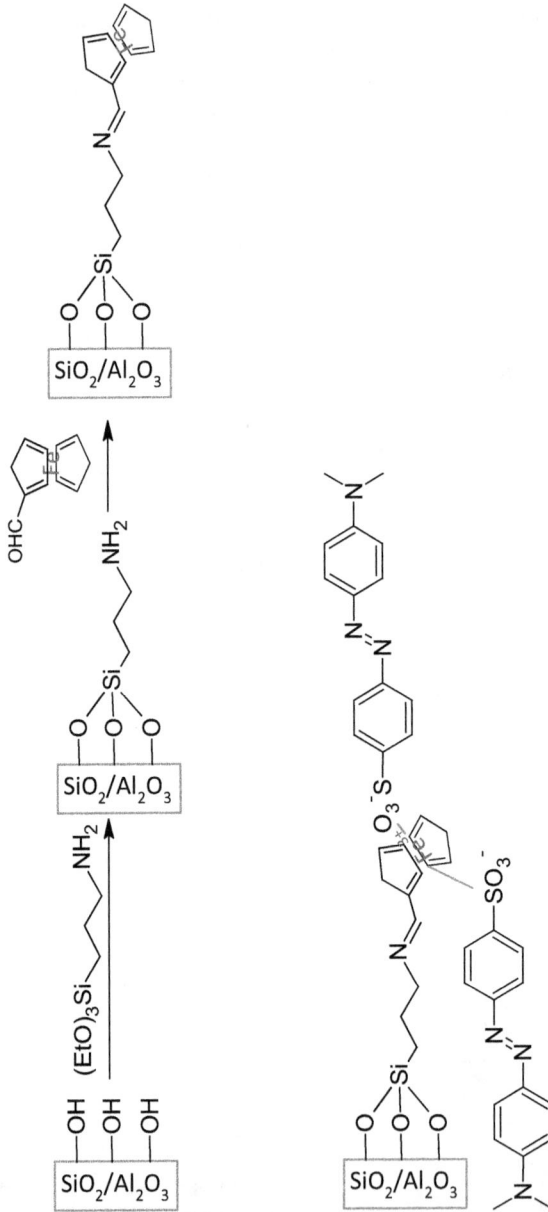

FIGURE 13.15 Immobilization procedure of the ferrocene on the SiO₂–Al₂O₃ mixed-oxides and the proposal mechanism for the heterogenization of MO on the Si/Al-Fe.

Source: (Arshadi et al., 2013)

analyzed or recycled. This was the case when MnOx-decorated MgAl layered double oxide (M-LDO) was performed via merging of memory effect and anion intercalation (Peng et al., 2016). To verify the hypothetic formation mechanism, calcination temperature and time for methyl orange (MO) adsorption were investigated. For the role of MnOx, the MnOx component in MgAl-LDO might act as an adsorbent for the removal of MO, and the active sites increase after decoration with MnOx because of the increase of the specific surface area. The possible adsorption mechanism is driven by electrostatic attraction because the layers charge of LDH is positive, and MO is an anionic dye–containing sulfonate group. It was found by calculation that the maximum capacity was 555.55 mg/g, a high value compared to other research. Moreover, the used adsorbent could be reused after regeneration for five cycles with an adsorption yield of 94%.

Another way of reusing MO after its removal from water is its adsorption and ferrocene is a reagent of choice as, when decorating nano-organo SiO_2–Al_2O_3, it can be easily removed from contaminated water, and thus be analyzed. This adsorbent was prepared by refluxing 3-aminopropyl-triethoxysilane in dry dichloromethane and then adding FCA in dry methanol (Figure 13.15). Then, by different adsorption measurement under various pH, it was found that the possible mechanism of MO adsorption may be considered as the strong electrostatic interaction between the positive active site of the adsorbent and the negative charge of MO (Arshadi et al., 2013).

In the same way, an immobilized Mn-nanoparticle can be used for MO adsorption (Figure 13.16). EPR and XPS of the Mn ions revealed that most of the covalently bonded active sites of the nanoadsorbent were in the form of Mn(III) ions, which were found to be effective as adsorbing MO ions from solution. This adsorption was endothermic and followed a pseudo–second-order kinetic model (Arshadi et al., 2014).

Adsorption can also be performed by ferric oxide–biochar nanocomposites derived from pulp and paper sludge. Indeed, if the use of synthetic adsorbent is attracting by its effectiveness and its low cost, in an One Health approach, the use of biobased materials is crucial, especially if they are ago-processing wastes.

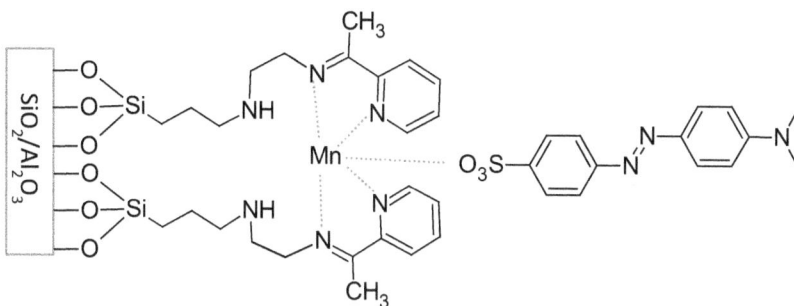

FIGURE 13.16   proposed mechanism for the heterogenization of MO by Mn@Si/Al.

Source: (Arshadi et al., 2014)

Biochar (BC) is one emerging adsorbent formed via pyrolysis of biomaterials and by synthesizing ferric oxide—biochar nanocomposites it was found that free energy calculations showed the feasibility of MO adsorption and that the efficiency of the adsorbents was in the order: Fe_2O_3–BC > BC (Chaukura et al., 2017).

13.3.2.2 Reactive Black 5

The sensing, adsorption and degradation of MO is thus possible using metallic nanoparticles and nanomaterials. However, other azobenzenic dyes, possessing several azo groups, can be identified as problematic for environment. This is the case of Reactive Black 5.

Reactive Black 5 (RB 5) is a bis-azobenzenic compound (Figure 13.17) mainly used in fabrics, textiles and leather products, but also known as a sensitizer, which may cause an allergic skin reaction, and may cause allergy or asthma symptoms or breathing difficulties if inhaled.

In this purpose, a type II 1D-ZnO/BiVO$_4$ heterojunction photocatalyst has been synthesized via a combination of hydrothermal and wet-chemical reactions (Chang et al., 2020). During the degradation process, it was found that h$^+$, O$_2^-$, and OH played decisive roles in the photocatalytic system. Moreover, the percentage of RB 5 degradation was excellent over the reuse cycles (92.0%, 91.3%, 88.7%, 87.4% and 87%). This reuse capability is clearly in favor of a green process for depolluting water by dye degradation.

Again, as for the degradation of MO, magnetic nanoparticles are of great interest in the RB 5 removal from wastewater, and particularly iron oxide nanoparticles (Chang and Shih, 2018). Indeed, magnetite nanoparticles (Fe_3O_4) can be used to activate peroxymonosulfate in the degradation of RB 5 from aqueous solutions (Fadaei et al., 2021). If the presence of chloride, carbonate, nitrate, and bicarbonate in the solution reduced the rate of RB 5 degradation, its degradation compound were

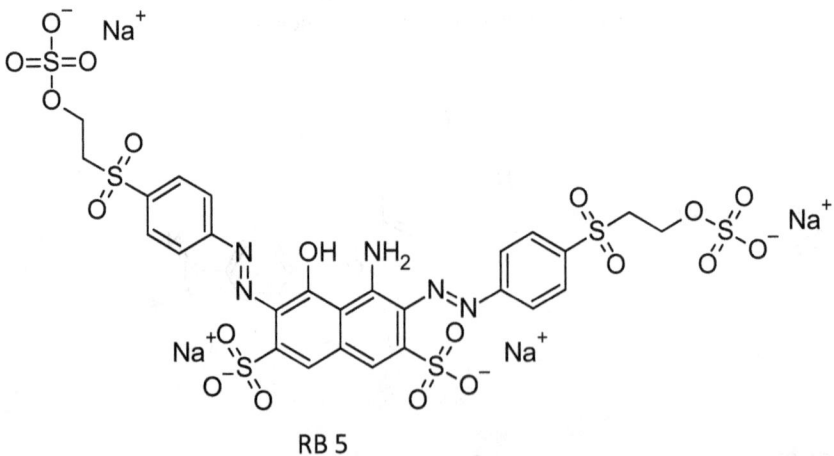

RB 5

FIGURE 13.17   Reactive Black 5 structure.

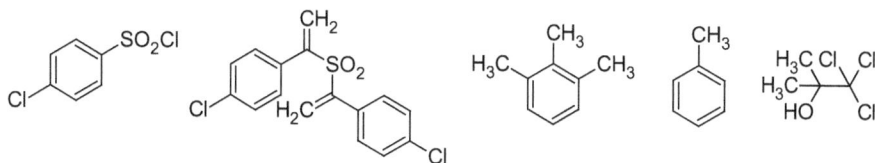

FIGURE 13.18   Degradation compounds from RB 5.

Source: (Fadaei et al., 2021)

clearly identified as 1-chloro-4-ethenylsulfonylbenzene, 4-chlorobenzenesulfonyl chloride, 1,2,3-trimethylbenzene. Further oxidation of RB 5 solution decomposes its products into toluene and 1,1,1-trichloro-2-methyl-2-propanol (Figure 13.18). In addition to these compounds, gas and small ions were identified such as CO_2, SO_4^{2-}, NO_3^-, NH_4^+.

Magnetic nanoparticles can also be coated by TiO_2 to combine the great performance of titanium oxide, plus the magnetic capacity for removal nanoparticles from environment. This $Fe_3O_4@SiO_2@TiO_2$ photocatalyst in combination with UV irradiation led to a RB 5 degradation of 91% after 60 minutes (Lucas et al., 2013). Moreover, it is noticeable that, thanks to the magnetic component, the decrease of RB 5 concentration in water can be followed by spectrophotometry over UV illumination time.

When palladium is added to iron nanoparticles, it leads to great capability for degrading azo dyes. Particularly, Fe/Pd nanoparticles were used to degrade RB 5. These nanoparticles were prepared by a simple successive reduction method and found to be ferromagnetic (Ms = 111.45 emu/g), which still facilitates its magnetic separation from the solution. The results showed that 0.5 g/L, Fe/Pd nanoparticles can completely decolorize 20 mg/L dye solution at 25°C (Samiee et al., 2016).

These examples showed that sensing, removing and degrading azo dyes from dye-polluted water are possible thanks to nanoparticles. Indeed, titanium oxides, zinc oxides, iron oxides, or coating or composites demonstrated their efficiency against MO and RB 5 molecules, taken as pollutant or simply as models for the medium tested. However, other dyes have attracted attention to chemists, such as polyaromatic compounds, known as toluidine or methylene blue, and rhodamine. The following part of this chapter will thus treat of the sensing and removal of such dyes from polluted solutions.

13.3.3   TOLUIDINE AND METHYLENE BLUE, RHODAMINE B ENVIRONMENTAL REMEDIATION

13.3.3.1   Toluidine Blue

Toluidine blue (TB) is generally used as histological dye or in chemistry, as it can give, for example, access to the number of carboxylate moieties on a surface when their dosing have to be performed. However, TB has carcinogenic effects (Redman et al., 1992) due to its phenothiazine structure (Figure 13.19).

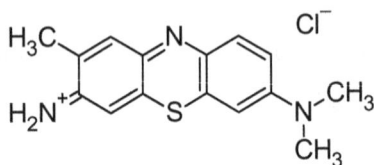

FIGURE 13.19   Toluidine blue structure.

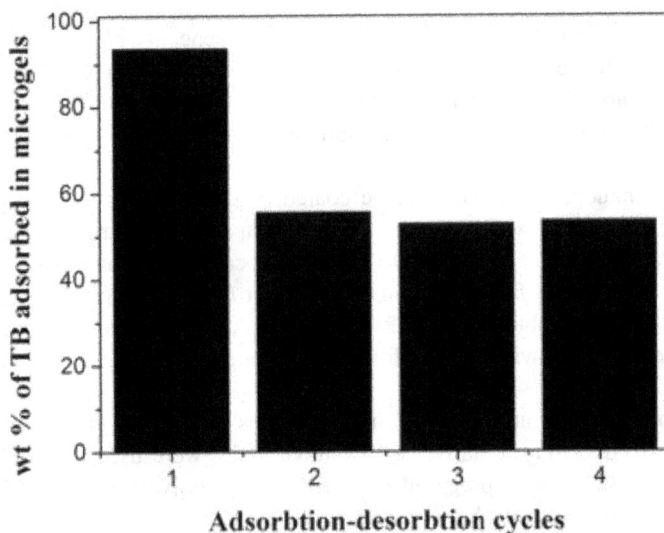

FIGURE 13.20   Recyclability of polymer coated magnetic clusters MC-pAMPS.

Source: (Craciunescu et al., 2017; reused with permission. Copyright [2017] Elsevier)

Its adsorption using a magnetic cluster is of great interest in terms of recycling, and Craciunescu et al. (2017) developed a magnetic hybrid system of magnetite nanoparticles covered with poly (2-acrylamido-2-methyl-1-propanesulfonic acid) (pAMPS) or polydopamine (pDPA) layers. This recyclable system was able to adsorb more than 90 wt% of TB during the first run (Figure 13.20), and around 50 wt% during the three following runs (Craciunescu et al., 2017).

And more importantly for the recovering of pollutants, high magnetization magnetic clusters coated with polymers will allow magnetic separation technology.

13.3.3.2   Methylene Blue

Methylene blue—prepared for the first time in 1876 by Heinrich Caro—is well-known as both medicine and tissue-dye whose action is based on redox properties. This redox molecule can also be used as a model molecule for dye degradation (Figure 13.21).

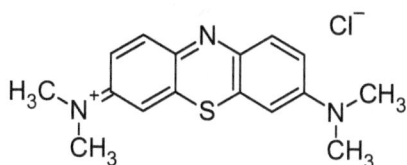

FIGURE 13.21 Methylene blue structure.

FIGURE 13.22 Photo of magnetic beads.

Source: (Bée et al., 2017; reused with permission. Copyright [2017] Elsevier)

In this context, as for TB, adsorbing cationic dyes such as methylene blue (MB) can be of great interest before trying to degrade them. That is why magnetic composite material was prepared, characterized and used as a magsorbent for the removal of a cationic dye, MB, from aqueous solutions. This material was composed of magnetic nanoparticles and clay encapsulated in cross-linked chitosan beads. To understand the adsorption mechanism, the effect of pH on adsorption was reported by the authors on MO and MB. For MO, the optimum range of pH was 3–5. To explain this fact, as chitosan is protonated at this pH range, there is a ion-exchange mechanism controlled by Coulombic interaction. In the case of the cationic MB, adsorption presents a maximum at a 8.5 pH and higher, due to the negative sites of magnetic nanoparticles. A decrease of pH value makes MB adsorption less important due to both the decrease of negative charges of the magnetic nanoparticles and the progressive protonation of chitosan (Bée et al., 2017). This kind of pH-dependent adsorption could be of real interest in the case of selective reuse of the adsorbed compounds. Indeed, having circular economy in mind, the desorption and reuse of these dyes could influence positively their pouring into the environment (Figure 13.22).

Another adsorbent is obviously the use of cyclodextrin cavities (Figure 13.23). Cyclodextrins are well-known not only for their capability to include in their cavity hydrophobic compounds and especially azobenzenes (Gerasimowicz and Wojcik, 1982; Shibusawa et al., 1998; Suzuki et al., 1994; Wu et al., 2004; Zhang et al., 2006), but also—while connected to magnetic nanoparticles—they can lead to an efficient adsorption/desorption device for MB (Badruddoza et al., 2010).

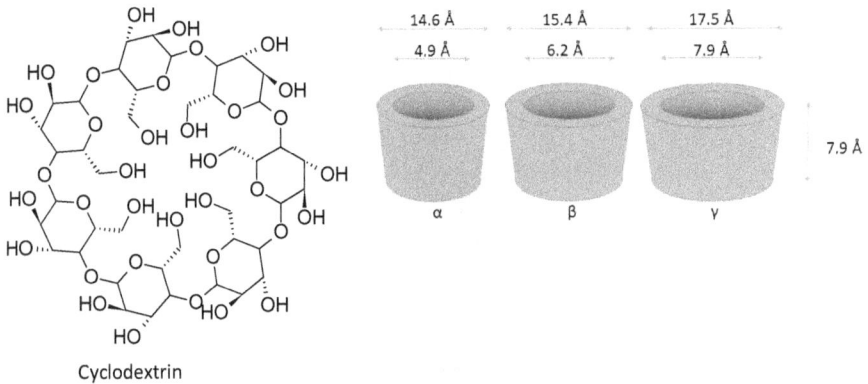

Cyclodextrin

FIGURE 13.23 Cyclodextrin structure and their various dimensions.

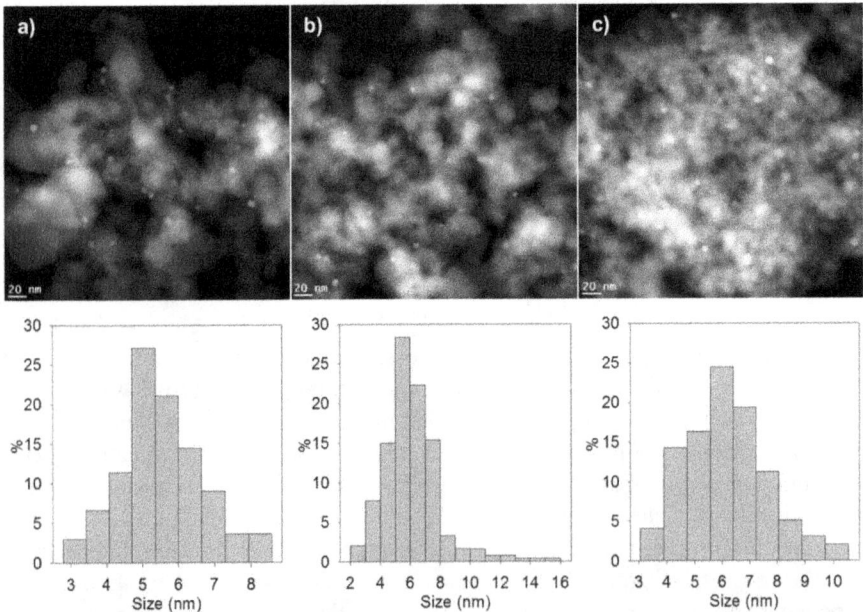

FIGURE 13.24 HAADF-STEM (High-angle annular dark-field) images of the Au/N-TiO₂ particles synthetized and their corresponding AuNPs size distributions: a) Au/N-P25; b) Au/N-P90; and c) Au/N-VP.

Source: (Luna et al., 2019b; reused with permission. Copyright [2019] Elsevier)

TiO$_2$ is also well known for its capability in depolluting, but its performance is limited by its low absorption in the visible range and this drawback can be overcome by titania doping with gold nanoparticles (AuNPs). The deposition-precipitation method using urea (a nitrogen source for doping TiO$_2$) has been employed in the preparation of well dispersed gold supported catalysts with a high yield and a small particle size, and a porous silica matrix with large surface area allowed the promotion of the photoactivity in comparison with a non-porous matrix (Luna et al., 2019b).

As seen in Figure 13.24, gold nanoparticles had size mainly from 4–8 nm but slight differences in the size distributions were observed depending on the TiO$_2$ used. The rate constants of MB photobleaching were compared between blanks, P25 TiO$_2$ (plus N-P25, plus Au/N-P25) and the same with P90 and VP titanium oxide. The best results were obtained for Au/N-P25 (3.94 h^{-1}), followed by Au/N-P90 (3.962 h^{-1}) and Au/N-VP (3.09 h^{-1}), all better than blank (0.05 h^{-1}) or without gold.

13.3.3.3   Rhodamine B

Rhodamine B (RhB), as MB or TB, is also an organic cationic dye. It is used as a tracer dye in water to determine volumes, flow rates and directions of flow and transport. Rhodamine B (Figure 13.25) is also used in microbiology as a fluorescent histological stain.

Two-dimensional (2D) nanosheets are ideal for dimensionally confined transport phenomenon investigation owing to specific surface atomic configuration. More than the real aim of degrading rhodamine B, 2D ZnO mesoporous single-crystal nanosheets (ZnO-MSN) proved to be selective of cationic molecules degradation under sunlight illumination over neutral or anionic molecules (Liu et al., 2016). Indeed, some tests of RhB (cationic), MO (anionic) and phenol (neutral) dyes were performed. It was observed that compared to the presence of single pollutant dyes, the degradation rate of cationic RhB in a mixture was much higher than those of anionic MO and neutral phenol. This was not explained by the mechanism of degradation but by the adsorption capabilities of RhB on ZnO-MSN. The sensitivity of MO over photogenerated charges may explain its degradation rate. Compared to ZnO (commercial) nanoparticles (ZnO-CNP, ZnO-CP), it was clearly demonstrated that the mesoporous nanosheets were the most selective over normalized surface adsorption (Figure 13.26).

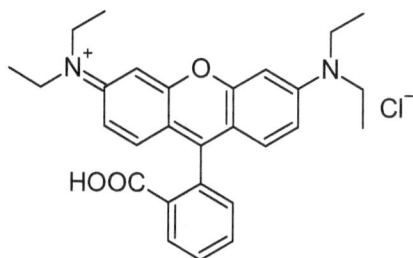

FIGURE 13.25   Rhodamine B structure.

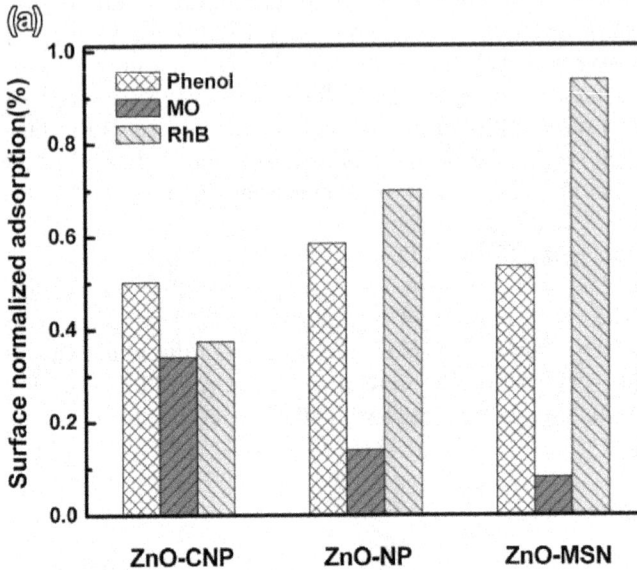

FIGURE 13.26 Surface normalized adsorption photocatalytic rates of ZnO-MSN, ZnO-CNP and ZnO-NP toward the mixed molecules.

Source: (Liu et al., 2016; reused with permission. Copyright [2016] Elsevier)

13.3.4 MULTI DYES DEGRADATION

Being able to functionalize common commercially available civil engineering materials for organic depollution is also really important in a vision of One Health. That is why ZnO nanowires were grown on the surface of tiling and concrete (Le Pivert et al., 2019). These functionalized commercial substrates were tested against three organic dyes for their photocatalytic efficiency (MO, MB and Acid Red 14–AR 14, Figure 13.27). AR 14 is mainly used as food or hair coloring agent.

It is really noticeable that in this case, photodegradation of these dyes (and especial MO which was found to be the hardest dye to degrade) even after four photocycles, the degrading yield was still quantitative or near quantitative. However, even if the reaction between these dyes and the hydroxyl radical generated is the most probable hypothesis mechanism involved, the authors did not seem to have detected the degradation products from the dyes.

Other dyes mixtures including those sent by textile industries have been successfully treated using silver nanoparticles (Gola et al., 2021). Another example can be found for the degradation of biological dyes. Indeed, aqueous leaf extract *C. papaya* has been used for the synthesis of AgNPs (Pa-AgNPs), whereby the results showed effective degradation ability of the nanoparticles with 90% and 83% removal for Blue CP and Yellow 3RS (Figure 13.28), respectively, at 50 mg/l of dye concentration (Jain et al., 2020).

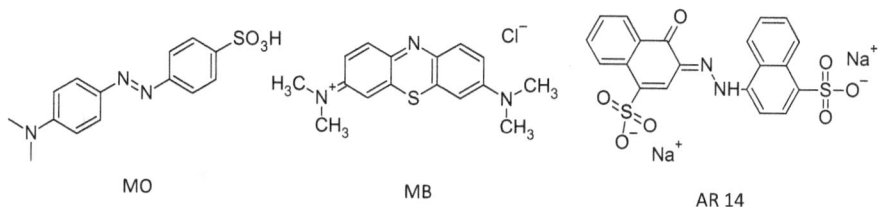

FIGURE 13.27 MO, MB and AR 14 structures.

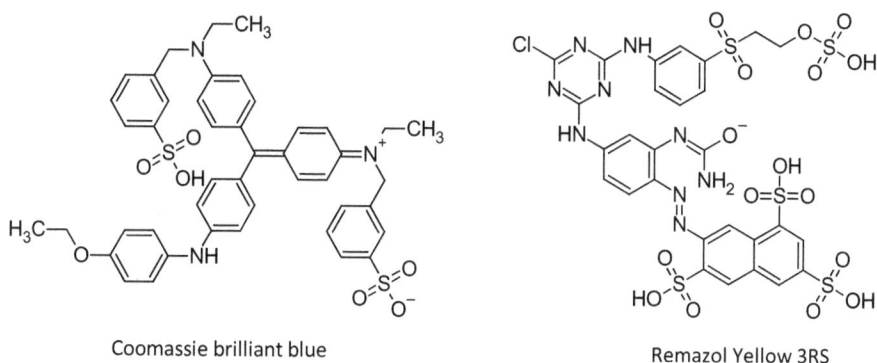

Coomassie brilliant blue

Remazol Yellow 3RS

FIGURE 13.28 Blue CP and Yellow 3RS structures.

It is interesting to notice that, here again, the molecular structures of the blue and yellow dyes are really different (azo dye for the yellow dye, not for the blue one) but that they possess both sulf(on)ate groups at their chain end.

Again, by monitoring changes in dye concentration by UV-visual spectrophotometry, rhodamine B (RhB), MB and 4-nitrophenol dyes were degraded thanks to $Ag_3PO_4/g-C_3N_4$ composites. The morphological features of heterojunction photocatalysts were specially well correlated to their photocatalytic activities regarding dyes pollutants (Deonikar et al., 2019).

13.4 CONCLUSION

To conclude, nanoparticles or nanomaterial proved to be exceptional sensors for dyes polluting water. Different dyes were tested such as diazo compounds, polyaromatic ones, biological-active dyes and even industrial mixtures. Not only nanomaterials proved to be able to detect the polluting dyes, by the use of spectrophotometric or colorimetric changing over time, but also they were able to degrade or adsorb them. Thanks to these nanomaterials, coupled with light or microwaves, their degradation compounds were detected by mass spectrometry, which showed complex mixture of products, and even by studying the gas emitted while decomposing. The adsorption of the problematic dyes led to a great valuable process knowing that these dyes can be reused after desorption, in a meaning of circular economy.

REFERENCES

Abd Ellah, N.H., Gad, S.F., Muhammad, K., Batiha, G.E., Hetta, H.F., 2020. Nanomedicine as a promising approach for diagnosis, treatment and prophylaxis against COVID-19. Nanomedicine (London, England) 15, 2085–2102. https://doi.org/10.2217/nnm-2020-0247

Alasvand, N., Urbanska, A.M., Rahmati, M., Saeidifar, M., Gungor-Ozkerim, P.S., Sefat, F., Rajadas, J., Mozafari, M., Grumezescu, A.M., 2017. Chapter 13: Therapeutic nanoparticles for targeted delivery of anticancer drugs, in: Multifunctional Systems for Combined Delivery, Biosensing and Diagnostics. Elsevier, Cambridge, USA, pp. 245–259. https://doi.org/10.1016/B978-0-323-52725-5.00013-7

Albrecht, M.A., Evans, C.W., Raston, C.L., 2006. Green chemistry and the health implications of nanoparticles. Green Chemistry 8, 417–432. https://doi.org/10.1039/B517131H

Aldewachi, H., Chalati, T., Woodroofe, M.N., Bricklebank, N., Sharrack, B., Gardiner, P., 2018. Gold nanoparticle-based colorimetric biosensors. Nanoscale 10, 18–33. https://doi.org/10.1039/c7nr06367a

Al-Qaradawi, S., Salman, S.R., 2002. Photocatalytic degradation of methyl orange as a model compound. Journal of Photochemistry and Photobiology A: Chemistry 148, 161–168. https://doi.org/10.1016/S1010-6030(02)00086-2

Ambartsumyan, O., Gribanyov, D., Kukushkin, V., Kopylov, A., Zavyalova, E., 2020. SERS-based biosensors for virus determination with oligonucleotides as recognition elements. International Journal of Molecular Sciences 21, 3373. https://doi.org/10.3390/ijms21093373

ANSES, 2021. Dioxyde de titane | ANSES—Agence nationale de sécurité sanitaire de l'alimentation, de l'environnement et du travail [WWW Document]. www.anses.fr/fr/content/dioxyde-de-titane (accessed 1.27.22).

Arquer, F.P.G. de, Dmitri, V.T., Victor, I.K., Yasuhiko, A., Manfred, B., Edward, H.S., 2021. Semiconductor quantum dots: Technological progress and future challenges. Science 373, eaaz8541. https://doi.org/10.1126/science.aaz8541

Arshadi, M., Salimi Vahid, F., Salvacion, J.W.L., Soleymanzadeh, M., 2013. A practical organometallic decorated nano-size SiO_2—Al_2O_3 mixed-oxides for methyl orange removal from aqueous solution. Applied Surface Science 280, 726–736. https://doi.org/10.1016/j.apsusc.2013.05.052

Arshadi, M., Salimi Vahid, F., Salvacion, J.W.L., Soleymanzadeh, M., 2014. Adsorption studies of methyl orange on an immobilized Mn-nanoparticle: Kinetic and thermodynamic. RSC Adv. 4, 16005–16017. https://doi.org/10.1039/C3RA47756H

Auta, M., Hameed, B.H., 2011. Preparation of waste tea activated carbon using potassium acetate as an activating agent for adsorption of Acid Blue 25 dye. Chemical Engineering Journal 171, 502–509. https://doi.org/10.1016/j.cej.2011.04.017

Badruddoza, A.Z.M., Hazel, G.S.S., Hidajat, K., Uddin, M.S., 2010. Synthesis of carboxymethyl-β-cyclodextrin conjugated magnetic nano-adsorbent for removal of methylene blue. Colloids and Surfaces A: Physicochemical and Engineering Aspects 367, 85–95. https://doi.org/10.1016/j.colsurfa.2010.06.018

Baig, N., Kammakakam, I., Falath, W., 2021. Nanomaterials: A review of synthesis methods, properties, recent progress, and challenges. Materials Advances 2, 1821–1871. https://doi.org/10.1039/d0ma00807a

Baiocchi, C., Brussino, M.C., Pramauro, E., Prevot, A.B., Palmisano, L., Marci, G., 2002. Characterization of methyl orange and its photocatalytic degradation products by HPLC/UV—VIS diode array and atmospheric pressure ionization quadrupole ion trap mass spectrometry. International Journal of Mass Spectrometry 214, 247–256. https://doi.org/10.1016/S1387-3806(01)00590-5

Banat, I.M., Nigam, P., Singh, D., Marchant, R., 1996. Microbial decolorization of textile-dyecontaining effluents: A review. Bioresource Technology 58, 217–227. https://doi.org/10.1016/S0960-8524(96)00113-7

Bée, A., Obeid, L., Mbolantenaina, R., Welschbillig, M., Talbot, D., 2017. Magnetic chitosan/clay beads: A magsorbent for the removal of cationic dye from water. Journal of Magnetism and Magnetic Materials 421, 59–64. https://doi.org/10.1016/j.jmmm.2016.07.022

Bokov, D., Turki Jalil, A., Chupradit, S., Suksatan, W., Javed Ansari, M., Shewael, I.H., Valiev, G.H., Kianfar, E., 2021. Nanomaterial by sol-gel method: Synthesis and application. Advances in Materials Science and Engineering 2021, 5102014. https://doi.org/10.1155/2021/5102014

Byzova, N.A., Zherdev, A.V., Khlebtsov, B.N., Burov, A.M., Khlebtsov, N.G., Dzantiev, B.B., 2020. Advantages of highly spherical gold nanoparticles as labels for lateral flow immunoassay. Sensors (Basel, Switzerland) 20, 3608. https://doi.org/10.3390/s20123608

Cabaj, J., Sołoducho, J., 2014. Nano-sized elements in electrochemical biosensors. Materials Sciences and Applications 5, 752–766. https://doi.org/10.4236/msa.2014.510076

Calia, A., Lettieri, M., Masieri, M., Pal, S., Licciulli, A., Arima, V., 2017. Limestones coated with photocatalytic TiO2 to enhance building surface with self-cleaning and depolluting abilities. Journal of Cleaner Production 165, 1036–1047. https://doi.org/10.1016/j.jclepro.2017.07.193

Chang, J.S., Phuan, Y.W., Chong, M.N., Ocon, J.D., 2020. Exploration of a novel Type II 1D-ZnO nanorods/BiVO4 heterojunction photocatalyst for water depollution. Journal of Industrial and Engineering Chemistry 83, 303–314. https://doi.org/10.1016/j.jiec.2019.12.002

Chang, M., Shih, Y., 2018. Synthesis and application of magnetic iron oxide nanoparticles on the removal of Reactive Black 5: Reaction mechanism, temperature and pH effects. Journal of Environmental Management 224, 235–242. https://doi.org/10.1016/j.jenvman.2018.07.021

Chaukura, N., Murimba, E.C., Gwenzi, W., 2017. Synthesis, characterisation and methyl orange adsorption capacity of ferric oxide—biochar nano-composites derived from pulp and paper sludge. Appl Water Sci 7, 2175–2186. https://doi.org/10.1007/s13201-016-0392-5

Chen, C., Liu, J., Liu, P., Yu, B., 2011. Investigation of photocatalytic degradation of methyl orange by using nano-sized ZnO catalysts. Advances in Chemical Engineering and Science 1, 9–14. https://doi.org/10.4236/aces.2011.11002

Chen, X., Leng, Y., Hao, L., Duan, H., Yuan, J., Zhang, W., Huang, X., Xiong, Y., 2020. Self-assembled colloidal gold superparticles to enhance the sensitivity of lateral flow immunoassays with sandwich format. Theranostics 10, 3737–3748. https://doi.org/10.7150/thno.42364

Chern, M., Kays, J.C., Bhuckory, S., Dennis, A.M., 2019. Sensing with photoluminescent semiconductor quantum dots. Methods and Applications in Fluorescence 7, 012005–012005. https://doi.org/10.1088/2050-6120/aaf6f8

Cho, I.-H., Kim, D.H., Park, S., 2020. Electrochemical biosensors: Perspective on functional nanomaterials for on-site analysis. Biomaterials Research 24, 6. https://doi.org/10.1186/s40824-019-0181-y

Chung, K.T., Fulk, G.E., Andrews, A.W., 1981. Mutagenicity testing of some commonly used dyes. Applied and Environmental Microbiology. https://doi.org/10.1128/aem.42.4.641-648.1981

Craciunescu, I., Petran, A., Liebscher, J., Vekas, L., Turcu, R., 2017. Synthesis and characterization of size-controlled magnetic clusters functionalized with polymer layer for wastewater depollution. Materials Chemistry and Physics 185, 91–97. https://doi.org/10.1016/j.matchemphys.2016.10.009

Cushing, B.L., Kolesnichenko, V.L., O'Connor, C.J., 2004. Recent advances in the liquid-phase syntheses of inorganic nanoparticles. Chemical Reviews 104, 3893–3946. https://doi.org/10.1021/cr030027b

Deonikar, V.G., Koteshwara Reddy, K., Chung, W.-J., Kim, H., 2019. Facile synthesis of Ag3PO4/g-C3N4 composites in various solvent systems with tuned morphologies and their efficient photocatalytic activity for multi-dye degradation. Journal of Photochemistry and Photobiology A: Chemistry 368, 168–181. https://doi.org/10.1016/j.jphotochem.2018.09.034

Devi, S.M., Nivetha, A., Prabha, I., 2019. Superparamagnetic properties and significant applications of iron oxide nanoparticles for astonishing efficacy: A review. Journal of Superconductivity and Novel Magnetism 32, 127–144. https://doi.org/10.1007/s10948-018-4929-8

Drumond Chequer, F.M., de Oliveira, G.A.R., Anastacio Ferraz, E.R., Carvalho, J., Boldrin Zanoni, M.V., de Oliveir, D.P., 2013. Textile dyes: Dyeing process and environmental impact, in: Gunay, M. (Ed.), Eco-Friendly Textile Dyeing and Finishing. InTech, London, UK. https://doi.org/10.5772/53659

El-Ansary, A., Faddah, L.M., 2010. Nanoparticles as biochemical sensors. Nanotechnol Sci Appl 3, 65–76. https://doi.org/10.2147/NSA.S8199

Erathodiyil, N., Ying, J.Y., 2011. Functionalization of inorganic nanoparticles for bioimaging applications. Accounts of Chemical Research 44, 925–935. https://doi.org/10.1021/ar2000327

Estelrich, J., Sánchez-Martín, M., Busquets, M., 2015. Nanoparticles in magnetic resonance imaging: From simple to dual contrast agents. Int J Nanomedicine 10, 1727–1741. https://doi.org/10.2147/IJN.S76501

Fabian, K., Marianne, E.M., Jan, G., Twan, L., 2014. Nanoparticles for imaging: Top or flop? Radiology 273, 10–28. https://doi.org/10.1148/radiol.14131520

Fadaei, S., Noorisepehr, M., Pourzamani, H., Salari, M., Moradnia, M., Darvishmotevalli, M., Mengelizadeh, N., 2021. Heterogeneous activation of peroxymonosulfate with Fe3O4 magnetic nanoparticles for degradation of Reactive Black 5: Batch and column study. Journal of Environmental Chemical Engineering 9, 105414. https://doi.org/10.1016/j.jece.2021.105414

Fanou, G.D., Traore, M., Yao, B.K., Kanaev, A., Chhor, K., 2021. Photocatalytic activity of TiO2—P25@n-TiO2@HAP composite films for air depollution. Environ Sci Pollut Res 28, 21326–21333. https://doi.org/10.1007/s11356-020-11924-4

Farjadian, F., Ghasemi, A., Gohari, O., Roointan, A., Karimi, M., Hamblin, M.R., 2018. Nanopharmaceuticals and nanomedicines currently on the market: Challenges and opportunities. Nanomedicine 14, 93–126. https://doi.org/10.2217/nnm-2018-0120

Feynman, R.P., 1960. There's plenty of room at the bottom. Engineering and Science 23, 22–36.

Folli, A., Pade, C., Hansen, T.B., De Marco, T., Macphee, D.E., 2012. TiO2 photocatalysis in cementitious systems: Insights into self-cleaning and depollution chemistry. Cement and Concrete Research 42, 539–548. https://doi.org/10.1016/j.cemconres.2011.12.001

Gadekar, V., Borade, Y., Kannaujia, S., Rajpoot, K., Anup, N., Tambe, V., Kalia, K., Tekade, R.K., 2021. Nanomedicines accessible in the market for clinical interventions. Journal of Controlled Release 330, 372–397. https://doi.org/10.1016/j.jconrel.2020.12.034

Gao, Y., Zhou, Y., Chandrawati, R., 2020. Metal and metal oxide nanoparticles to Enhance the Performance of Enzyme-Linked Immunosorbent Assay (ELISA). ACS Applied Nano Materials 3, 1–21. https://doi.org/10.1021/acsanm.9b02003

Gerasimowicz, W.V., Wojcik, J.F., 1982. Azo dye-α-cyclodextrin adduct formation. Bioorganic Chemistry 11, 420–427. https://doi.org/10.1016/0045-2068(82)90033-5

Gibbs, E.P.J., 2014. The evolution of One Health: A decade of progress and challenges for the future. Veterinary Record 174, 85–91. https://doi.org/10.1136/vr.g143

Goel, S., England, C.G., Chen, F., Cai, W., 2017. Positron emission tomography and nanotechnology: A dynamic duo for cancer theranostics. Advanced Drug Delivery Reviews 113, 157–176. https://doi.org/10.1016/j.addr.2016.08.001

Gola, D., kriti, A., Bhatt, N., Bajpai, M., Singh, A., Arya, A., Chauhan, N., Srivastava, S.K., Tyagi, P.K., Agrawal, Y., 2021. Silver nanoparticles for enhanced dye degradation. Current Research in Green and Sustainable Chemistry 4, 100132. https://doi.org/10.1016/j.crgsc.2021.100132

Gossard, A., Lepeytre, C., 2017. An innovative green process for the depollution of Cr(VI)-contaminated surfaces using TiO2-based photocatalytic gels. Journal of Environmental Chemical Engineering 5, 5573–5580. https://doi.org/10.1016/j.jece.2017.10.026

Guerrini, L., Alvarez-Puebla, R.A., Pazos-Perez, N., 2018. Surface modifications of nanoparticles for stability in biological fluids. Materials 11, 1154.

Guo, P., Wang, X., Guo, H., 2009. TiO2/Na-HZSM-5 nano-composite photocatalyst: Reversible adsorption by acid sites promotes photocatalytic decomposition of methyl orange. Applied Catalysis B: Environmental 90, 677–687. https://doi.org/10.1016/j.apcatb.2009.04.028

Han, X., Xu, K., Taratula, O., Farsad, K., 2019. Applications of nanoparticles in biomedical imaging. Nanoscale 11, 799–819. https://doi.org/10.1039/c8nr07769j

Hotze, E.M., Phenrat, T., Lowry, G.V., 2010. Nanoparticle aggregation: Challenges to understanding transport and reactivity in the environment. Journal of Environmental Quality 39, 1909–1924. https://doi.org/10.2134/jeq2009.0462

Hu, K.H., Hu, X.G., Xu, Y.F., Sun, J.D., 2010. Synthesis of nano-MoS2/TiO2 composite and its catalytic degradation effect on methyl orange. J Mater Sci 45, 2640–2648. https://doi.org/10.1007/s10853-010-4242-9

Huang, J., J. Xie, et al., 2010. HSA coated MnO nanoparticles with prominent MRI contrast for tumor imaging. Chemical Communications 46(36): 6684–6686.

Jain, A., Ahmad, F., Gola, D., Malik, A., Chauhan, N., Dey, P., Tyagi, P.K., 2020. Multi dye degradation and antibacterial potential of Papaya leaf derived silver nanoparticles. Environmental Nanotechnology, Monitoring & Management 14, 100337. https://doi.org/10.1016/j.enmm.2020.100337

Jason-Moller, L., Murphy, M., Bruno, J., 2006. Overview of biacore systems and their applications. Current Protocols in Protein Science 45, 19.13.1–19.13.14. https://doi.org/10.1002/0471140864.ps1913s45

Kahraman, M., Mullen, E.R., Korkmaz, A., Wachsmann-Hogiu, S., 2017. Fundamentals and applications of SERS-based bioanalytical sensing. Nanophotonics 6, 831–852. https://doi.org/10.1515/nanoph-2016-0174

Kang, B., Opatz, T., Landfester, K., Wurm, F.R., 2015. Carbohydrate nanocarriers in biomedical applications: Functionalization and construction. Chemical Society Reviews 44, 8301–8325. https://doi.org/10.1039/c5cs00092k

Kirk, R.E., Othmer, D.F., Kroschwitz, J.I., Howe-Grant, M., 1993. Encyclopedia of Chemical Technology, Fourth. ed. J. Wiley & Sons, New York.

Koczkur, K.M., Mourdikoudis, S., Polavarapu, L., Skrabalak, S.E., 2015. Polyvinylpyrrolidone (PVP) in nanoparticle synthesis. Dalton Transactions 44, 17883–17905. https://doi.org/10.1039/c5dt02964c

Krenkel, P. 2012. Water Quality Management. Elsevier, Burlington, e-Book.

Kreyling, W.G., Semmler-Behnke, M., Möller, W., 2006. Health implications of nanoparticles. J Nanopart Res 8, 543–562. https://doi.org/10.1007/s11051-005-9068-z

Kumar, N., Kumbhat, S., 2016. Unique properties, in: Kumar, N., Kumbhat, S. (Eds.), Essentials in Nanoscience and Nanotechnology. John Wiley & Sons, Inc., Weinheim, Germany, pp. 326–360. https://doi.org/10.1002/9781119096122.ch8

Kumar, R., Kumar, G., Umar, A., 2013. ZnO nano-mushrooms for photocatalytic degradation of methyl orange. Materials Letters 97, 100–103. https://doi.org/10.1016/j.matlet.2013.01.044

LaGrow, A.P., Besenhard, M.O., Hodzic, A., Sergides, A., Bogart, L.K., Gavriilidis, A., Thanh, N.T.K., 2019. Unravelling the growth mechanism of the co-precipitation of iron oxide nanoparticles with the aid of synchrotron X-Ray diffraction in solution. Nanoscale 11, 6620–6628. https://doi.org/10.1039/c9nr00531e

Lakshmipriya, T., Gopinath, S.C.B., Gang, F., 2021. 1—Introduction to nanoparticles and analytical devices, in: Nanoparticles in Analytical and Medical Devices. Elsevier, Cambridge, USA, pp. 1–29. https://doi.org/10.1016/B978-0-12-821163-2.00001-7

Lee, D.-E., Koo, H., Sun, I.-C., Ryu, J.H., Kim, K., Kwon, I.C., 2012. Multifunctional nanoparticles for multimodal imaging and theragnosis. Chemical Society Reviews 41, 2656–2672. https://doi.org/10.1039/c2cs15261d

Le Pivert, M., Poupart, R., Capochichi-Gnambodoe, M., Martin, N., Leprince-Wang, Y., 2019. Direct growth of ZnO nanowires on civil engineering materials: Smart materials for supported photodegradation. Microsyst Nanoeng 5, 1–7. https://doi.org/10.1038/s41378-019-0102-1

Li, J., Wu, Q., Wu, J., 2015. Synthesis of nanoparticles via solvothermal and hydrothermal methods, in: Aliofkhazraei, M. (Ed.), Handbook of Nanoparticles. Springer International Publishing, Cham, pp. 1–28. https://doi.org/10.1007/978-3-319-13188-7_17-1

Li, W., Li, D., Lin, Y., Wang, P., Chen, W., Fu, X., Shao, Y., 2012. Evidence for the active species involved in the photodegradation process of methyl orange on TiO2. J. Phys. Chem. C 116, 3552–3560. https://doi.org/10.1021/jp209661d

Li, X., Zhang, X.-N., Li, X.-D., Chang, J., 2016. Multimodality imaging in nanomedicine and nanotheranostics. Cancer Biology & Medicine 13, 339–348. https://doi.org/10.20892/j.issn.2095-3941.2016.0055

Li, Y., Li, X., Li, J., Yin, J., 2006. Photocatalytic degradation of methyl orange by TiO2-coated activated carbon and kinetic study. Water Research 40, 1119–1126. https://doi.org/10.1016/j.watres.2005.12.042

Liao, D.L., Badour, C.A., Liao, B.Q., 2008. Preparation of nanosized TiO2/ZnO composite catalyst and its photocatalytic activity for degradation of methyl orange. Journal of Photochemistry and Photobiology A: Chemistry 194, 11–19. https://doi.org/10.1016/j.jphotochem.2007.07.008

Liu, J., Hu, Z.-Y., Peng, Y., Huang, H.-W., Li, Y., Wu, M., Ke, X.-X., Tendeloo, G.V., Su, B.-L., 2016. 2D ZnO mesoporous single-crystal nanosheets with exposed {0001} polar facets for the depollution of cationic dye molecules by highly selective adsorption and photocatalytic decomposition. Applied Catalysis B: Environmental 181, 138–145. https://doi.org/10.1016/j.apcatb.2015.07.054

Liu, S., Han, M.-Y., 2010. Silica-coated metal nanoparticles. Chemistry: An Asian Journal 5, 36–45. https://doi.org/10.1002/asia.200900228

Lozsan, A., GarcÃa-Sucre, M., Urbina-Villalba, G., 2005. Steric interaction between spherical colloidal particles. Physical Review E 72, 061405. https://doi.org/10.1103/PhysRevE.72.061405

Lucas, M.S., Tavares, P.B., Peres, J.A., Faria, J.L., Rocha, M., Pereira, C., Freire, C., 2013. Photocatalytic degradation of reactive black 5 with TiO2-coated magnetic nanoparticles. Catalysis Today, Selected Contributions of the 7th European Meeting on Solar Chemistry and Photocatalysis: Environmental Applications (Spea 7) 209, 116–121. https://doi.org/10.1016/j.cattod.2012.10.024

Luna, M., Gatica, J.M., Vidal, H., Mosquera, M.J., 2019a. Au-TiO2/SiO2 photocatalysts with NOx depolluting activity: Influence of gold particle size and loading. Chemical Engineering Journal 368, 417–427. https://doi.org/10.1016/j.cej.2019.02.167

Luna, M., Gatica, J.M., Vidal, H., Mosquera, M.J., 2019b. One-pot synthesis of Au/N-TiO2 photocatalysts for environmental applications: Enhancement of dyes and NOx photodegradation. Powder Technology 355, 793–807. https://doi.org/10.1016/j.powtec.2019.07.102

Marais, B., Crawford, J., Iredell, J., Ward, M., Simpson, S., Gilbert, L., Griffiths, P., Kamradt-Scott, A., Colagiuri, R., Jones, C., Sorrell, T., 2012. One world, one health: Beyond the millennium development goals. The Lancet 380, 805–806. https://doi.org/10.1016/S0140-6736(12)61450-0

Medintz, I.L., Mattoussi, H., 2009. Quantum dot-based resonance energy transfer and its growing application in biology. Physical Chemistry Chemical Physics 11, 17–45. https://doi.org/10.1039/b813919a

Michalet, X., Pinaud, F.F., Bentolila, L.A., Tsay, J.M., Doose, S., Li, J.J., Sundaresan, G., Wu, A.M., Gambhir, S.S., Weiss, S., 2005. Quantum dots for live cells, in vivo imaging, and diagnostics. Science (New York, N.Y.) 307, 538–544. https://doi.org/10.1126/science.1104274

Mirkin, C.A., Letsinger, R.L., Mucic, R.C., Storhoff, J.J., 1996. A DNA-based method for rationally assembling nanoparticles into macroscopic materials. Nature 382, 607–609. https://doi.org/10.1038/382607a0

Modi, S., Prajapati, R., Inwati, G.K., Deepa, N., Tirth, V., Yadav, V.K., Yadav, K.K., Islam, S., Gupta, P., Kim, D.-H., Jeon, B.-H., 2021. Recent trends in fascinating applications of nanotechnology in allied health sciences. Crystals 12, 39.

Mujawar, M.A., Gohel, H., Bhardwaj, S.K., Srinivasan, S., Hickman, N., Kaushik, A., 2020. Nano-enabled biosensing systems for intelligent healthcare: Towards COVID-19 management. Materials Today Chemistry 17, 100306. https://doi.org/10.1016/j.mtchem.2020.100306

Nag, A., Mukhopadhyay, S.C., 2019. Nanoparticles-based flexible wearable sensors for health monitoring applications, in: Kumar, C.S.S.R. (Ed.), Nanotechnology Characterization Tools for Environment, Health, and Safety. Springer, Berlin, Heidelberg, pp. 245–284. https://doi.org/10.1007/978-3-662-59600-5_9

Nam, J., Won, N., Bang, J., Jin, H., Park, J., Jung, Sungwook, Jung, Sanghwa, Park, Y., Kim, S., 2013. Surface engineering of inorganic nanoparticles for imaging and therapy. Advanced Drug Delivery Reviews 65, 622–648. https://doi.org/10.1016/j.addr.2012.08.015

Orlov, A.V., Khodakova, J.A., Nikitin, M.P., Shepelyakovskaya, A.O., Brovko, F.A., Laman, A.G., Grishin, E.V., Nikitin, P.I., 2013. Magnetic immunoassay for detection of staphylococcal toxins in complex media. Analytical Chemistry 85, 1154–1163. https://doi.org/10.1021/ac303075b

Peng, B., Zhao, F., Ping, J., Ying, Y., 2020. Recent advances in nanomaterial-enabled wearable sensors: Material synthesis, sensor design, and personal health monitoring. Small 16, 2002681. https://doi.org/10.1002/smll.202002681

Peng, H.H., Chen, J., Jiang, D.Y., Guo, X.L., Chen, H., Zhang, Y.X., 2016. Merging of memory effect and anion intercalation: MnOx-decorated MgAl-LDO as a high-performance nano-adsorbent for the removal of methyl orange. Dalton Trans. 45, 10530–10538. https://doi.org/10.1039/C6DT00335D

Perkin, E., Gubala, V., 2017. Nanomaterials: On the brink of revolution? Or the endless pursuit of something unattainable? Chemistry International 39, 10–13. https://doi.org/10.1515/ci-2017-0206

Pineiro, Y., Gonzalez Gamez, M., de Castro Alves, L., Arnosa Prieto, A., GarcÃa Acevedo, P., Seco GudiÃ±a, R., Puig, J., Teijeiro, C., YÃ¡Ã±ez Vilar, S., Rivas, J., 2020. Hybrid nanostructured magnetite nanoparticles: From bio-detection and theragnostics to regenerative medicine. Magnetochemistry 6, 4.

Pingarron, J.M., Yanez-Sedeao, P., Gonzalez-Cortes, A., 2008. Gold nanoparticle-based electrochemical biosensors. Electrochimica Acta 53, 5848–5866. https://doi.org/10.1016/j.electacta.2008.03.005

Pirzada, M., Altintas, Z., 2019. Nanomaterials for healthcare biosensing applications. Sensors 19, 5311.

Polte, J., 2015. Fundamental growth principles of colloidal metal nanoparticles: A new perspective. CrystEngComm 17, 6809–6830. https://doi.org/10.1039/c5ce01014d

Prosposito, P., Burratti, L., Venditti, I., 2020. Silver nanoparticles as colorimetric sensors for water pollutants. Chemosensors 8, 26. https://doi.org/10.3390/chemosensors8020026

Rawat, R.S., 2015. Dense plasma focus - From alternative fusion source to versatile high energy density plasma source for plasma nanotechnology. J Phys Conf Ser 591, 012021. https://doi.org/10.1088/1742-6596/591/1/012021

Redman, R.S., Krasnow, S.H., Sniffen, R.A., 1992. Evaluation of the carcinogenic potential of toluidine blue O in the hamster cheek pouch. Oral Surgery, Oral Medicine, Oral Pathology 74, 473–480. https://doi.org/10.1016/0030-4220(92)90299-6

Robinson, T.P., Bu, D.P., Carrique-Mas, J., Fèvre, E.M., Gilbert, M., Grace, D., Hay, S.I., Jiwakanon, J., Kakkar, M., Kariuki, S., Laxminarayan, R., Lubroth, J., Magnusson, U., Thi Ngoc, P., Van Boeckel, T.P., Woolhouse, M.E.J., 2016. Antibiotic resistance is the quintessential One Health issue. Transactions of the Royal Society of Tropical Medicine and Hygiene 110, 377–380. https://doi.org/10.1093/trstmh/trw048

Said, A.E.-A.A., Aly, A.A.M., El-Wahab, M.M.A., Soliman, S.A., El-Hafez, A.A.A., Helmey, V., Goda, M.N., 2013. Application of modified bagasse as a biosorbent for reactive dyes removal from industrial wastewater. Journal of Water Resource and Protection 5, 10–17. https://doi.org/10.4236/jwarp.2013.57A003

Salamanca-Buentello, F., Daar, A.S., 2021. Nanotechnology, equity and global health. Nature Nanotechnology 16, 358–361. https://doi.org/10.1038/s41565-021-00899-z

Samiee, S., Goharshadi, E.K., Nancarrow, P., 2016. Successful degradation of reactive black 5 by engineered Fe/Pd nanoparticles: Mechanism and kinetics aspects. Journal of the Taiwan Institute of Chemical Engineers 67, 406–417. https://doi.org/10.1016/j.jtice.2016.07.012

Sapsford, K.E., Algar, W.R., Berti, L., Gemmill, K.B., Casey, B.J., Oh, E., Stewart, M.H., Medintz, I.L., 2013. Functionalizing nanoparticles with biological molecules: Developing chemistries that facilitate nanotechnology. Chemical Reviews 113, 1904–2074. https://doi.org/10.1021/cr300143v

Sarfraz, N., Khan, I., 2021. Plasmonic gold nanoparticles (AuNPs): Properties, synthesis and their advanced energy, environmental and biomedical applications. Chemistry an Asian Journal 16, 720–742. https://doi.org/10.1002/asia.202001202

Schubert, J., Chanana, M., 2019. Coating matters: Review on colloidal stability of nanoparticles with biocompatible coatings in biological media, living cells and organisms. Current Medicinal Chemistry 25, 4553–4586. https://doi.org/10.2174/0929867325666180601101859

Sengupta, A., Sarkar, C.K., 2015. Introduction to Nano: Basics to Nanoscience and Nanotechnology, Engineering Materials. Springer, Berlin, Heidelberg. https://doi.org/10.1007/978-3-662-47314-6

Shen, Z., Zhou, H., Chen, H., Xu, H., Feng, C., Zhou, X., 2018. Synthesis of nano-zinc oxide loaded on mesoporous silica by coordination effect and its photocatalytic degradation property of methyl orange. Nanomaterials 8, 317. https://doi.org/10.3390/nano8050317

Shibusawa, T., Okamoto, J., Abe, K., Sakata, K., Ito, Y., 1998. Inclusion of azo disperse dyes by cyclodextrins at dyeing temperature. Dyes and Pigments 36, 79–91. https://doi.org/10.1016/S0143-7208(96)00118-0

Siddique, F., Mirzaei, A., Gonzalez-Cortes, S., Slocombe, D., Al-Megren, H.A., Xiao, T., Rafiq, M.A., Edwards, P.P., 2022. Sustainable chemical processing of flowing wastewater through microwave energy. Chemosphere 287, 132035. https://doi.org/10.1016/j.chemosphere.2021.132035

Sperling, R.A., Parak, W.J., 2011. Surface modification, functionalization and bioconjugation of colloidal inorganic nanoparticles. Philosophical Transactions of the Royal Society A: Mathematical, Physical and Engineering Sciences 368, 1333–1383. https://doi.org/10.1098/rsta.2009.0273

Stephen, Z.R., Kievit, F.M., Zhang, M., 2011. Magnetite nanoparticles for medical MR imaging. Materials Today 14, 330–338. https://doi.org/10.1016/S1369-7021(11)70163-8

Su, S.S., Chang, I., 2017. Review of production routes of nanomaterials, in: Brabazon, D., Pellicer, E., Zivic, F., Sort, J., Dolors BarÃ³, M., Grujovic, N., Choy, K.-L. (Eds.), Commercialization of Nanotechnologies: A Case Study Approach. Springer International Publishing, Cham, pp. 15–29. https://doi.org/10.1007/978-3-319-56979-6_2

Subasioglu, T., Bilkay, I.S., 2009. Determination of biosorption conditions of Methyl Orange by *Humicola fuscoatra*. JSIR 68(12), December.

Suzuki, M., Ohmori, H., Kajtar, M., Szejtli, J., Vikmon, M., 1994. The association of inclusion complexes of cyclodextrins with azo dyes. J Incl Phenom Macrocycl Chem 18, 255–264. https://doi.org/10.1007/BF00708732

Szunerits, S., Spadavecchia, J., Boukherroub, R., 2014. Surface plasmon resonance: Signal amplification using colloidal gold nanoparticles for enhanced sensitivity. Reviews in Analytical Chemistry 33, 153–164. https://doi.org/10.1515/revac-2014-0011

Taniguchi, N., 1974. On the basic concept of nanotechnology. Proceedings of the International Conference on Production Engineering 18–23.

Thompson, L.H., Doraiswamy, L.K., 1999. Sonochemistry: Science and engineering. Ind. Eng. Chem. Res. 38, 1215–1249. https://doi.org/10.1021/ie9804172

Trofimovaite, R., Parlett, C.M.A., Kumar, S., Frattini, L., Isaacs, M.A., Wilson, K., Olivi, L., Coulson, B., Debgupta, J., Douthwaite, R.E., Lee, A.F., 2018. Single atom Cu(I) promoted mesoporous titanias for photocatalytic Methyl Orange depollution and H2 production. Applied Catalysis B: Environmental 232, 501–511. https://doi.org/10.1016/j.apcatb.2018.03.078

Vreeland, E.C., Watt, J., Schober, G.B., Hance, B.G., Austin, M.J., Price, A.D., Fellows, B.D., Monson, T.C., Hudak, N.S., Maldonado-Camargo, L., Bohorquez, A.C., Rinaldi, C., Huber, D.L., 2015. Enhanced nanoparticle size control by Extending LaMer's Mechanism. Chemistry of Materials 27, 6059–6066. https://doi.org/10.1021/acs.chemmater.5b02510

Wang, H., Niu, J., Long, X., He, Y., 2008. Sonophotocatalytic degradation of methyl orange by nano-sized Ag/TiO2 particles in aqueous solutions. Ultrasonics Sonochemistry 15, 386–392. https://doi.org/10.1016/j.ultsonch.2007.09.011

Wegner, K.D., Jin, Z., LindÃ©n, S., Jennings, T.L., Hildebrandt, N., 2013. Quantum-dot-based FÃ¶rster resonance energy transfer immunoassay for sensitive clinical diagnostics of low-volume serum samples. ACS Nano 7, 7411–7419. https://doi.org/10.1021/nn403253y

Weiss, C., Carriere, M., Fusco, L., Capua, I., Regla-Nava, J.A., Pasquali, M., Scott, J.A., Vitale, F., Unal, M.A., Mattevi, C., Bedognetti, D., MerkoÃ§i, A., Tasciotti, E., Yilmazer, A., Gogotsi, Y., Stellacci, F., Delogu, L.G., 2020. Toward nanotechnology-enabled approaches against the COVID-19 pandemic. ACS Nano 14, 6383–6406. https://doi.org/10.1021/acsnano.0c03697

World Health Organization, "One Health" [WWW Document], 2017. www.who.int/newsroom/q-a-detail/one-health (accessed 9.7.21).

Wolfbeis, O.S., 2015. An overview of nanoparticles commonly used in fluorescent bioimaging. Chemical Society Reviews 44, 4743–4768. https://doi.org/10.1039/c4cs00392f

Wu, M., Yuguchi, Y., Kumagai, T., Endo, T., Hirotsu, T., 2004. Nano-complex formation of cyclodextrin and azobenzene using supercritical carbon dioxide. Chem. Commun. 1288–1289. https://doi.org/10.1039/B400289J

Wuithschick, M., Birnbaum, A., Witte, S., Sztucki, M., Vainio, U., Pinna, N., Rademann, K., Emmerling, F., Kraehnert, R., Polte, J., 2015. Turkevich in new robes: Key questions answered for the most common gold nanoparticle synthesis. ACS Nano 9, 7052–7071. https://doi.org/10.1021/acsnano.5b01579

Zhang, D.E., Xie, Q., Zhang, X.B., Li, S.Z., Han, G.Q., Ying, A.L., Tong, Z.W., 2010. Simple method for fabrication of CeO2 trigon-like structures and application for methyl-orange sensors. Micro & Nano Letters 5, 58–62. https://doi.org/10.1049/mnl.2009.0104

Zhang, Huarong, Chen, G., Wang, L., Ding, L., Tian, Y., Jin, W., Zhang, Hanqi, 2006. Study on the inclusion complexes of cyclodextrin and sulphonated azo dyes by electrospray ionization mass spectrometry. International Journal of Mass Spectrometry 252, 1–10. https://doi.org/10.1016/j.ijms.2006.01.021

Zyoud, A., Zu'bi, A., Helal, M.H.S., Park, D., Campet, G., Hilal, H.S., 2015. Optimizing photo-mineralization of aqueous methyl orange by nano-ZnO catalyst under simulated natural conditions. J Environ Health Sci Engineer 13, 46. https://doi.org/10.1186/s40201-015-0204-0

14 Implications of the Use of Functional Nanomaterials for the Environment and Health
Risk Assessment and Challenges

Magdalena Luty-Błocho,
Manuel Varon Hoyos, and Volker Hessel

CONTENTS

DOI: 10.1201/9781003263852-14

14.1 INTRODUCTION

A sensor is a device allowing for qualification and/or quantification of a target analyte in a working environment (air, waste stream, water, human body, etc.). Measured
amount of the target analyte is converted into a signal which can be read by an
observer or by an instrument. There are many different types of sensors. Based on
their detection targets, as well as detection mechanisms, sensors can be categorized
as mechanical, electromagnetic, thermal, optical, acoustic, radiation and chemical
sensors [1]. The trend in developing of future sensors is moving toward sensors with
better sensitivity and selectivity, faster response time, and those which are more portable, remotely capable, uncomplicated to use, long lasting and cheap in production
and operation. More advanced sensor design also tends to be multifunctional, i.e. in
addition to its main sensing function, it keeps track of many environmental factors
such as temperature, humidity, time, geographic location, etc. Better, faster and more
compact sensors offer powerful tools to cover all interesting areas and working environments. Recently, however, most interesting are sensors based on functional nanomaterials (FNs) due to their unique physicochemical properties and surface ability
to multiplying their basic properties, depending on the working environment or,
on the contrary, focusing on the one selected effect/aspect. According to Su et al. [2]
there are five categories of nanomaterials (NMs): carbon-based nanomaterials, metal
nanoparticles, silica-based nanomaterials, semiconductor quantum dots and other
nanomaterials (Figure 14.1). Compared to micro-sized materials, NMs are characterized by a higher surface-to-volume ratio, usually leading to an enhancement of
optical, mechanical, electrical, structural and magnetic properties [3]. Therefore,
the use of materials at the nanodimensional scale provides several improvements in
terms of analytical features including sensitivity, rapidity of response, selectivity and
robustness [4]. On the other hand, FNs are a group of advanced materials designed
and synthesized for specific functions, and therefore, they can be applied for different purposes. Sensors based on nanomaterials (NMs) are divided into four group
as is shown schematically in Figure 14.2, these are: biochemical, chemical, electrochemical and fluorescent [5]. The development of FNs-based sensors, especially
used in the context of environmental protection, is a topic that is gaining more and
more importance. Taking into account the Web of Science database, the number of
articles published in this field in 2021 alone (search words: sensors, functional nanomaterials) is more than six times higher than the number of articles published in 2010
(Figure 14.3). The great interest in sensors is the result of an increased awareness of

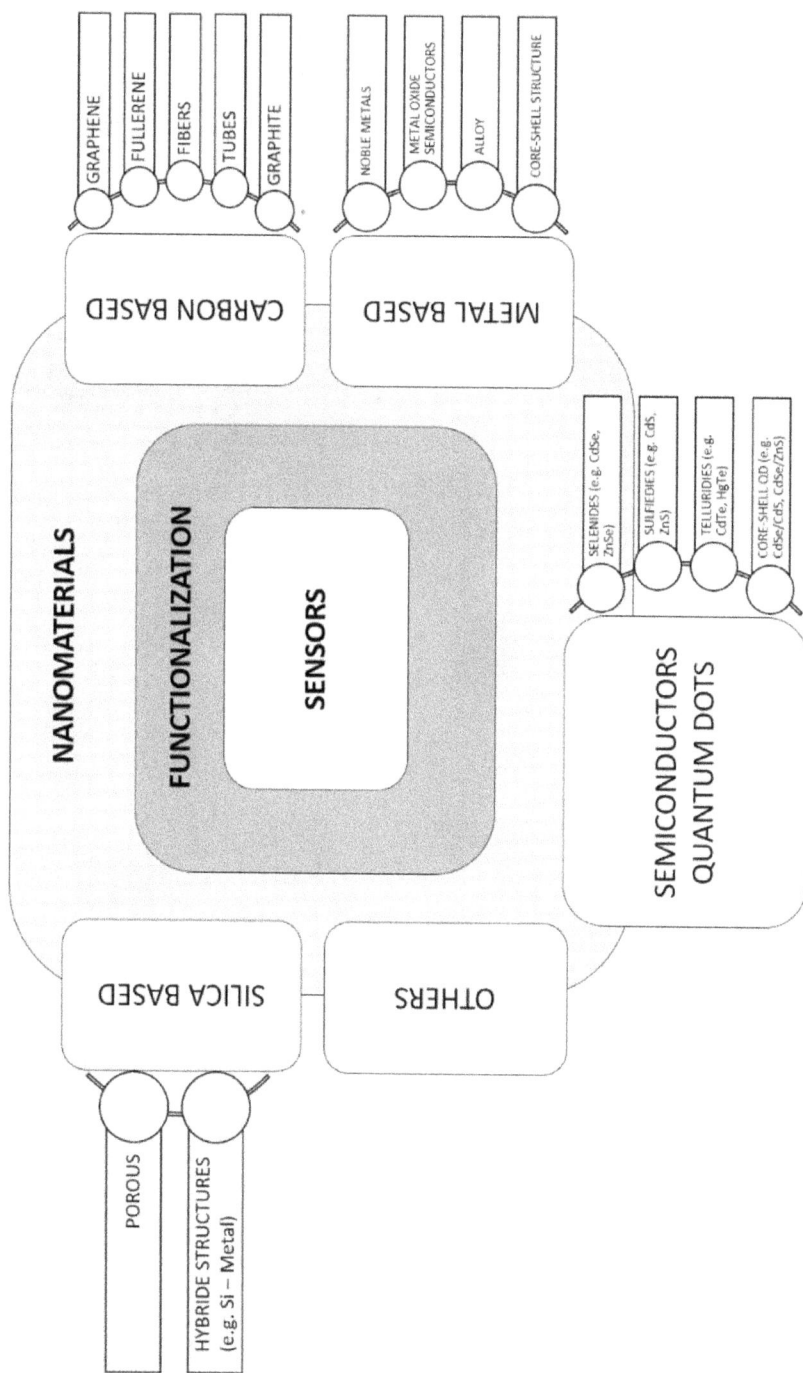

FIGURE 14.1 Classification of nanomaterials.

Source: (Su et al.)

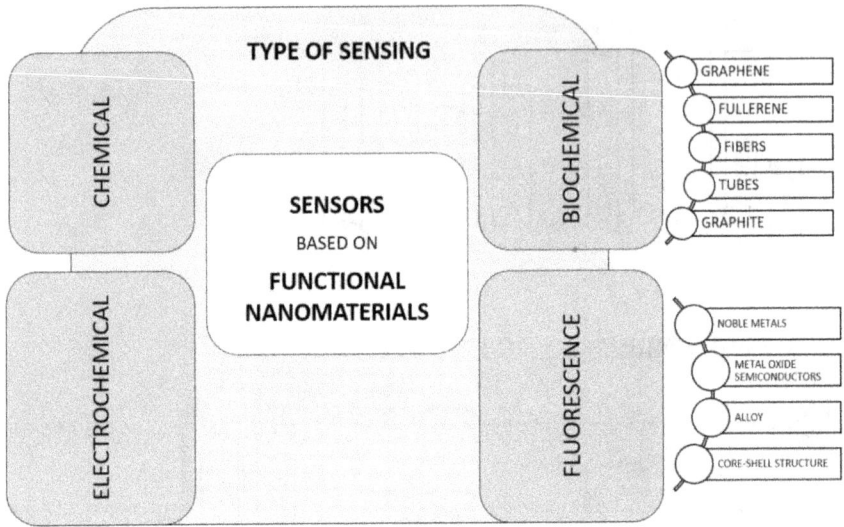

FIGURE 14.2   Types of sensors based on nanomaterials.

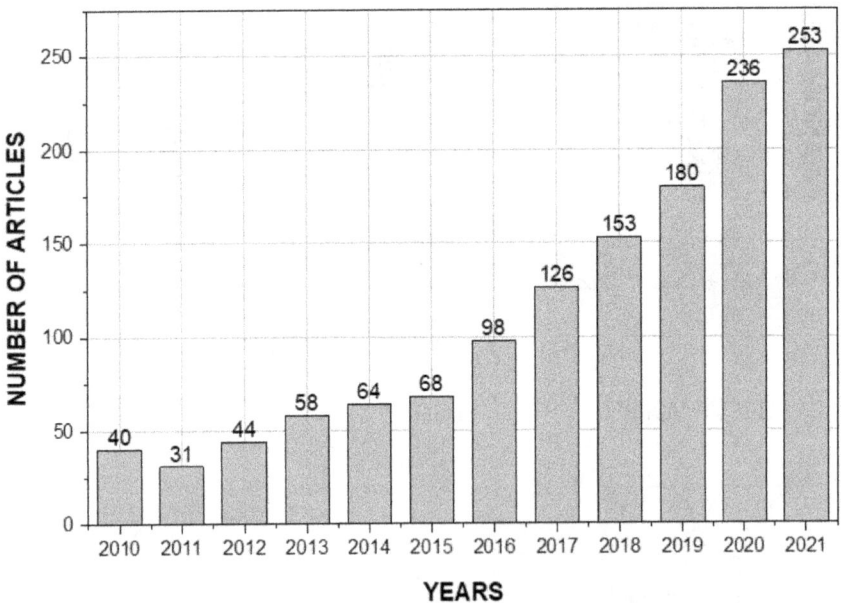

FIGURE 14.3   Number of published articles per year related to sensors and functional nanomaterials.

Source: (Web of Science)

the environment around the world—not only in relation to the health of people, but also of ecosystems. Therefore, to the extent that it seeks to protect the environment from contamination, it is also necessary to have at hand the appropriate technological methodologies and tools to monitor it, such as sensors.

Sensors are among the fastest-growing applications of NMs due to their huge demands in numerous industries, including communication, medicine, transportation, agriculture, energy, materials and manufacturing, consumer products, and households [6] (see Figure 14.4). Sensors can be divided into several classes, including mechanical, thermal, optical, magnetic, gas, chemical, and biological [7]. Nanosensors technology is developing rapidly and is the focus of a global research effort. Particularly noteworthy is the progress of nanosensors in health, military and security, environmental monitoring, detection of heavy metals, and food and agriculture, among other sectors [6]. Various NMs—such as carbon nanotubes, gold nanoparticles, silicon nanowires and quantum dots—have been extensively explored in detecting and measuring toxic metal ions, toxic gases, pesticides and hazardous industrial chemicals with high sensitivity, selectivity and simplicity [2]. Thus, there is a significantly important use of nanosensors in evaluation, detection and environmental monitoring [8]. Due to the characteristics and advantages of nanosensors of contaminants and biological and synthetic compounds present in food, water and living beings [9], it is now possible to check the levels of environmental contaminants, evaluate the effects of their presence in the environment and take appropriate measures to maintain essential environmental resources based on increasingly accurate information.

This chapter focuses on risk assessment and challenges related to the use of FNs for nanosensor fabrication. Difficulties in proper classification of these materials is related to the high rate of new functional materials development. The large amount of compounds and new production techniques make difficult not only the control and proper classification to proper groups, but also determining how harmful or environment friendly a compound is. In addition, it is difficult to carry out responsible tests, which allow not only to have an overview of this type of material in a short time, but also to establish whether or not they are safe for people, animals and the environment.

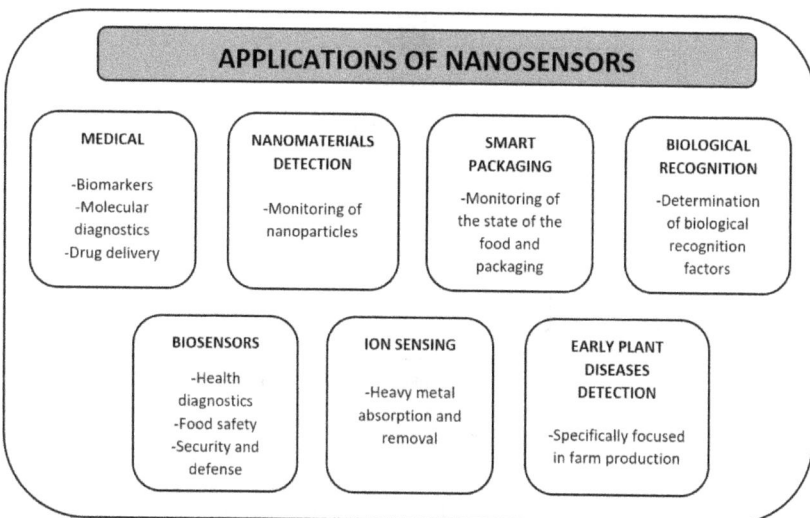

FIGURE 14.4 Common application of nanomaterials-based sensors.

Furthermore, this issue tends to be complex due to the fact that the proper classification process for sensors based on FNs is based on their nature. To this group of sensors belong those that contain a certain form of nanomaterial (carbon, silica, metallic nanoparticles, etc.) and have a surface area, whose properties can be easily multiplied and modified through their functional compounds. Surface functionalization is usually achieved with the presence of different types of organic and inorganic compounds. These compounds, on the one hand, are responsible for the unique properties of NMs and, on the other hand, are considered to be able to generate toxicity.

Therefore, in this context, it is not possible to classify all sensors correctly. Said classification can rather be made based on the synergistic effect of the properties of the compounds in the environment, which in fact may not be precise enough. It can also be assumed that the surface compounds could match the properties of the base NMs and effectively reduce the negative effects of their use.

14.2 STATE OF THE ART OF EU REGULATION RELATED TO USING FUNCTIONAL NANOMATERIALS IN SENSING

European regulations related to sensors are clear when stating that their proper classification should be done basing on the nature of FNs. When analyzing the environmental impacts of these materials, not only must the morphology and properties of the surfaces of these materials be taken into account, but also the interaction between them and the environment throughout their life cycle.

The life cycle of the sensors has as neuralgic stages their production and their use, however, in order to make a better approximation to the environmental impacts that their existence implies, the end of life is also taken into account, either whether these can be neutralized, eliminated or reused (see Figure 14.5). It should be noted that the

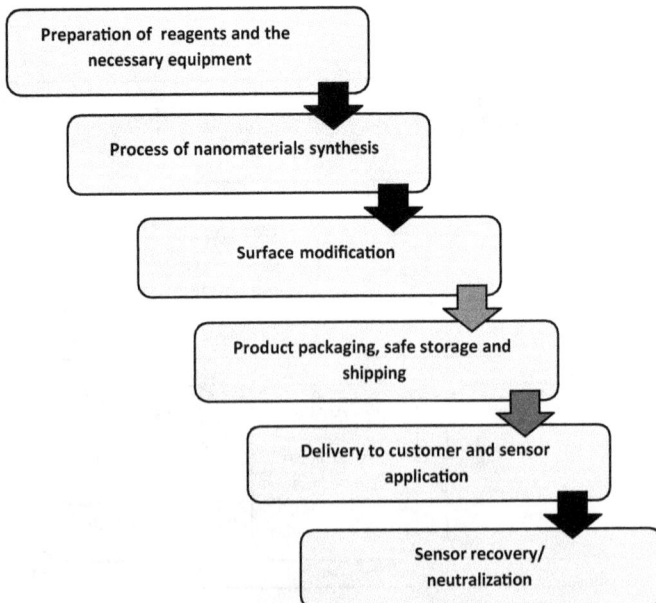

FIGURE 14.5 The sensor's life cycle.

impacts of exposure to nanomaterials of people who work or have some type of close contact with them during the life cycle of nanosensors is an issue that is becoming increasingly important.

As previously mentioned, the process of the proper classification of sensors is complicated and it is associated with the large amount of materials created. Basically, European Union (EU) regulations give an information about nanomaterials, which are a base compound as described in this book about sensors [10].

14.3 RISK ASSESSMENT

Risk assessment (RA) is a technique for proactively identifying and addressing risks in all settings [11]. In general terms, it is a careful examination of what could cause harm to people, in order to enable precautions to be taken to prevent injury and ill health effects [12]. RA includes three basic steps: (1) risk identification; (2) risk analysis; and (3) risk evaluation [13].

1. *Risk identification* is the process used to find, recognize and describe hazards that may affect the achievement of objectives. The starting point for proper hazard identification is information on accidents/hazardous events that have occurred in the past. Furthermore, experience and information from different countries can be useful when identifying risks and can be discussed with a wide range of stakeholders and employers, as well as appropriate data analysis from the private and industrial sector, etc. The good quality of these information is important in the process of proper risk identification.

2. *Risk analysis* is a process used to understand the nature, sources and causes of identified hazards and to estimate the level of risk associated with their occurrence. Risk analysis is also used to study impacts and consequences. Moreover, it is used for examining and verifying the existing controls. Analysis techniques can be qualitative, quantitative or a combination of these, depending on the circumstances and intended use. Quantitative and qualitative risk analysis methods and those with characteristics of the first two types described in the sublist (a, b) which follows are widely used for different purposes and with their respective strengths and weaknesses [14].

    a. **Quantitative Risk Analysis**
    Quantitative risk analysis attempts to estimate the risk in form of the probability (or frequency) of a loss and evaluates such probabilities to make decisions and communicate the results. This kind of analysis is the preferred approach when adequate field data, test data and other types of evidence to estimate the probability (or frequency) of the magnitude of losses.

    b. **Qualitative Risk Analysis**
    This type of analysis is perhaps the most widely used, just because it is simple and quick to perform. In this type of analysis, the potential loss is qualitatively estimated using linguistic scales such as low, medium and high. Qualitative risk analysis does not need to rely on actual data and probabilistic treatment of such data; the analysis is far simpler and easier to use and understand, but is extremely subjective.

  c. **Mixed Quantitative-Qualitative Analysis**

Risk analysis may use a mix of qualitative and quantitative analyses. This mix can happen in two ways: the frequency or potential for loss is measured qualitatively, but the magnitude of the loss (consequence) is measured quantitatively, or vice versa. Further, it is possible that both the frequency and magnitude of the loss are measured quantitatively, but with the policy setting and decision-making part of the analysis relying on quantitative methods.

3. *Risk evaluation* is the process used to compare the results of a risk analysis with risk criteria to determine whether a certain level of risk is acceptable, tolerable or unacceptable. Decisions should take into account of the wider context and the actual and perceived consequences to external and internal stakeholders. The outcome of risk evaluation should be recorded, communicated and then validated at appropriate levels of an organization. Nevertheless, RA does not only apply to organizations.

Moreover, many activities, projects and products can be considered to have a life cycle starting from initial concept and definition through realization to a final completion which might include decommissioning and disposal of hardware. RA can be applied at all stages of the life cycle and is usually applied many times with different levels of detail to assist in the decisions that need to be made at each phase [15].

Likewise, it is important to highlight that the assessment of risks in a life cycle may require the application of different techniques, depending on the characteristics of each stage of the cycle and/or the needs that exist.

14.3.1 Risk Assessment Techniques

There are numerous RA techniques whose application depends on aspects such as the complexity of the problem to be analyzed; the level of uncertainty and the quantity and quality of the information available; the availability of material, economic and human resources; and the need for get measurable results. Based on the aforementioned, it is possible to establish six groups of RA techniques (Table 14.1), which will be explained as follows [15].

1. **Look-Up Methods**

This group is made up of checklists and preliminary hazard analysis. In general terms, these are simple methods that do not require too many resources and whose results are eminently qualitative.

2. **Supporting Methods**

These methods are human reliability analysis (HRA), Delphi technique, SWIFT structured ("what-if?") and structured interview and brainstorming. With the exception of the HRA, the remaining methods generate results of a qualitative nature.

3. **Scenario Analysis Methods**

Several methods like root cause analysis, toxicological risk assessment, business impact analysis, event tree analysis and cause/consequence analysis are used for RA within the framework of probable future scenarios or events.

4. **Function Analysis Methods**

   Techniques such as FMEA (failure mode and effect analysis), HAZOP (hazard and operability studies) and HACCP (hazard analysis and critical control points) are based, in general terms, on the identification of errors and/or failures in a system that may entail a certain level of risk.

5. **Controls Assessment**

   Through methodologies like LOPA (layers of protection analysis) and bow-tie analysis, it is possible to review the controls that must be made regarding the risk factors and their possible effects.

6. **Statistical Methods**

   Methodologies such as Markov analysis, Monte-Carlo analysis and Bayesian analysis are based on probabilistic calculations because these techniques allow to work with systems that are too complex due to the effects of uncertainty.

14.3.2 Risk Assessment Applied to Nanomaterials

Questions have been raised about the applicability of the general chemical RA approach in the specific case of NMs. Most scientists and stakeholders assume that the current standard methods are in principle suitable, but point out that experimental aspects and practical guidelines need specific adaptations [16]. Although nanoforms may have the same chemical composition as bulk materials, the differences that exist between these two types of materials in terms of physical and chemical properties (size, shape, charge, etc.) mean that there is an enormous variety of NMs, which is why the evaluation of the risks of using these can become a difficult task to achieve.

Therefore, in order to deal with the complexity involved in nanomaterials RA, progress is being made in adopting new approaches, such as modification of tools such as (Q)SAR (quantitative structure–activity relationship), clustering, extrapolation, and high-throughput screening/testing for NMs. For successful applicability of such new approaches it is crucial that sufficient good-quality nanospecific information becomes available [17].

14.4 NANOMATERIALS IN THE ENVIRONMENT

In the first instance, NMs can be classified into two types: natural and synthetic. Among the synthetics, incidental NMs and engineering NMs can be distinguished (Figure 14.6). Natural nanomaterials (NNMs) are generally formed through different biogeochemical processes, such as natural chemical weathering processes, volcanic eruptions, aeroplasma, lightning and flash-pyrolysis, among others [18]. Synthetic nanomaterials (SNMs) are produced by anthropogenic activities (such as energy, telecommunications, computing, agrichemicals, and personal care products sectors), both intentional and unintentional [19]. Although NNMs are present in low amounts compared to NNMs, these NMs are considered a threat to the environment and are called pollutants [20].

As NMs may not be detectable after discharge into the environment, they can cause various types of environmental problems if a remediation plan is not secured

TABLE 14.1
Risk Assessment Techniques [15]

Type of Risk Assessment Technique	Technique	Brief Description	Can Provide Quantitative Output
Look-up methods	Checklists	Provides a listing of typical uncertainties which need to be considered.	No
	Preliminary hazard analysis	Its objective is to identify the hazards and hazardous situations and events that can cause harm for a given activity, facility or system.	No
Supporting methods	Structured interview and brainstorming	A means of collecting a broad set of ideas and evaluation, ranking them by a team.	No
	Delphi technique	A means of combining expert opinions that may support the source and influence identification, probability and consequence estimation and risk evaluation.	No
	SWIFT structured "what-if?"	A system for prompting a team to identify risks. Normally used within a facilitated Workshop.	No
	Human reliability analysis (HRA)	It deals with the impact of humans on system performance and can be used to evaluate human error influences on the system.	Yes
Scenario analysis	Root cause analysis (single loss analysis)	Seeks to understand contributory causes of a loss and how the system or process can be improved to avoid such future losses.	No
	Scenario analysis	Scenario analysis: possible future scenarios are identified through imagination or extrapolation.	No
	Toxicological risk assessment	Hazards are identified and analyzed, and possible pathways by which a specified target might be exposed to the hazard are identified.	Yes
	Business impact analysis	Provides an analysis of how key disruption risks could affect an organization's operations.	No
	Fault tree analysis	Starts with the undesired event (top event) and determines all the ways in which it could occur.	Yes

Category	Method	Description	
	Event tree analysis	Using inductive reasoning to translate probabilities of different initiating events into possible outcomes.	Yes
	Cause/consequence analysis	A combination of fault and event tree analysis that allows inclusion of time delays.	Yes
	Cause-and-effect analysis	An effect can have a number of contributing factors which may be grouped into different categories.	No
Function analysis	FMEA (failure mode and effect analysis) and FMECA	A technique which identifies failure modes and mechanisms, and their effects.	Yes
	Reliability-centered maintenance	Identifies the necessary policies to manage failures so as to efficiently and effectively achieve the successful operation of all types of equipment.	Yes
	Sneak analysis (Sneak circuit analysis)	A methodology for identifying design errors. A sneak condition is a latent hardware, software, or integrated condition that may cause an unwanted event to occur.	No
	HAZOP (hazard and operability studies)	A general process of risk identification to define possible deviations from the expected or intended performance.	No
	HACCP (hazard analysis and critical control points)	Preventive system for assuring product quality, reliability and safety of processes.	No
Controls assessment	LOPA (layers of protection analysis)	It allows controls and their effectiveness to be evaluated.	Yes
	Bowtie analysis	A simple diagrammatic way of describing and analyzing the pathways of a risk.	Yes
Statistical methods	Markov analysis	It is commonly used in the analysis of repairable complex systems that can exist in multiple states.	Yes
	Monte-Carlo analysis	It is used to establish the aggregate variation in a system resulting from variations in the system.	Yes
	Bayesian analysis	A statistical procedure which utilizes prior distribution data to assess the probability of the result.	Yes

Engineered NMs

-Ceramic nanoparticles
-Metal   oxides   and   carbon
nanoparticles

Incidental NMs

-Nanoplastics
-Paint and pigments
-Metal   oxides   and   carbon
nanoparticles
-Degraded products in landfills or
biosolids or natural weathering

Natural NMs

-Humic substance, biomolecules, viruses
-Nanoparticles by microbes
-Ceramic nanoparticles
-Metal oxides and carbon nanoparticles
-Degraded products in landfills or biosolids
or natural weathering

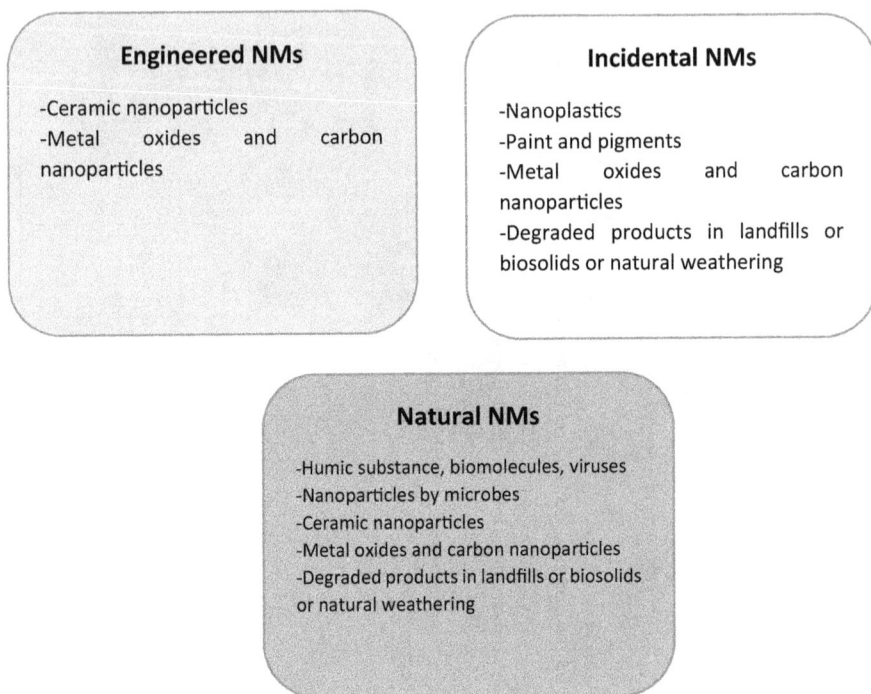

FIGURE 14.6   Different classes of natural and anthropogenic nanomaterials [20].

[21]. Once NMs are discharged, they accumulate in different environmental matrices, for example, air, water, soil, and sediments [22].

14.4.1 NATURAL NANOMATERIALS IN THE ENVIRONMENT

NNMs from volcanic ash, forest fires, lightning, or formed in the outer space can easily be present in the air or can quickly spread to different surface water sources (lakes, rivers, sea, and oceans), as well, by winds and rain. Precipitation can deposit air-suspended NMs into ground and surface water sources [19]. Other forms of NMs can also be suspended in air and carried across the planet with the wind [23].

The toxicity of NNMs mainly arises from their dissolution and release of metal ions, which is prevalent under anoxic conditions and depends on the kinetics and mechanism of dissolution under environmentally relevant oxic/anoxic conditions [19]. However, NNMs can also serve as essential nutrient sources; for example, volcanic ash NMs serves as nutrients for phytoplankton but can also increase toxicity levels [24].

The rapid advancement of analytical techniques to measure NMs makes it possible to understand the diverse role of NMs in climate change, microbial-soil-rhizosphere interfaces and the different cycles of nature. Currently, however, there is a substantial gap in knowledge about the chemical composition of NNMs and the effects of their life cycle on natural systems [19].

14.4.2 SYNTHETIC NANOMATERIALS IN THE ENVIRONMENT

Synthetic NMs, both incidental NMs (INMs) and engineered NMS (ENMs), can potentially be released to the atmosphere at the source of production and can remain suspended for a prolonged period [19]. Road traffic exhaust, combustion, explosion, and oxidation of atmospheric gases contribute to the production of synthetic NMs released into the atmosphere [25]. In turn, waste incineration plants generate ENM emissions that largely depend on the composition of said waste [26].

Synthetic NMs suspended in air can follow a similar pathway as followed by NNMs and end up in various soil and water sources [27]. Urban runoff can have high volumes of synthetic NMs, which can be tracked in stormwater and end up in wastewater treatment facilities, which can further infiltrate groundwater, and surface runoff can pollute surface water sources [28]. Waste and biosolids can also contain ENMs, which can easily leach from landfills and reach surface water or infiltrate to groundwater [19].

On the other hand, the fate of NMs in the aquatic system is thus affected by various processes, such as accumulation, disaggregation, diffusion, interaction with other components (and aquatic organisms), biological degradation (aerobic and anaerobic) and abiotic degradation (including photolysis and hydrolysis) [29]. Likewise, ENMs can enter soils through different sources and pathways, such as the use of fertilizers and plant protection products, biosolids, sewage water and floodplains [30]. Previous studies have reported that plants can take up and translocate NMs from soil, which may influence the germination and plant growth [31, 32]).

14.5 NANOMATERIALS' EFFECTS ON THE ENVIRONMENT

14.5.1 POSITIVE EFFECTS

The application of the advances of nanotechnology in a growing number of activities generates great benefits for society and the environment (Figure 14.7). This field of science is helping to decrease the human footprint on the environment by providing more efficient and energy saving innovations; development of precision manufacturing needs nanotechnological devices, because these may lead to the generation of less waste and reduce the requirement for large industrial plants [21].

Furthermore, nanotechnology has been used to make more efficient and environmentally friendly batteries [33]. Additionally, by breaking down oil into biodegradable compounds, NMs may play an important role cleaning up oil spills [34].

14.5.2 NEGATIVE EFFECTS

As the environmental impacts of NMs cannot be clearly diagnosed and there are too many variables to account for, it is very difficult to reach any conclusion about the ecological effects and environmental stability of NMs [21]. Nevertheless, scientists have investigated the ecological effects of Ag_2S nanoparticles. Accordingly, it was found that plants (e.g., dicotyledonous cucumber [*Cucumis sativus*] and monocotyledonous wheat [*Triticum aestivum* L.]) may uptake nanosilver if it is available in soil

- Help build lighter and stronger structures
- Boost chemical reactions
- Allows carbon dioxide isolation
- Help produce low resistant conductors
- Help produce more efficient solar cells
- Self-cleaning
- Precision manufacturing
- Purification, filtration and oxidation
- Disaster management

Positive Impacts

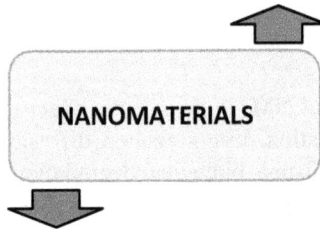

NANOMATERIALS

Negative Impacts

- Difficult to identify
- Plant intake vis root
- Synthetization of NMs
- Toxicological effects
- Dust cloud formation

FIGURE 14.7 Positive and negative impacts of nanomaterials on the environment [21].

[35]. The exact mechanism of toxicity associated with nanoparticles is yet unclear. However, a few groups claimed that the toxicity of nanoparticles is majorly associated with the dissolved material or metals leached from the nanoparticles [36].

The presence of NMs was also demonstrated to exert low- to high-toxicity impacts on aquatic life [21]. The carbon nanotubes and their byproducts were also reported to exhibit toxic effects on marine organisms, e.g., *Thalassiosira pseudonana*, *Tigriopus japonicas* and *Oryzias melastigma* [37]. Moreover, the NMs exhibited toxic effects on several aquatic microorganisms, e.g., *Escherichia coli* and *Aeromonas hydrophila* [38]. NMs are also reported to exert negative impact on the life of soil organisms and on the air system [21]. Likewise, the NMs played an important role in the formation of dust clouds after being released into the environment [39].

Overall, NMs exerted negative effects on environment by complicating the different environmental systems. The NMs not only affected the life of aquatic and terrestrial systems but also degraded the quality of air. In other words, NMs can affect all the environmental media, i.e., air, water, and land [21].

14.6 HUMAN HEALTH IMPACTS OF NANOMATERIALS EXPOSURE

The need for ENMs is thriving among consumers, as well as in commercial products such as food additives, water purification, soil cleaning, sunscreen, biocides, supplements, shampoos, agriculture, energy production, feed, veterinary drugs, packaging and information technology [40]. Just as the effects of NMs on the environment are becoming known, we are also beginning to understand how human health can be affected by these materials.

14.6.1 EXPOSURE PATHWAYS TO HUMANS

Studies on the impact of NNMs on human health are lacking, and maximum research has focused on ENMs [41]. The exposure of ENMs to humans develops via several pathways such as inhalation, ingestion, dermal penetration, injection and eye contact. This last pathway is often an overlooked potential hazard [42].

- **Inhalation**
  Inhalation is the main pathway of exposure to humans; consequently, the existence of NMs in the air poses a substantial health risk [43]. Urban air can contain 10,000–50,000 nanoparticles per cubic centimeter [44]. These estimated limits can increase multi-fold in NMs processing and production industries, which can be a significant concern for workplace safety [45]. The air quality index in urban areas can act as a possible indicator of suspended NMs, which can be a cause for a variety of cancers [19]. Inhalation of NMs in some cases may result in chronic inflammation in the lung tissue causing an inflammatory cascade, immune cell recruitment, and oxidative stress that can cause tissue injury [46]. Aerodynamic size, size and shape of NMs are key aspects when considering where an inhaled material will deposit in the respiratory tract, which includes the extra thoracic, upper bronchial, lower bronchial and alveolar regions [47].
- **Ingestion**
  The gastrointestinal (GI) tract offers a route for entry by NMs into the body following intentional consumption, leaching from food containers, deposition onto food, or secondary exposure from inhalation [47]. For example, the food coloring agent E171 consists of TiO_2 NMs at concentrations from 1–5 µg mg^{-1} and may be used in sweets [48]. Likewise, polymeric NMs— used for drug delivery and generally considered to be less toxic—are often also used in medication. There are also few studies on the effects of long-term exposure from these NMs [49]. The increasing use of nanofertilizers and nanopesticides in food production implies that these contaminants may bioaccumulate in soils and food crops, leading to a potential source of exposure when applied or ingested [50].
- **Dermal Penetration**
  Skin exposure may be the result of deposition of an airborne NMs, unintentional contact, or intentional application via NMs-containing personal care

products [47]. ENMs such as TiO_2, ZnO and silver nanoparticles are estimated to be present in personal care products, including baby products
[48, 51]. Moreover, there is an ongoing debate on the potential for the passage of NMs through the skin barrier [19]. As skin pores are small, it is evi-
dent that smaller particles, specifically those < 4 nm, can cross over easily [52].

* **Injection**
  Direct injection of NMs is used for biomedical applications for drug delivery [53]. Medical use of ENMs has provided encouraging results to fight
  diseases, but may also lead to unwanted consequences [54].

* **Ocular Exposure**
  Although studies evaluating the risks posed by NMs to the eyes are limited,
  a number of investigations have been undertaken [47]. For example, the
  effect of 20 nm and 80 nm gold NPs on mouse retinas over a 72-hour period
  demonstrated a significant increase in oxidative stress [55].

14.6.2 HUMAN HEALTH IMPACTS

Risk factors in humans are governed by the level of exposure, routes of exposure and
the size, type, distribution, reactivity and shape of ENMs [41]. Therefore, the ultra-
small size together with the geometry of the NMs may result in a higher probability
of the material entering the human body, translocating to different regions other than
the portal of entry and promoting adverse interactions at the level of organs, tissues
and cells (Figure 14.8) [47].

The most common ENMs affect human health according to numerous previous
studies, as evidenced in the following list.

* **Silver Nanoparticles (AgNPs)**
  Prior investigations have demonstrated that oral exposure to AgNPs is able
  to direct their transference to numerous areas—for example, to the spleen,
  lungs, bone marrow, kidneys, liver, parathyroid, thyroid, brain, skin, eyes,
  heart, muscles, small intestine, stomach, prostate, tongue, blood, teeth, duo-
  denum and pancreas [41, 56]. Likewise, an in vivo study with rats proved
  that AgNPs are able to be passed on to the offspring, along with the oral
  administration of AgNPs in dosages higher than 100 mg/kg of body weight/
  day, which might produce oxidative stress in hepatic tissue in the time of
  pregnancy [57]. It has also been proven that AgNPs cause deoxyribonucleic
  acid (DNA) destruction in TK6 cells at a concentration of as little as 5 μg/
  ml [58].

* **Carbon Nanotubes (CNTs) and Graphene**
  There have been concerns about safety issues involving CNTs because of
  their insolubility in the lungs [41]. Likewise, is estimated that CNTs produce the same detrimental consequence as asbestos, such as mesothelioma,
  cancer, pulmonary inflammation and fibrosis [59]. It has been found that
  single-walled CNT could produce scratches and interstitial infection in rats
  within 7–90 days [60]. Numerous studies have mentioned that longer CNT
  produced more significant toxicity than shorter ones. More specifically, is

estimated that long multi-walled CNT induced inflammation, along with genotoxicity [61]. However, other studies indicate that short CNTs cause lysosomal destruction, which in turn results in mitochondrial damage and oxidative stress [62]. Graphene nanosheets have been recommended as adjuncts for tissue engineering and have also been applied in photothermal cancer therapy in conjunction with drug delivery [41]. It was found that graphene oxide nanosheets drive cytotoxicity by interrupting the cell membrane at the initial contact of the cell with the graphene oxide. This allows to conclude that graphene oxide nanosheets are more toxic than multi-walled CNT [63].

- **Silica Nanoparticles (SiNPs)**
  SiNPs have been shown to produce epigenetic transformations in human bronchial epithelial cells as well as transformations in keratinocyte cell lines (made up of skin cells) when exposed to 15-nm SiNPs [64]. Likewise, a mutagenic response to SiNPs 7.172 nm and 7.652 nm in size was notified for mouse lymphoma cell lines at 100 µg/ml and 150 µg/ml [65]. It has also been proven that SiNPs of size 20–30 nm generated structural transformations to human hemoglobin, producing heme displacement and deterioration of the heme protein [66]. Likewise, it has been shown that SiNPs demonstrated toxicity to the immune system [67]. Despite these discoveries, SiNPs have demonstrated lower toxicity than alternative nanomaterials [41].

- **Titanium Dioxide Nanoparticles (TiO$_2$ NPs)**
  Exposure of TiO$_2$ NPs cause DNA damage in the human fetus [68] and on human embryonic kidney cells (HEK293), along with mouse embryonic fibroblast cells (NIH/3T3) at 1,000 µg/ml (21 nm and 50 nm) [65]. Higher inflammation in the liver when female mice exposed to 25–80 nm of TiO$_2$ NPs near 5 g/kg was reported, as well [69]. Mice exposed to TiO$_2$ NPs in lesser dosages demonstrated tremors, lethargy and loss of appetite, which gradually ended. At a high dosage, these mice revealed intense indications of lethargy, anorexia, tremors, body weight loss and diarrhea [70]. The International Agency for Research on Cancer (IARC) distinguished pigment grade (lower than 2.5 µm), as well as ultrafine (lower than 100 µm) TiO$_2$ NPs as potential carcinogens, considering they could induce inhaling-tract cancer in rats; likewise, titanium dioxide nanofilaments and nanorods may exhibit significant cytotoxicity to epithelial cells [41].

- **Gold Nanoparticles (AuNPs)**
  The AuNPs are widely applied in many medical industries; however, they are known to affect human embryonic stem cells, primarily due to their size [71]. Another study based on human exposure to gold nanorods (39 nm long, 18 nm wide), nanospheres (6.3 nm) and nanostars (215 nm) found that gold nanostars are more toxic to human fetal osteoblasts and pancreatic duct cells [72].

In general terms, it should also be noted that simplifying the size limits of different NMs to endorse toxicity is intricate, as there are presently no standardized toxicity proceedings for scientists to correlate various outcomes [41].

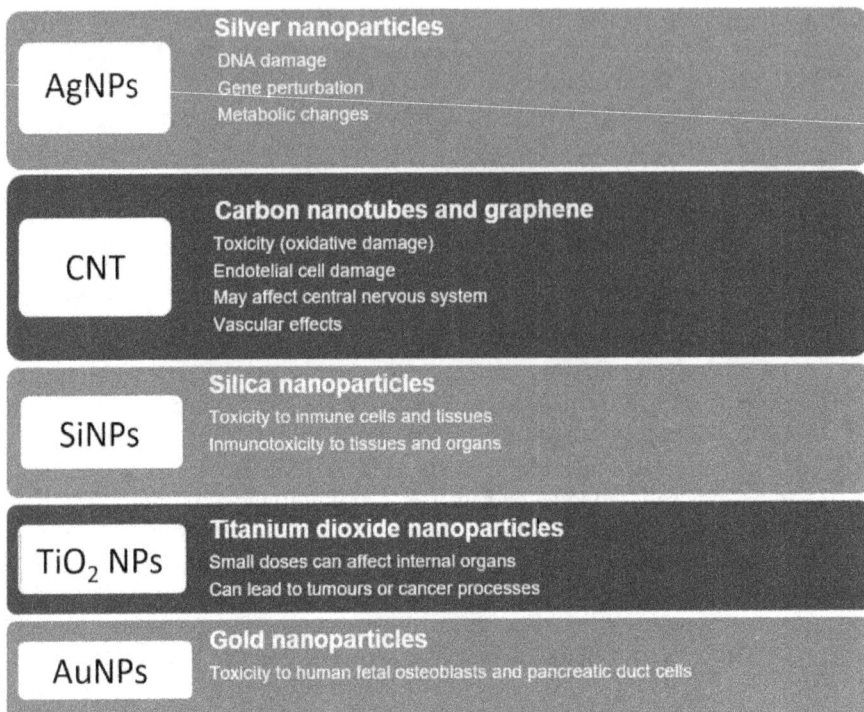

AgNPs
Silver nanoparticles
DNA damage
Gene perturbation
Metabolic changes

CNT
Carbon nanotubes and graphene
Toxicity (oxidative damage)
Endotelial cell damage
May affect central nervous system
Vascular effects

SiNPs
Silica nanoparticles
Toxicity to inmune cells and tissues
Inmunotoxicity to tissues and organs

TiO$_2$ NPs
Titanium dioxide nanoparticles
Small doses can affect internal organs
Can lead to tumours or cancer processes

AuNPs
Gold nanoparticles
Toxicity to human fetal osteoblasts and pancreatic duct cells

FIGURE 14.8   Most common impacts of ENMs on human health.

14.7 RISK ASSESSMENT OF FUNCTIONAL NANOMATERIALS

14.7.1 REGULATORY FRAMEWORK FOR NANOMATERIALS

Worldwide, regulatory authorities carefully observe recent developments in this area, striving to find a balance between consumer safety and the interests of the industry. The regulation of NMs is following different approaches, for example the one-time reporting of discrete forms by the U.S. Environmental Protection Agency (EPA) [73] or the Canadian Section 71 reporting of NMs [74]. Several Asian countries have found their own definitions of NMs, starting with China [75] releasing nanostandards in 2004 followed by Japan and South Korea [76, 77]. In Brazil, the Chamber of Deputies presented the bill n. 6741/2013 proposing the creation of the national nanotechnology policy [78]. Since then, nanotechnology has played an emerging role in Brazilian Health Surveillance Agency (ANVISA).

In Europe, the most important regulatory framework adopted to improve the protection of human health and the environment from the risks that can be posed by chemicals is known as REACH (registration, evaluation, authorization and restriction of chemicals) and entered into force on 1 June 2007. Despite the aforementioned, NMs fall within the scope of different pieces of EU legislation, some of which address NMs explicitly using a variety of terms, e.g. nanomaterial, engineered nanomaterial or nanoform [79]. Likewise, the definitions used in REACH, which addresses industrial chemicals—and in the sector-specific regulation, and in national inventories—are each different, awaiting harmonization [80].

New information requirements for substances in the nanoform that are subject to registration under REACH entered into force on 1 January 2020 as laid down in the amended Annexes of REACH [81]. The most profound change is that REACH Annex VI now includes a legally binding definition of nanomaterial, referred to as nanoform in the Annex, that, in turn, is based on the 2011 European Commission recommendation [82].

Likewise, the nano-information requirements in REACH address nanospecific minimum characterization information including the substance identification, physicochemical characterization, chemical safety assessment, and downstream user information. There is also the possibility to define sets of similar nanoforms, for which hazard assessment, exposure assessment and risk assessment can be performed jointly [81].

On the other hand, test guidelines (TGs) have been developed at the level of the Organization for Economic Cooperation and Development (OECD) aimed at the adoption of regulatory tests of chemical products and published under the Mutual Acceptance of Data (MAD) agreement, which is an essential component in the international harmonization of approaches to chemical safety through regulatory recognition of these test guidelines [83]. The development of TGs has progressed for NMs addressing physicochemical properties, effects on biotic systems, environmental fate and behavior, and health effects [81]. In particular, three TGs specifically addressing NMs have been adopted: a new TG on "Dispersion Stability of Nanomaterials in Simulated Environmental Media" (TG318) and adaptation of two TGs on Subacute Inhalation Toxicity: 28-Day Study/90-Day Study (TG412 and TG413). Several new TGs for physicochemical properties are under development.

14.7.2 Frameworks of Risk Assessment of Nanomaterials

The basic components of RA of chemicals are hazard and exposure assessments, dose-response estimation, risk characterization, and accounting for uncertainty in the overall assessment [84]. Nevertheless, many of the tools, test protocols and guidelines for determination and assessment of physicochemical properties, fate, exposures and effects used for conventional chemicals need modifications when applied to (the regulatory) evaluation of NMs [85].

Therefore, there have been ongoing concerns expressed by all stakeholders that acceptable regulatory data and methods for assessing the environmental, health and safety (EHS) of NMs are not fully available [86, 87].

In general terms, there are two types of RA frameworks applied to NMs that differ mainly in the way they are addressed; therefore, some of them are mainly aimed at obtaining information for regulatory filing, and others are mainly aimed at informing decision-making in the innovation chain [84].

The early users of many frameworks are mostly manufacturers and developers looking to efficiently direct the creation of safe and sustainable products or materials, and collect the information needed for regulatory approval. The frameworks are also useful for risk assessors in government agencies considering the need for additional information or risk reduction measures, and for prioritizing those applications and situations that should be considered first or most [84].

14.7.2.1 Risk Assessment Strategies Mainly Directed toward Regulatory Submission

- **NanoRiskCat**
  This method can support companies and regulators in their first-tier assessment and communication on what they know about the hazard and exposure potential of consumer products containing NMs. Input is given on how the potential for exposure, human health hazards and environmental hazards can be assessed [88].
- **DF4nanoGrouping Framework**
  This framework focuses on hazards to human health through inhalation. Therefore, this framework is an efficient strategy to classify NMs that could be subject to hazard assessment without further testing [89, 90].
- **MARINA Risk Assessment Strategy**
  The aim of this tool is developing a flexible and efficient approach for data collection and RA. The generated information should be sufficient to assess the risks of NMs [91].
- **Nanomaterial Categorization for Assessing Risk Potential**
  This is a prioritization methodology to target materials of high concern that need additional scrutiny, while material categories that pose the least risk can receive expedited review [92].
- **Test Strategy for Assessing the Risks of Nanomaterials in the Environment**
  This is a conceptual framework that aims to develop a test and RA strategy for NMs which specifically addresses environmental fate and effects [93].
- **A Strategy toward Grouping and Read-Across**
  This is a general strategy that comprises testing strategies for NMs that are in compliance with REACH. The aim of this framework is to develop testing strategies for NMs in order to characterize the potential risks to human health and the environment [94].
- **Risk Assessment and Grouping Strategy Based on Clouds of Predefined Test Strategies**
  This framework is intended to describe the need and outline of a RA framework for nanomaterial identification and grouping, using 'clouds' [95].
- **NANoREG Nanospecific Approach for Risk Assessment**
  This approach provides alternative ways to address the RA of NMs, by prioritizing those applications with the highest potential health risk [17].
- **Risk Banding Framework**
  This is a screening-level assessment framework for inhalation routes which looks for the development of a scientifically based risk banding tool by combining information on deposition of particles in the respiratory tract, lung burden and clearance, diffusion through lung mucus layer, translocation and cellular uptake, and local and systemic toxicity [96].

- **NANoREG Framework for the Safety Assessment of Nanomaterials**
  This framework is aimed to analyze the applicability of the current EU regulatory framework on NMs and give concrete, practical direction to industry and regulatory authorities on how to address NMs in a legislative context, with focus on REACH [97].
- **Sustainable Nanotechnology Decision Support System (SUNDS)**
  This is a conceptual decision framework in which various tools and exposure models have been integrated. It considers links to risk governance and risk management approaches [98].

14.7.2.2 Risk Assessment Strategies Mainly Directed toward the Innovation Chain

- **LICARA NanoSCAN**
  This is a generally applicable risk comparison framework for SMEs that provides a qualitative evaluation of the potential benefits and risks of a new or existing nano-enabled products. This tool is preferably to be used at an early stage in the innovation chain with the aim to facilitate the development of sustainable and competitive nano-enabled products [99].
- **Alternatives Assessments for Nanomaterials**
  In this case, the aim is to assess the overall applicability of alternatives assessment methods for NMs and to outline recommendations. Alternatives assessment for nanomaterials is complicated by the sheer number of nanomaterials possible. Although science may not (yet) be in the position to predict or explain nanotoxicology, science may be (more) ready for making better and safer choices [100].
- **NANoREG D6.04**
  This is a generally applicable risk screening framework intended to screen for indicators of potential risks during early stages of an innovation process. Is based on six key risk potentials: solubility/dissolution rate, stability of coating, accumulation, genotoxicity, inflammation and ecotoxicity [101].

14.8 CHALLENGES CONCERNING RISK ASSESSMENT OF NANOMATERIALS

Some questions concerning the safety and sustainability of nanotechnology applications in consumer and industrial products remain open, and they become even more challenging when addressing new generations of NMs. Therefore, is necessary to promote the development of new smart NMs that are safe and sustainable [102].

Safety of a nanomaterial implies that: (1) the materials are non-hazardous for humans and the environment; (2) the production should aim to eliminate risks at the workplace and to eliminate waste; and (3) exposure during use of the product should be avoided, and efficient recycling and disposal routes should be available [97, 103].

Taking into account the aforementioned, it is necessary to mention certain challenging issues that must be addressed to improve both the knowledge and the RA of NMs.

14.8.1 REGULATORY FRAMEWORK

Regulators need to be aware of upcoming new materials, technologies and innovations from the early stages of the innovation process on to understand whether the existing legislation addresses all relevant aspects regarding their human and environmental safety sustainably [102]. The lack of appropriate RA and regulation to address safety concerns is considered among the possible barriers to the implementation of smart NMs in agriculture, which is a just now emerging technology [104]. Likewise, risk frameworks need to include the use of life-cycle thinking, systems thinking and probabilistic analysis (e.g. engineering concepts, system dynamics models, fault trees, event trees) and move from a deterministic to a probabilistic interpretation of risk. Additionally, the concept of risk would need to be broadened to address the social and ethical impacts that the new risks of active nanostructures may pose [105].

14.8.2 SPECIFIC ISSUES RELATED TO RISK ASSESSMENT

Several important nanospecific issues should be considered because NMs show particle-specific behavior [84]. First of all, different nanoforms can display different behavior both in their fate/toxicokinetics and hazard, and thus in their risk. As long as different nanoforms are not addressed, this can be considered a regulatory gap which requires further research and policy considerations [84].

Another important issue is regarding life cycle and exposure. It is known that physicochemical properties of NMs relevant for their potential risk may change during their life cycle [97, 106]. Further investigation may be relevant, as these life cycle changes may lead to more dangerous NMs than their pristine counterparts [84].

On the other hand, NMs tend to agglomerate, which can have serious impacts on factors such as dilution, internalized dose, and application of the assessment [107], which may lead to underestimation of risk [107, 108]. Information on internalized dose is thus considered highly relevant, though still technically challenging [84].

Having relevant information regarding bioaccumulation of NMs in tissues—as their elimination from them can be very slow (i.e., this can take years) [84]—is an important matter, as well. Linking toxicokinetic data, including information on accumulation and elimination, to physicochemical properties of the nanomaterial and to their dissolution rate in physiologically relevant media may make this issue easier to handle in the future [84].

Finally, it is also necessary to advance in the possibility of extrapolating information related to exposure routes to NMs (taking into account the particularities of these) in the development of functional tests (guides and testing guidelines) internationally accepted and specific to the RA of NMs and in improving the availability of publications with risk assessments of specific NMs [84].

14.8.3 IMPROVEMENT OF THE FEASIBILITY OF RISK ASSESSMENT
OF NANOMATERIALS

The standardization of test protocols and assurance of high quality data (including sufficient quality controls), unification of dose metrics for linking in vitro and in vivo tests, good-quality physicochemical characterization and exposure and/or (chronic) hazard studies and improvement of available data related to in silico approaches will allow to increase the feasibility to perform RA of NMs in practice [84].

14.8.4 UNCERTAINTY AND EFFICIENCY IN RISK ASSESSMENT FRAMEWORKS

It is anticipated that a major hurdle in building and implementing efficient RA frameworks in product and substance/material regulations will be the choices and decisions that will be made that will lead to uncertainty in the number of false negatives, i.e. in the quantity of NMs present on the market despite the fact that their existence implies unacceptable risks [84].

Another challenge for RA of NMs is the urgent need for efficient RA with a focus on potential chronic effects, limited availability of good-quality information, and the high cost and time efforts needed to increase this information [84].

On the other hand, to enable decisions to be made about efficient RA frameworks in the near future, or application of (parts of) RA frameworks in existing regulations, international agreement is required on practical cut-off and trigger values, while realizing that these can only be scientifically founded to a limited extent [84].

It is necessary to improve the way in which information is obtained for the RA of NMs, as well. One way in which progress can be made in this direction is by grouping and read-across approaches [109]; another one is stratifying the information needs according to the anticipated potential for hazards and risks in order to focus more on NMs with the greatest potential [84].

14.9 CONCLUSIONS

The trend in developing of future sensors is moving toward sensors with better sensitivity and selectivity, faster response time, more portability, more remote capability and greater ease of use, and those which are long lasting and cheap to produce and operate. In consideration of these needs, the creation and use of nanosensors has facilitated the improvement of numerous activities related to medicine, commerce, agriculture and environmental protection, among other sectors. Therefore, the creation of more functional nanomaterials (FNs) will allow higher levels of efficiency in nanosensors to be achieved; however, the risks inherent to their life cycle must also be reduced, since the physicochemical characteristics of nanomaterials (NMs) can lead to negative impacts in ecosystems and on human health.

NMs can be of either natural or synthetic origin. Although synthetic NMs are not as abundant as natural ones, they are considered pollutants. Synthetic NMs, however, can greatly contribute to improving the efficiency of energy use, to the introduction of best industrial practices, as well as to minimizing the emission of waste and pollutants, among other advantages. On the other hand, overall, NMs exerted negative effects on the environment by complicating the different environmental systems.

In relation to human health, NMs can enter the body either through inhalation, ingestion, skin contact, injections or eye contact. As far as it is known at present, there is a risk that the presence of NMs in the human body implies effects at the cellular level such as DNA damage, oxidative stress, mutations, carcinogenic processes and toxicity to the immune system, among other possible effects.

On the other hand, risk assessment (RA) is a recurring activity when thinking about a product's design, production, consumption and end of life. The application of this approach to the case of FNs is pertinent, although their particular characteristics must be taken into account. Therefore, although nanoforms may have the same chemical composition as bulk materials, the differences that exist between these two types of materials in terms of physical and chemical properties imply an enormous variety of NMs, which can make RA a difficult task to achieve.

There are numerous RA techniques whose application depends on aspects such as the complexity of the problem, the level of uncertainty and the quantity and quality of the information available, the availability of resources and the need for achieving measurable results. Although at the global level, there are some regulatory frameworks that already address topics related to NMs, these regulatory efforts are still at an early stage. A similar situation occurs with RA frameworks applicable to nanomaterials.

In view of this, the RA concerning nanomaterials faces challenges such as the structuring of one or more regulatory reference frameworks that can be applied to adequately address the risks. Likewise, it is necessary to deepen the research that allows filling the gaps that exist regarding aspects such as behaviors, life cycles, routes of exposure and toxicity. Also required are the standardization of protocols and techniques that allow improving the reliability of the results, and the implementation of strategies that allow the efficient management of uncertainty in RA frameworks.

REFERENCES

[1] Hau Wang H. (2010). Chapter 9: Flexible chemical sensors. Pages 247–273. In: Sun Y., Rogers J.A. (eds.), *Micro and Nano Technologies, Semiconductor Nanomaterials for Flexible Technologies*. William Andrew Publishing, Oxford. https://doi.org/10.1016/B978-1-4377-7823-6.00009-X.

[2] Su S., Wu W., Gao J., Lu J., Fan C. (2012). Nanomaterials-based sensors for applications in environmental monitoring. *Journal of Materials Chemistry*, 22(35), 18101–18110.

[3] Lombardi M. (2012). Nanostructured materials for sensing. In: Bhushan B. (eds.), *Encyclopedia of Nanotechnology*. Springer, Dordrecht. https://doi-org.ezproxy.utp.edu.co/10.1007/978-90-481-9751-4_254.

[4] Arduini F., Cinti S., Scognamiglio V., Moscone D. (2020). 13—Nanomaterial-based sensors. Pages 329–359. In: Hussain C.M. (ed.), *Handbook of Nanomaterials in Analytical Chemistry*. Elsevier, Amsterdam. https://doi.org/10.1016/B978-0-12-816699-4.00013-X.

[5] Ali N., Bilal M., Khan A., Ali F., Khan H., Khan H.A., Iqbal H.M.N. (2021). Chapter 4—Fabrication strategies for functionalized nanomaterials. Pages 55–95. In: Tahir M.B., Sagir M., Asiri A.M. (eds.), *Nanomaterials: Synthesis, Characterization, Hazards and Safety*. Elsevier, Amsterdam. https://doi.org/10.1016/B978-0-12-823823-3.00010-0.

[6] Adam T., Gopinath S.C.B. (2022). Nanosensors: Recent perspectives on attainments and future promise of downstream applications. *Process Biochemistry*, 117, 153–173. https://doi.org/10.1016/j.procbio.2022.03.024.

[7] Tuantranont A. (2012). Nanomaterials for sensing applications: Introduction and perspective. In: Tuantranont A. (ed.), *Applications of Nanomaterials in Sensors and Diagnostics*. Springer Series on Chemical Sensors and Biosensors, vol. 14. Springer, Berlin, Heidelberg. https://doi-org.ezproxy.utp.edu.co/10.1007/5346_2012_41.

[8] Chakraborty U., Kaur G., Chaudhary G.R. (2021). Development of environmental nanosensors for detection monitoring and assessment. In: Kumar R., Kumar R., Kaur G. (eds.), *New Frontiers of Nanomaterials in Environmental Science*. Springer, Singapore. https://doi-org.ezproxy.utp.edu.co/10.1007/978-981-15-9239-3_5.

[9] Singh K. (2020). Nanosensors for food safety and environmental monitoring. In: Thangadurai D., Sangeetha J., Prasad R. (eds.), *Nanotechnology for Food, Agriculture, and Environment*. Nanotechnology in the Life Sciences. Springer, Cham. https://doi-org.ezproxy.utp.edu.co/10.1007/978-3-030-31938-0_4.

[10] European Commission. (2013). Guidance on the protection of the health and safety of workers from the potential risks related to nanomaterials at work. Employment, Social Affairs and Inclusion.

[11] The Tavistock and Portman. NHS Foundation Trust. (2018). Conducting a Risk Assessment Procedure. https://trustsrv-io-tavistock-tenant-mediabucket-jxlat5oi107p.s3.amazonaws.com/media/documents/procedure-conducting-risk-assessment.pdf?X-Amz-Algorithm=AWS4-HMAC-SHA256&X-Amz-Credential=AKIASPFGSFA5MV75PZDY%2F20220617%2Feu-west-2%2Fs3%2Faws4_request&X-Amz-Date=20220617T132829Z&X-Amz-Expires=3600&X-Amz-SignedHeaders=host&X-Amz-Signature=1993957806e3040705120 3ba877ab11e1740a7d3599db17537b67ed6d12eb032.

[12] HSENI (Health and Safety Executive for Northern Ireland). (2017). Five Steps to Risk Assessment. www.hseni.gov.uk/sites/hseni.gov.uk/files/publications/%5Bcurrent-domain%3Amachine-name%5D/five-steps-to-risk-assessment-2017.pdf.

[13] International Organization for Standardization (ISO). (2018). ISO 31000. Risk Management-Guidelines. Second Edition 2018–02.

[14] Modarres M. (2006). *Risk Analysis in Engineering: Techniques, Tools, and Trends*. CRC Press.

[15] International Organization for Standardization (ISO). (2009). IEC 31010:2009—Risk management.

[16] Jahnel J. (2015). Conceptual questions and challenges associated with the traditional risk assessment paradigm for nanomaterials. *Nanoethics*, 9, 261–276. https://doi-org.ezproxy.utp.edu.co/10.1007/s11569-015-0235-0.

[17] Dekkers S., Oomen A., Bleeker E., Vandebriel R., Micheletti C., Cabellos J., Janer G., Fuentes N., Vázquez-Campos S., Borges T., João Silva M., Prina-Mello A., Movia D., Nesslany F., Ribeiro A., Leite P., Groenewold M., Cassee F., Sips A., Dijkzeul A., Teunenbroek T., Wijnhoven S. (2016). Towards a nanospecific approach for risk assessment. *Regulatory Toxicology and Pharmacology*, 80, 46–59. https://doi.org/10.1016/j.yrtph.2016.05.037.

[18] Courty M.A., Allue E., Henry A. (2020). Forming mechanisms of vitrified charcoals in archaeological firing-assemblages. *J. Archaeol. Sci. Rep.*, 30, 102215. https://doi.org/10.1016/j.jasrep.2020.102215.

[19] Malakar A., Kanel S., Ray C., Snow D., Nadagouda M. (2021). Nanomaterials in the environment, human exposure pathway, and health effects: A review. *Science of the Total Environment*, 759, 143470. https://doi.org/10.1016/j.scitotenv.2020.143470.

[20] Bundschuh M., Filser J., Lüderwald S., McKee M.S., Metreveli G., Schaumann G.E., Schulz R., Wagner S. (2018). Nanoparticles in the environment: Where do we come from, where do we go to? *Environ. Sci. Eur.*, 30, 6. https://doi.org/10.1186/s12302-018-0132-6.

[21] Kabir E., Kumar V., Kim K., Yip A., Sohn J.R. (2018). Environmental impacts of nanomaterials. *Journal of Environmental Management*, 225, 261–271. https://doi.org/10.1016/j.jenvman.2018.07.087.

[22] Iavicoli I., Leso V., Ricciardi W., Hodson L.L., Hoover M.D. (2014). Opportunities and challenges of nanotechnology in the green economy. *Environmental Health*, 13(1), 1–11.

[23] Ermolin M.S., Fedotov P.S., Malik N.A., Karandashev V.K. (2018). Nanoparticles of volcanic ash as a carrier for toxic elements on the global scale. *Chemosphere*, 200, 16–22.

[24] Maters E.C., Delmelle P., Bonneville S. (2016). Atmospheric processing of volcanic glass: Effects on iron solubility and redox speciation. *Environmental Science & Technology*, 50(10), 5033–5040.

[25] John A.C., Küpper M., Manders-Groot A.M., Debray B., Lacome J.M., Kuhlbusch T.A. (2017). Emissions and possible environmental implication of engineered nanomaterials (ENMs) in the atmosphere. *Atmosphere*, 8(5), 84.

[26] Ounoughene G., Le Bihan O., Chivas-Joly C., Motzkus C., Longuet C., Debray B., Lopez-Cuesta J.M. (2015). Behavior and fate of halloysite nanotubes (HNTs) when incinerating PA6/HNTs nanocomposite. *Environmental Science & Technology*, 49(9), 5450–5457.

[27] Malakar A., Snow D.D. (2020). Nanoparticles as inorganic pollutant in water. In *Inorganic Pollutants in Water*. Elsevier, Oxford.

[28] Wang J., Nabi M.M., Mohanty S.K., Afrooz A.N., Cantando E., Aich N., Baalousha M. (2020). Detection and quantification of engineered particles in urban runoff. *Chemosphere*, 248, 126070.

[29] Vale G., Mehennaoui K., Cambier S., Libralato G., Jomini S., Domingos R.F. (2016). Manufactured nanoparticles in the aquatic environment-biochemical responses on freshwater organisms: A critical overview. *Aquatic Toxicology*, 170, 162–174.

[30] Batley G.E., Kirby J.K., McLaughlin M.J. (2013). Fate and risks of nanomaterials in aquatic and terrestrial environments. *Accounts of Chemical Research*, 46(3), 854–862.

[31] Khodakovskaya, M., Dervishi E., Mahmood M., Xu Y., Li Z., Watanab F., Biris A.S. (2009). Carbon nanotubes are able to penetrate plant seed coat and dramatically affect seed germination and plant growth. *ACS Nano*, 3(10), 3221–3227.

[32] Hong J., Peralta-Videa J.R., Rico C., Sahi S., Viveros M.N., Bartonjo J., Gardea-Torresdey J.L. (2014). Evidence of translocation and physiological impacts of foliar applied CeO_2 nanoparticles on cucumber (Cucumis sativus) plants. *Environmental Science & Technology*, 48(8), 4376–4385.

[33] Sun Q., Bijelić M., Djurišić A.B., Suchomski C., Liu X., Xie M., Popović J. (2017). Graphene-oxide-wrapped ZnMn2O4 as a high performance lithium-ion battery anode. *Nanotechnology*, 28(45), 455401.

[34] Daza E.A., Misra S.K., Scott J., Tripathi I., Promisel C., Sharma B.K., Pan D. (2017). Multi-shell nano-carboscavengers for petroleum spill remediation. *Scientific Reports*, 7(1), 1–15.

[35] Wang P., Lombi E., Sun S., Scheckel K.G., Malysheva A., McKenna B.A., Kopittke P.M. (2017). Characterizing the uptake, accumulation and toxicity of silver sulfide nanoparticles in plants. *Environmental Science: Nano*, 4(2), 448–460.

[36] Chen X., O'Halloran J., Jansen M.A. (2016). The toxicity of zinc oxide nanoparticles to Lemna minor (L.) is predominantly caused by dissolved Zn. *Aquatic Toxicology*, 174, 46–53.

[37] Kwok K.W., Leung K.M., Flahaut E., Cheng J., Cheng S.H. (2010). Chronic toxicity of double-walled carbon nanotubes to three marine organisms: Influence of different dispersion methods. *Nanomedicine*, 5(6), 951–961.

[38] Tong T., Wilke C.M., Wu J., Binh C.T.T., Kelly J.J., Gaillard J.F., Gray K.A. (2015). Combined toxicity of nano-ZnO and nano-TiO2: From single-to multinanomaterial systems. *Environmental Science & Technology*, 49(13), 8113–8123.

[39] Turkevich L.A., Dastidar A.G., Hachmeister Z., Lim M. (2015). Potential explosion hazard of carbonaceous nanoparticles: Explosion parameters of selected materials. *Journal of Hazardous Materials*, 295, 97–103.

[40] Martirosyan A., Schneider Y.J. (2014). Engineered nanomaterials in food: Implications for food safety and consumer health. *Int J Environ Res Public Health* 11, 5720–5750. https://doi-org.ezproxy.utp.edu.co/10.3390/ijerph110605720.

[41] Asmatulu E., Andalib M.N., Subeshan B., Abedin F. (2022). Impact of nanomaterials on human health: A review. *Environ Chem Lett*. https://doi-org.ezproxy.utp.edu. co/10.1007/s10311-022-01430-z.

[42] Zhu S., Gong L., Li Y., Xu H., Gu Z., Zhao Y. (2019). Safety assessment of nanomaterials to eyes: An important but neglected issue. *Adv Sci* (Weinheim, Baden-Wurttemberg, Germany), 6, 1802289.

[43] Helland A., Scheringer M., Siegrist M., Kastenholz H.G., Wiek A., Scholz R.W. (2008). Risk assessment of engineered nanomaterials: A survey of industrial approaches. *Environmental Science & Technology*, 42(2), 640–646.

[44] Oberbek P., Kozikowski P., Czarnecka K., Sobiech P., Jakubiak S., Jankowski, T. (2019). Inhalation exposure to various nanoparticles in work environment: Contextual information and results of measurements. *Journal of Nanoparticle Research*, 21(11), 1–24.

[45] Geiser M., Jeannet N., Fierz M., Burtscher H. (2017). Evaluating adverse effects of inhaled nanoparticles by realistic in vitro technology. *Nanomaterials* (Basel), Feb 22;7(2), 49. doi: 10.3390/nano7020049.

[46] Cheresh P., Kim S.-J., Tulasiram S., Kamp D.W. (2013). Oxidative stress and pulmonary fibrosis. *Biochim Biophys Acta*, 1832, 1028–1040.

[47] Evans S.J., Vecchiarelli P.M., Clift M.J.D., Doak S.H., Lead J.R. (2021). Overview of nanotoxicology in humans and the environment: Developments, challenges and impacts. In: Lead J.R., Doak S.H., Clift M.J. (eds.), *Nanotoxicology in Humans and the Environment: Molecular and Integrative Toxicology*. Springer, Cham. https://doi-org. ezproxy.utp.edu.co/10.1007/978-3-030-79808-6_1.

[48] De Matteis V. (2017). Exposure to inorganic nanoparticles: Routes of entry, immune response, biodistribution and in vitro/in vivo toxicity evaluation. *Toxics*, 5(4), 29.

[49] Jesus S., Schmutz M., Som C., Borchard G., Wick P., Borges, O. (2019). Hazard assessment of polymeric nanobiomaterials for drug delivery: What can we learn from literature so far. *Frontiers in Bioengineering and Biotechnology*, 7, 261.

[50] Iavicoli I., Leso V., Beezhold D.H., Shvedova A.A. (2017). Nanotechnology in agriculture: Opportunities, toxicological implications, and occupational risks. *Toxicology and Applied Pharmacology*, 329, 96–111.

[51] Ding G., Zhang N., Wang C., Li X., Zhang J., Li W., Yang Z. (2018). Effect of the size on the aggregation and sedimentation of graphene oxide in seawaters with different salinities. *Journal of Nanoparticle Research*, 20(11), 1–10.

[52] Filon F.L., Mauro M., Adami G., Bovenzi M., Crosera, M. (2015). Nanoparticles skin absorption: New aspects for a safety profile evaluation. *Regulatory Toxicology and Pharmacology*, 72(2), 310–322.

[53] Hashem N.M., Sallam S.M. (2020). Reproductive performance of goats treated with free gonadorelin or nanoconjugated gonadorelin at estrus. *Domest. Anim. Endocrinol.*, 71, 106390. doi: 10.1016/j.domaniend.2019.106390.

[54] Huang Y.W., Cambre M., Lee H.J. (2017). The toxicity of nanoparticles depends on multiple molecular and physicochemical mechanisms. *International Journal of Molecular Sciences*, 18(12), 2702.

[55] Söderstjerna E., Bauer P., Cedervall T., Abdshill H., Johansson F., Johansson U.E. (2014). Silver and gold nanoparticles exposure to *In Vitro* cultured retina—studies on

nanoparticle internalization, apoptosis, oxidative stress, glial- and microglial activity. *PLoS One*, 9(8), e105359. https://doi.org/10.1371/journal.pone.0105359.

[56] Silva-León S., Fernández-Luqueño F., López-Valdez F. (2016). Silver nanoparticles (AgNP) in the environment: A review of potential risks on human and *Environmental Health*. *Water, Air, & Soil Pollution*, 227, 1–20. doi:10.1007/s11270-016-3022-9.

[57] Gaillet S., Rouanet J.M. (2015). Silver nanoparticles: Their potential toxic effects after oral exposure and underlying mechanisms-a review. *Food Chem Toxicol*, 77, 58–63. https://doi-org.ezproxy.utp.edu.co/10.1016/j.fct.2014.12.019.

[58] Watson C., Ge J., Cohen J., Pyrgiotakis G., Engelward B.P., Demokritou P. (2014). High-throughput screening platform for engineered nanoparticle-mediated genotoxicity using CometChip technology. *ACS Nano*, 8(3), 2118–2133.

[59] Sousa S.P.B., Peixoto T., Santos R.M., Lopes A., Paiva M.D.C., Marques A.T. (2020). Health and safety concerns related to CNT and graphene products, and related composites. *J. Compos. Sci.*, 4, 106. https://doi.org/10.3390/jcs4030106.

[60] Mangum J.B., Turpin E.A., Antao-Menezes A., Cesta M.F., Bermudez E., Bonner J.C. (2006). Single-walled carbon nanotube (SWCNT)-induced interstitial fibrosis in the lungs of rats is associated with increased levels of PDGF mRNA and the formation of unique intercellular carbon structures that bridge alveolar macrophages in situ. *Particle and Fibre Toxicology*, 3(1), 1–13.

[61] Sharma M., Nikota J., Halappanavar S., Castranova V., Rothen-Rutishauser B., Clippinger A.J. (2016). Predicting pulmonary fibrosis in humans after exposure to multi-walled carbon nanotubes (MWCNTs). *Archives of Toxicology*, 90(7), 1605–1622.

[62] Bhattacharya K., Andón F.T., El-Sayed R., Fadeel B. (2013). Mechanisms of carbon nanotube-induced toxicity: Focus on pulmonary inflammation. *Adv Drug Deliv Rev*, 65, 2087–2097. https://doi-org.ezproxy.utp.edu.co/10.1016/j.addr.2013.05.012.

[63] Hu W., Peng C., Lv M., Li X., Zhang Y., Chen N., Huang Q. (2011). Protein corona-mediated mitigation of cytotoxicity of graphene oxide. *ACS Nano*, 5(5), 3693–3700.

[64] Mebert A.M., Baglole C.J., Desimone M.F., Maysinger D. (2017). Nanoengineered silica: Properties, applications and toxicity. *Food Chem Toxicol*, 109, 753–770. https://doi-org.ezproxy.utp.edu.co/10.1016/j.fct.2017.05.054.

[65] Demir E., Castranova V. (2016). Genotoxic effects of synthetic amorphous silica nanoparticles in the mouse lymphoma assay. *Toxicol Rep*, 3, 807–815. https://doi-org.ezproxy.utp.edu.co/10.1016/j.toxrep.2016.10.006.

[66] Azimipour S., Ghaedi S., Mehrabi Z., Ghasemzadeh S.A., Heshmati M., Barikrow N., Falahati M. (2018). Heme degradation and iron release of hemoglobin and oxidative stress of lymphocyte cells in the presence of silica nanoparticles. *International Journal of Biological Macromolecules*, 118, 800–807.

[67] Liangjiao C., Yiyuan K., Hongbing G., Jia L., Wenchao Z., Yanli Z., Longquan S. (2019). The current understanding of immunotoxicity induced by silica nanoparticles. *Nanomedicine*, 14(10), 1227–1229.

[68] Trouiller B., Reliene R., Westbrook A., Solaiman P., Schiestl R.H. (2009). Titanium dioxide nanoparticles induce DNA damage and genetic instability in vivo in mice. *Cancer Research*, 69(22), 8784–8789.

[69] Wang L., Yan L., Liu J., Chen C., Zhao Y. (2018). Quantification of nanomaterial/nanomedicine trafficking in vivo. *Anal. Chem*, 90(1), 589–614.

[70] Chen J., Dong X., Zhao J., Tang G. (2009) In vivo acute toxicity of titanium dioxide nanoparticles to mice after intraperitioneal injection. *J Appl Toxicol*, 29, 330–337. https://doi-org.ezproxy.utp.edu.co/10.1002/jat.1414.

[71] De Berardis B., Marchetti M., Risuglia A., Ietto F., Fanizza C., Superti F. (2021). Correction to: Exposure to airborne gold nanoparticles: A review of current toxicological data on the respiratory tract. *Journal of Nanoparticle Research*, 23(2), 1–2.

[72] Steckiewicz K.P., Barcinska E., Malankowska A., Zauszkiewicz-Pawlak A., Nowaczyk G., Zaleska-Medynska A., Inkielewicz-Stepniak I. (2019). Impact of gold nanoparticles shape on their cytotoxicity against human osteoblast and osteosarcoma in in vitro model: Evaluation of the safety of use and anti-cancer potential. *Journal of Materials Science: Materials in Medicine*, 30(2), 1–15.

[73] U.S. Environmental Protection Agency (EPA). (2017). Fed. Regist., 82.

[74] Morin D. (2015). Section 71 of the Canadian Environmental Protection Act, 1999 (CEPA 1999) *Canada Gazette*, Part, I p. 149.

[75] SAC. (2004). GB/T 19619–2004 Terminology for Nanomaterials.

[76] MHLV. (2009). Notification on present preventive measures for the prevention of exposure at workplaces manufacturing and handling nanomaterials. LSB Notification No.0331013. March 31st, 2009. https://www.jniosh.johas.go.jp/publication/doc/houkoku/nano/files/mhlw/Notification_0331013_en.pdf

[77] KATS. (2009). KS A 6202 Guidance to safe handling of manufactured nanomaterials in workplace/industry. https://kssn.net/search/stddetail.do?itemNo=K001010103390

[78] Brazilian Chamber of Deputies. (2013). Bill n. 6741/2013 amending law no. 9605 (1998).

[79] Rauscher H., Rasmussen K., Sokull-Klüttgen B. (2017). Regulatory aspects of nanomaterials in the EU Chem. *Ing. Tech.*, 89, 224–231.

[80] Ministère de l'Environnement de l'Énergie et de la Mer (2015). Éléments issus des déclarations des substances à l'état nanoparticulaire. Rapport d'e etude. https://www.ecologie.gouv.fr/sites/default/files/2015-12%20-%20Rapport%20R-nano%202015.pdf

[81] Tschiche H., Bierkandt F., Creutzenberg O., Fessard V., Franz R., Giese B., Greiner R., Haas K., Haase A., Hartwig A., Hund-Rinke K., Iden P., Kromer C., Loeschner K., Mutz D., Rakow A., Rasmussen K., Rauscher H., Richter H., Schoon J., Schmid O., Som C., Tovar G., Westerhoff P., Wohlleben W., Luch A., Laux P. (2022). Environmental considerations and current status of grouping and regulation of engineered nanomaterials. *Environmental Nanotechnology, Monitoring & Management*, 18, 100707. https://doi.org/10.1016/j.enmm.2022.100707.

[82] Carlander D., Skentelbery C. (2021). EU regulations and nanotechnology innovation. In: Lead J.R., Doak S.H., Clift M.J. (eds.), *Nanotoxicology in Humans and the Environment: Molecular and Integrative Toxicology*. Springer, Cham. https://doi-org.ezproxy.utp.edu.co/10.1007/978-3-030-79808-6_8.

[83] OECD. (1981). Decision of the Council Concerning the Mutual Acceptance of Data in the Assessment of Chemicals—C(81)30/FINAL. https://www.oecd.org/env/ehs/2017640.pdf.

[84] Oomen A., Steinhäuser K., Bleeker E., Broekhuizen F., Sips A., Dekkers S., Wijnhoven S., Sayre P. (2018). Risk assessment frameworks for nanomaterials: Scope, link to regulations, applicability, and outline for future directions in view of needed increase in efficiency, *NanoImpact*, 9, 1–13. https://doi.org/10.1016/j.impact.2017.09.001.

[85] Sayre P., Steinhäuser K. (2016). ProSafe—Roadmap for members of task force when reviewing data, protocols, reports and guidance notes for regulatory relevance. Final version.

[86] NAS. (2012). A Research Strategy for Environmental, Health, and Safety Aspects of Engineered Nanomaterials. National Academies Press, Washington, DC 231 pages. www.nap.edu/catalog/13347/a-research-strategy-for-environmental-health-and-safety-aspects-of-engineered-nanomaterials.

[87] OECD. (2016). Future Challenges Related to the Safety of Manufactured Nanomaterials: Report From the Special Session No. 75. www.oecd.org/officialdocuments/publicdisplaydocumentpdf/?cote=env/jm/mono(2012)14&doclanguage=en.

[88] Hansen F.S., Alstrup Jensen K., Baun A. (2014). NanoRiskCat: A conceptual tool for categorization and communication of exposure potentials and hazards of nanomaterials in consumer products. *J. Nanopart. Res.*, 16, 2195. doi: 10.1007/s11051–013–2195–z.

[89] Arts J.H., Hadi M., Irfan M.A., Keene A.M., Kreiling R., Lyon D., Maier M., Michel K., Petry T., Sauer U.G., Warheit D., Wiench K., Wohlleben W., Landsiedel R. (2015). A decision-making framework for the grouping and testing of nanomaterials (DF4nanoGrouping). *Regul. Toxicol. Pharmacol.*, 71, S1–27.

[90] Arts J.H., Irfan M.A., Keene A.M., Kreiling R., Lyon D., Maier M., Michel K., Neubauer N., Petry T., Sauer U.G., Warheit D., Wiench K., Wohlleben W., Landsiedel R. (2016). Case studies putting the decision-making framework for the grouping and testing of nanomaterials (DF4nanoGrouping) into practice Regul. *Toxicol. Pharmacol.*, 76, 234–261.

[91] Bos P.M., Gottardo S., Scott-Fordsmand J.J., van Tongeren M., Semenzin E., Fernandes T.F., Hristozov D., Hund-Rinke K., Hunt N., Irfan M.A., Landsiedel R., Peijnenburg W.J., Sanchez Jimenez A., van Kesteren P.C., Oomen A.G. (2015). The MARINA risk assessment strategy: A flexible strategy for efficient information collection and risk assessment of nanomaterials. *Int. J. Environ. Res. Public Health*, 12, 15007–15021.

[92] Godwin H., Nameth C., Avery D., Bergeson L.L., Bernard D., Beryt E., Boyes W., Brown S., Clippinger A.J., Cohen Y., Doa M., Hendren C.O., Holden P., Houck K., Kane A.B., Klaessig F., Kodas T., Landsiedel R., Lynch I., Malloy T., Miller M.B., Muller J., Oberdorster G., Petersen E.J., Pleus R.C., Sayre P., Stone V., Sullivan K.M., Tentschert J., Wallis P., Nel A.E. (2015). Nanomaterial categorization for assessing risk potential to facilitate regulatory decision-making. *ACS Nano*, 9(4), 3409–3417.

[93] Hund-Rinke K., Herrchen M., Schlich K., Schwirn K., Völker D. (2015). Test strategy for assessing the risks of nanomaterials in the environment considering general regulatory proceduresEnviron. *Sci. Eur.*, 27, 24. http://dx.doi.org/10.1186/s12302-015-0053-6.

[94] Sellers K., Deleebeeck N.M.E., Messiaen M., Jackson M., Bleeker E.A.J., Sijm D.T.H.M., van Broekhuizen F.A. (2015). Grouping nanomaterials: A strategy towards grouping and read-across RIVM Report 2015–0061, National Institute for Public Health and the Environment (RIVM) & Arcadis, Bilthoven, The Netherlands. www.rivm.nl/bibliotheek/rapporten/2015-0061.html.

[95] Walser T., Studer C. (2015). Sameness: The regulatory crux with nanomaterial identity and grouping schemes for hazard assessment *Regul. Toxicol. Pharmacol.*, 72(3), 569–571. doi: 10.1016/j.yrtph.2015.05.031.

[96] Oosterwijk M.T., Feber M.L., Burello E. (2016). Proposal for a risk banding framework for inhaled low aspect ratio nanoparticles based on physicochemical properties. *Nanotoxicology*, 10(6), 780–793. doi: 10.3109/17435390.2015.1132344.

[97] Gottardo S., Alessandrelli M., Amenta V., Atluri R., Barberio G., Bekker C., Bergonzo P., Bleeker E., Booth A.M., Borges T., Buttol P., Carlander D., Castelli S., Chevillard S., Clavaguera S., Dekkers S., Delpivo C., Di Prospero Fanghella P., Dusinska M., Einola J., Ekokoski E., Fito C., Gouveia H., Grall R., Hoehener K., Jantunen P., Johanson G., Laux P., Lehmann H.C., Leinonen R., Mech A., Micheletti C., Noorlander C., Olof-Mattsson M., Oomen A., Quiros Pesudo L., Letizia Polci M., Prina-Mello A., Rasmussen K., Rauscher H., Sanchez Jimenez A., Riego Sintes J., Scalbi S., Sergent J.-A., Stockmann-Juvala H., Simko M., Sips A., Suarez B., Sumrein A., van Tongeren M., Vázquez-Campos S., Vital N., Walser T., Wijnhoven S., Crutzen H. (2017). NANoREG framework for the safety assessment of nanomaterials. *EUR*, 28550 EN. doi: 10.2760/245972.

[98] Subramanian V., Semenzin E., Hristozov D., Zabeo A., Malsch I., McAlea E., Murphy F., Mullins M., van Harmelen T., Ligthart T., Linkov I., Marcomini A. (2016). Sustainable nanotechnology decision support system: Bridging risk management, sustainable innovation and risk governance. *J. Nanopart. Res.*, 18, 89. doi: 10.1007/s11051-016-3375-4.

[99] Van Harmelen T., Zondervan-van den Beuken E.K., Brouwer D.H., Kuijpers E., Fransman W., Buist H.B., Ligthart T.N., Hincapié I., Hischier R., Linkov I., Nowack B., Studer J., Hilty L., Som C. (2016). LICARA nanoSCAN: A tool for the self-assessment

of benefits and risks of nanoproducts. *Environ. Int.*, 91, 150–160. doi: 10.1016/j.envint. 2016.02.021.

[100] Hjorth R., Hansen S.F., Jacobs M., Tickner J., Ellenbecker M., Baun A. (2016). The applicability of chemical alternatives assessment for engineered nanomaterials Integr. *Environ. Assess. Manag.* doi: 10.1002/ieam.1762.

[101] NANoREG 6.04 Inventory of Existing Regulatory Accepted Toxicity Tests Applicable for Safety Screening of MNMs. (2016). http://rivm.nl/en/About_RIVM/International/ International_Projects/Completed/NANoREG/deliverables/NANoREG_D6_04_DR_ Inventory_of_existing_regulatory_accepted_toxicity_tests.org.

[102] Gottardo S., Mech A., Drbohlavová J., Małyska A., Bøwadt S., Riego Sintes J., Rauscher H. 2021. Towards safe and sustainable innovation in nanotechnology: State-of-play for smart nanomaterials. *NanoImpact*, 21, 100297. https://doi.org/10.1016/j.impact.2021.100297.

[103] Soeteman-Hernandez L.G., Apostolova M.D., Bekker C., Dekkers S., Grafström R.C., Groenewold M., Handzhiyski Y., Herbeck-Engel P., Hoehener K., Karagkiozaki V., Kelly S., Kraegeloh A., Logothetidis S., Micheletti C., Nymark P., Oomen A., Oosterwijk T., Rodríguez-LLopis I., Sabella S., Sanchez Jiménez A., Sips A.J.A.M., Suarez-Merino B., Tavernaro I., van Engelen J., Wijnhoven S.W.P., Noorlander C.W. (2019). Safe innovation approach: Towards an agile system for dealing with innovations. *Mater. Today Commun.*, 20, 100548.

[104] Lowry G.V., Avellan A., Gilbertson L.M. (2019). Opportunities and challenges for nanotechnology in the agri-tech revolution. *Nat. Nanotechnol.*, 14, 517–522.

[105] Kuzma J., Roberts J.P. (2016). Is adaptation or transformation needed? Active nanomaterials and risk analysis. *J. Nanopart. Res.*, 18, 1–18.

[106] OECD. (2012). Important Issues on Risk Assessment of Manufactured Nanomaterials, Series on the Safety of Manufactured Nanomaterials No. 33. www.oecd.org/official documents/publicdisplaydocumentpdf/?cote=env/jm/mono(2012)8&doclanguage=en.

[107] Lützhøft H.C.H., Hartmann N.B., Brinch A., Kjølholt J., Baun, A. (2015). Environmental effects of engineered nanomaterials: estimations of predicted no-effect concentrations (PNECs).

[108] van Kesteren P.C., Cubadda F., Bouwmeester H., van Eijkeren J.C., Dekkers S., de Jong W.H., Oomen A.G. (2014). Novel insights into the risk assessment of the nanomaterial synthetic amorphous silica, additive E551, in food. *Nanotoxicology*, 1–10. doi: 10.3109/17435390.2014.940408.

[109] ECHA/JRC/RIVM. (2016). Usage of (Eco)Toxicological Data for Bridging Data Gaps between and Grouping of Nanoforms of the Same Substance. Elements to Consider. https://echa.europa.eu/documents/10162/13630/eco_toxicological_for_bridging_ grouping_nanoforms_en.pdf.

Index

For Product Safety Concerns and Information please contact our EU
representative GPSR@taylorandfrancis.com
Taylor & Francis Verlag GmbH, Kaufingerstraße 24, 80331 München, Germany